Prehistoric Life

Dedication (from BSL)

To my parents, Burt Lieberman and Mimi Smith, and my professors, Niles Eldredge and Stephen Jay Gould (in memoriam).

Prehistoric Life

Evolution and the Fossil Record

Bruce S. Lieberman and Roger Kaesler

WILEY-BLACKWELL

A John Wiley & Sons, Ltd., Publication

This edition first published 2010, © 2010 by Bruce S. Lieberman and Roger Kaesler

Blackwell Publishing was acquired by John Wiley & Sons in February 2007. Blackwell's publishing program has been merged with Wiley's global Scientific, Technical and Medical business to form Wiley-Blackwell.

Registered office: John Wiley & Sons Ltd, The Atrium, Southern Gate, Chichester, West Sussex, PO19 8SQ, UK

Editorial offices: 9600 Garsington Road, Oxford, OX4 2DQ, UK
      The Atrium, Southern Gate, Chichester, West Sussex, PO19 8SQ, UK
      111 River Street, Hoboken, NJ 07030-5774, USA

For details of our global editorial offices, for customer services and for information about how to apply for permission to reuse the copyright material in this book please see our website at www.wiley.com/wiley-blackwell

Library of Congress Cataloguing-in-Publication Data
Lieberman, Bruce S.
  Prehistoric life : evolution and the fossil record / Bruce S. Lieberman and Roger Kaesler.
    p. cm.
  Includes index.
  ISBN 978-1-4443-3408-1 (hardcover)—ISBN 978-0-6320-4472-6 (pbk.) 1. Evolutionary paleobiology. 2. Animals, Fossil. 3. Evolution (Biology) 4. Life—Origin. I. Kaesler, Roger L. II. Title.
  QE721.2.E85L54 2010
  560—dc22

            2009038741

ISBN: 9780632044726 (paperback) and 9781444334081 (hardback)

A catalogue record for this book is available from the British Library.
Set in 11.5/13pt Plantin by Graphicraft Limited, Hong Kong
Printed and bound in Malaysia by Vivar Printing Sdn Bhd

1 2010

Contents

Preface, viii

1 **Introduction to Fossils, 1**
History, Science, and Historical Science
Time, Life, and Stratigraphy
What is a Fossil?
How do Fossils Form?
Conclusions: Fossils as Curious Stones
Additional Reading

2 **The Nature of the Fossil Record, 29**
Fossils in Sedimentary Rock
Taphonomy
Time Averaging
Mode of Growth
Colonial Organisms
Trace Fossils
Concluding Remarks
Additional Reading

3 **Organizing the Fossil Record, 52**
History of Ideas on Biological Classification
Applying Linnaeus' Hierarchy
What is a Species and How Does a Paleontologist
   Identify Them?
Conclusions: the Difference Between Inanimate
   Atoms and Living Things
Additional Reading

4 **Introduction to Evolution, 74**
Introduction
A Biological Definition of Evolution
The History of Evolutionary Thought
Science and Religion
Darwin and Wallace: Never Ask a Stranger to
   Present Your Paper at a Meeting You Cannot
   Attend
Natural Selection
Conclusions: Why was Natural Selection Not
   Endorsed at Once by Many Scientists?
Additional Reading

5 **Macroevolution, Progress, and
the History of Life, 100**
Introduction
Competition and Macroevolution

Does Evolution Happen Gradually or Episodically?
Natural Selection Operating Above and Below the
   Level of the Individual Organism
Progress and the History of Life
Conclusions: Patterns and Processes of Increasing
   Complexity
Additional Reading

6 **Extinctions: The Legacy of the
Fossil Record, 123**
Introduction
Contingency
Boundaries in the Geological Time Scale and
   the Nature of Extinction
The Cretaceous–Tertiary Mass Extinction
How has the Existence of Mass Extinctions
   Influenced the History of Life?
Were Most Extinctions Caused by Asteroid
   Impact?
The Permo-Triassic Mass Extinction—Causes and
   Consequences
The Ordovician–Silurian Mass Extinction
Other Mass Extinction Events: The Late Devonian
   and the End of the Triassic
Habitat Degradation and Mass Extinctions
The Sixth Great Mass Extinction: The Current
   Biodiversity Crisis
Conclusions: Lessons from the Past and Future
   Prospects for Humanity
Additional Reading

7 **Systematics and the Fossil Record, 150**
Introduction
Methods and Approaches in Systematics
The Growth of Molecular Biology and
   Improvements in DNA Sequencing Technology
The Spread of Computers and Computer
   Programs Used to Study Evolutionary
   Relationships
Systematics and How to go About Identifying
   Species in the Fossil Record
Systematics and its Relevance for Identifying
   Patterns of Mass Extinction
Systematics and the Meaning of Adaptations
Concluding Remarks
Additional Reading

8 **Principles of Growth and Form: Life, the Universe, and Gothic Cathedrals, 170**
Introduction
Galileo's Principle
Galileo's Principle and its Relevance to the Biology of Living Organisms
Galileo's Principle and Constraints on the Evolution of Large Body Size
Galileo's Principle and its Relevance to Medieval Architecture
Galileo's Principle and its Relevance to Cratering Density in our Solar System
Concluding Remarks
Additional Reading

9 **The Role of Fossils in the Genesis of Myths and Legends, 192**
Introduction
Paleontologist's Have Come from Many Different Walks of Life and Have Sported Many Different Hairdos
Paleontology in Ancient Greece
Native American Contributions to Paleontology
Concluding Remarks
Additional Reading

10 **Plate Tectonics and its Effects on Evolution, 208**
Introduction
Early Ideas on Continents in Motion: Continental Drift
Plate Tectonics: Continental Drift in a Different Guise and with a Valid Mechanism
The Evolutionary Implications of Plate Tectonics
Biogeography
Concluding Remarks
Additional Reading

11 **Life, Climate, and Geology, 227**
Introduction
Some of the Major Factors that Govern the Climate System
Examples of How Life has Influenced Climate: The Difference Between the Proterozoic and the Permian
Life Influencing Geology: the Form and Shape of Rivers and the Rocks they Leave Behind
Plants, Oxygen, and Coal: More Examples of Life Affecting the Atmosphere and Geology

How Geology Affects Climate: Considering How Plate Tectonic Changes have Contributed to Climate Changes Over the Last 60 Million Years
Concluding Remarks
Additional Reading

12 **Patterns and Processes of Precambrian Evolution, 253**
Introduction
The Earliest Evidence for Life in the Geological Record
The Time of Pond Scum and the Rise of Oxygen
For Billions of Years Organisms Have Been Modifying the Atmosphere and Their Environment
More Effects of Rising Oxygen Levels
The Evolution of the Eukaryotic Cell
Concluding Remarks
Additional Reading

13 **The Cambrian Radiation and Beyond: Understanding Biology's Big Bang, 273**
Introduction
Life Before the Cambrian Radiation
The Burgess Shale and the Cambrian Radiation
The Ordovician Radiation and Concluding Remarks
Additional Reading

14 **The Evolution and Extinction of Reefs Through Time: From the Precambrian to the Current Biodiversity Crisis, 299**
Introduction
Reef-Forming Organisms Today
Cnidarians and Outer Space
How Modern Cnidarian Corals Feed
Reefs Through Geological Time
Corals and the Biodiversity Crisis
Lessons from Human Effects on Modern Reefs
Concluding Remarks
Additional Reading

15 **Key Evolutionary Transitions: The Origins of Multicellularity and the Evolution of the Vertebrate Brain, 319**
Introduction
Origins of Multicellularity
The Evolution of the Vertebrate Brain

Trends in Brain Size Within Primates and
  Hominids
Concluding Remarks
Additional Reading

16 **Key Events in Vertebrate Evolution, 337**
Introduction
Cambrian Origins
Major Groups of Chordates
The Vertebrates
Lobe-Finned Aquatic Vertebrates
The Origins and Evolution of the Tetrapods
Concluding Remarks
Additional Reading

17 **Are We Alone in the Universe?, 359**
Introduction
What is the Potential that Humans will Encounter
  Extraterrestrial Civilizations?
Radio Waves and the Search for Extra-Terrestrial
  Intelligence

Possible Evidence for Life in a Martian Meteorite?
Concluding Remarks
Additional Reading

18 **Humanity: Origins and Prospects, 371**
Introduction
How do New Species Evolve—The Shift from
  Chimps to Hominids
Humans in a Changing Climate
The Current Biodiversity Crisis
Mapping a Course for Future Changes—Climate
  and Life
Concluding Remarks
Additional Reading

Index, 381

Color plate section between pp. 232 and pp. 233

**Companion website available at
www.wiley.com/go/lieberman/prehistoric**

Preface

Samuel Clemens, in the form of Huckleberry Finn, remarked that "if I'd a know'd what a trouble it was to make (this) book I never would a tackled it." Although I would not go quite that far, aspects of this sentiment ring true because I signed on to do this project with my colleague and mentor, Roger Kaesler. However, not too far into the project Roger became deathly ill with the disease that ultimately took his life. In spite of the fact that he passed away in 2006, certainly his stamp is present throughout the book. For instance, in addition to the contributions he made before he passed away, I endeavored to follow the topics and styling's that I think (and hope) he would have most preferred. In any event, the book's genesis was based on a class that both he and I taught (and I continue to teach) at the University of Kansas, the eponymous *Prehistoric Life*.

The fossil record represents a remarkable pageant of life: what historian of science Peter Bowler, paraphrasing the famous paleontologist William Diller Matthew, referred to as "Life's splendid drama." It also provides a unique window on the patterns and processes of evolution. This book focuses on both the patterns and the processes of evolution and how the history of life is intertwined with the geological and climatic history of this planet. Ultimately, what first inspired me to become a paleontologist was a trip via New York City subway to the American Museum of Natural History (AMNH) when I was three years old. There, accompanied by my parents, I saw the spectacular skeleton of *Tyrannosaurus rex*. My parents were my first teachers: they instilled in me the wonder of scientific discovery. Later teachers, especially Stephen Jay Gould, my undergraduate advisor at Harvard College, and Niles Eldredge, my PhD advisor at Columbia University and the AMNH, continued to inspire my interest in paleontology and the history of life. This book is dedicated to these four teachers.

My source of professional inspiration today is no longer solely based on dramatic ancient life forms (although I still find these fascinating) but also on how we can use the study of that life to comprehend evolution. Certainly people of all ages continue to be fascinated by spectacular examples of the life that has evolved on this planet, including dinosaurs. We as paleontologists in general, and book authors in particular, enjoy capitalizing on this phenomenon, and aspire to transfer that fascination to a broader understanding of evolutionary and geological processes. The particular audience aimed at here is college students taking an introductory course in the history of life.

There are several individuals who assisted with the completion of this project. First, I am especially grateful to my editor, Ian Francis, for his support. He helped shepherd this project not only through the good times but also through various travails big and small. Delia Sanford helped in many important ways, including coordinating the project. Camille Poire also assisted in this area, especially in 2009. Rosie Hayden and Harry Langford skillfully helped move the project into production. I am also very grateful to Tony Martin, Emory University, and several anonymous reviewers, who generously gave their time and advice by providing useful comments on earlier versions of this book. Finally, several individuals provided images used in the book. This was of significant assistance, and they are acknowledged in the relevant image captions.

BRUCE S. LIEBERMAN
University of Kansas

Chapter 1

Introduction to Fossils

Outline

- History, Science, and Historical Science
- Time, Life, and Stratigraphy
- What is a Fossil?
- How do Fossils Form?
- Conclusions: Fossils as Curious Stones
- Additional Reading

History, Science, and Historical Science

Paleontologists are those people who are fortunate enough to be able to study fossils for a living. George Gaylord Simpson was one of the most famous American paleontologists. He studied vertebrate fossils—those animals with backbones; among other things he specialized on the evolution of horses. In his work on the evolution of fossils through time, what we call the fossil record, Simpson was very much a historian—a historian of life on Earth.

Studying the History of Life

We often think of history as implying something distinctly human. For example, at a university the football or basketball program may have had a long history with many illustrious athletes associated with it, but because of the nature of universities (student's actually graduate in spite of some professor's best efforts), the teams change very year. The city in which the university is located also has a history,

Prehistoric Life: Evolution and the Fossil Record. 1st edition. By Bruce S. Lieberman and Roger Kaesler. Published 2010 by Blackwell Publishing.

and information on the changes that have taken place through time is available through archives of the local newspaper or perhaps made evident in plaques marking the sites of noteworthy past events. Civilizations have histories, too. Those that flourished in the ancient world are now extinct. Even their languages are no longer accessible to most of us, except perhaps in obscure academic corridors.

What Simpson knew and what every other paleontologist recognizes is that the Earth, too, has a history. Life has a history, and the history of the Earth and its life are topics worthy of our study and understanding. That is why we study fossils: to understand the long history of life. This history helps put what we see today, for instance, the current environment of the planet, the rich diversity of living organisms found in many different habitats, into that historical, evolutionary, and geological context. It also helps us to understand where our species comes from, and to make predictions about where we and the other organisms that occupy this planet are headed.

History

Simpson defined history as "configurational change through time, i.e., a sequence of real, individual but related events." These configurational changes have to do with changes in the state of the universe or any part of it through time. From our special point of view, those of interest to us are changes in the fossil record through time as organisms have evolved, been faced with environmental change, and coped—or failed to cope—with the forces of nature that at times in the past have caused the extinction of many kinds of organisms with which they shared their environments. Can we study in a scientific way these changes that have taken place through time? Is there such a thing as historical science? You can bet that the paleontologists think so. Their science is based on the idea of investigating the history of life in a scientific way.

Science

Simpson presented one of the most intriguing definitions of science that has ever been proposed. "Science" he wrote "is an exploration of the material universe that seeks natural, orderly relationships among observed phenomena and that is self-testing." His definition says it all. It captures the excitement of science and implies that the exploration is far from being completed. It makes the important point that science deals with the natural world (not the supernatural), and it captures the essence of science as a self-testing, self-correcting enterprise. This means that when a scientist presents an idea—an hypothesis—he or she no longer owns it, and he or she should feel no compulsion to defend it. Instead, all qualified scientists, including the one who proposed the idea in the first place, are obligated to test the idea and to seek to reject it. If they are unable to reject the idea, it may be accepted. The idea, however, is always subject to further testing and to possible rejection. This implies another important point about ideas in science. Ideas in science are never proved. Mathematicians prove their theorems, but scientists seek only to test the ideas and to reject them when it is possible to do so.

If an idea in science is broad enough in scope to explain many different kinds of phenomena and if it has been tested repeatedly and accepted, the idea can be regarded as a scientific theory or even as a law of science. Simpson defined a scientific law as "a recurrent, repeatable relationship between variables that is itself invariable to the extent that the factors affecting the relationship are explicit in the law." We shall come back to this idea later.

Historical science

Simpson defined historical science as "the determination of configurational sequences, their explanation, and the testing of such sequences and explanations." History typically deals with unique events. There has been only one American Civil War, only a single origin of life more than 3.5 billion years ago (3.5 Ga, for *giga annum*), only a single invasion of the land by primitive Devonian amphibians, only one ultimate extinction of the dinosaurs at the end of the Cretaceous. It is the historical aspects of paleontology that continue to fascinate so many people interested in the history of life.

Process and Pattern

The way scientists like paleontologists connect history with science is through the study of processes. Indeed, following Simpson's definition, a fundamental part of science is the search for orderly relationships: these are scientific processes that explain why certain phenomena occur. Further, the way scientists discover processes are from patterns: these patterns can be the results of an experiment or the distribution of species in the fossil record. That is, scientists seek to determine what processes are at work in nature from the patterns these processes produce—in the laboratory, in the stars, or in the fossil record preserved in ancient rock.

The way of going about this in the historical sciences, however, is different from the procedures used in the physical sciences, where research often takes place in the laboratory. Years ago, before we realized how dangerous to health metallic mercury can be, a standard experiment in beginning chemistry laboratory classes was to heat mercuric oxide (HgO, an orange powder) to about 600° C. at which temperature it decomposed into metallic mercury (Hg, a silvery liquid) and free oxygen O_2. A property of mercuric oxide is that at 600° C. it decomposes in this manner. It has always done so, and it will do so anywhere in the universe where conditions are appropriate because the laws of science are invariant in time and space. By conducting this experiment, the beginning chemistry student demonstrated the process of thermal decomposition by the pattern it produced: change of an orange powder into a silvery liquid and a colorless gas.

Experiments of this sort lie at the heart of the physical sciences. In historical science, however, all the experiments have already been run by nature in the distant past. The task of the historical scientist is to determine what experiments nature has conducted. It is much as if you were to venture into a chemistry laboratory at the end of the day, poke around in the sink and waste containers, and try to determine what experiments had occupied the students in the laboratory that day. You would still be determining process from pattern, but the pattern would be little blebs of mercury, the remains of students' playing about with free oxygen, and perhaps a broken test tube or two. The process that you would deduce is the same, but you would be establishing the historical, configurational fact that the chemistry course had just begun rather than making any profound discoveries about the chemical process of decomposition. Discovering the immanent properties and processes is the job of the physical scientist; applying the properties to understanding the process of history is what historical science is all about.

Fossils as a historical record

The early English scientist Robert Hooke (1635–1703) was a remarkable individual by any possible measure one could propose. Among his many accomplishments, he studied springs and invented the hairspring, which led later to the invention of wristwatches and to ship's chronometers, a discovery that made global navigation possible. While studying microscopically a thin slice of cork, he observed and named as cells the small compartments in the tissue, a term we still use for these structures today. We now know that all life is organized into cells, but it was Robert Hooke who started our thinking in this direction. From our special paleontological point of view, Hooke observed in 1688 that fossils could be used to record passing time. Evolution provides the process that explains the change in fossils through time that Hooke observed. It is interesting, therefore, that Hooke made this observation 171 years before Charles Darwin published his book *On the Origin of Species*.

As we shall see in some detail in the next section of this chapter, Hooke was absolutely correct. Fossils have recorded the chronicle of the history of life, and their study remains the best way we have of determining the past processes that have led to the patterns we see in the modern world.

Time's cycle and time's arrow

Conflicting adages abound. A common saying these days, "What goes around comes around," implies some sort of cyclical recurrence of events. The Greek philosopher Heraclitus (ca. 540 to ca. 480 BCE) said, on the other hand, "You could not step twice into the same rivers; for other waters are ever flowing on to you." Both are correct. Patterns in history and historical science can be thought of as comprising three components—trend, signal, and noise—the latter radio listeners refer to as static.

Trends, referred to as time's arrow by paleontologist and evolutionary biologist Stephen Jay Gould, are unidirectional changes, patterns caused by long-term processes. In history they include such things as the flow of rivers mentioned by Heraclitus, the progressive and now alarming increase of human population, increased speed of computers, the growing levels of carbon dioxide (CO_2) in the atmosphere, and since the beginning of the 20th century, the increased effectiveness of weaponry and the consequent increased potential lethality of wars. In paleontology trends include such patterns as increased complexity of the most complex forms of life, movement of vertebrates from an aquatic to a terrestrial mode of life, and increased diversity of life—the number of different kinds of organisms that have lived at any one time.

Note that we do not include evolutionary trends here. In the past, the prevailing concept of evolution was as a gradual change of the form of organisms directed over a long time. As we shall see in a later chapter, paleontologists no longer think evolution produces a pattern of gradual change.

Signals, referred to as time's cycle by Stephen Jay Gould, are events that recur in some more-or-less regular fashion. In our everyday lives, time's cycle is manifest as diurnal cycles, tides that ebb and flow, the phases of the Moon, and the changing seasons. In historical science signals from nature include such patterns as the waxing and waning of glaciers over intervals of tens of thousands of years and, some evidence suggests, the recurrence of events of mass extinction in which a large proportion of life is eradicated.

Now, what about noise? In this context noise is configuration, the very kind of thing with which the historical sciences deal most often. Noise includes such chance events as predation, extinction events that are not periodic or cyclical, and the evolution of new species from ancestral forms. Often a clearer understanding of events of the past can be gained if one is able to subtract out the trend and signal and focus on the noise as the stuff of history. It may seem a little funny to say so, but if we think of the history of life as being played on some sort of imaginary historical radio, as students of that history we are likely to be as interested in the static as in the music.

Time, Life, and Stratigraphy

Stratigraphy is the study of layered rock, most of which is conglomerate, sandstone, siltstone, shale, mudstone, limestone, or such evaporites as gypsum and rock salt. Geologists refer to all these types of rock as sedimentary rock, meaning that they were deposited by a fluid medium, either air or water. Paleontology is the study of fossils, nearly all of which occur in sedimentary rock. Because of the important roles organisms have played in the Earth's history and because fossils provide the only information we have about the history of life, you would be correct in supposing that stratigraphy, understanding geological time, and the history of life are all closely intertwined.

Some Principles

As is true of most historical science, only a very few principles underlie the study of layered rock and the fossils it contains. Three of these principles were recognized by Nicholas Steno (1631–1687), a Dane working in Florence, Italy, as physician to the Grand Duke of Tuscany. In 1667 Steno was one of the first to understand that fossils were the remains of past life that had been deposited with the sediment, although he supposed that fossils had been deposited by the biblical deluge or, in the case of fossil elephants, were the remains of Hannibal's army as it crossed the Alps to invade Rome in 218 BCE. Previously, many supposed that fossils grew in the rock under the influence of emanations from heaven or perhaps had been placed supernaturally into the rock as a test of religious faith.

Steno seems also to have been one of the first to understand that the Earth has a history that is worthy of study, and he applied his principles of stratigraphy in his attempts to understand the geological history of the area around Florence.

Superposition

If you are playing cards or tossing dirty laundry into a pile, you know that the first item played or tossed is the one at the bottom of the pile. As one moves up the stack or pile, the items were deposited more and more recently. This is all there is to the principle of superposition (Figure 1-1). Steno recognized that in any undisturbed sequence of sedimentary rock, the oldest is on the bottom and the youngest is on the top. It is not rocket science, but Steno's recognition of this fact grew from his recognition that the rocks have a history, and it allowed him to begin interpreting that history.

Original horizontality

Steno recognized a further principle—that sediment is deposited from a fluid medium—air or water—and that the upper surface of a layer of sediment should be horizontal if the layer has not been disturbed subsequently.

Figure 1.1 Principle of superposition. A layered sedimentary rock, where the unit labeled "1" is older than the unit labeled "2", and the unit labeled "2" is older than the unit labeled "3". Image from iStockPhoto.com © G. Clerk.

Original lateral continuity

Finally, Steno understood the principle that layers of sedimentary rock are laterally continuous so that a layer of rock cropping out on opposite sides of a river valley was once continuous across the valley. Today his principle tells us that layers of rock continue until they reach the edge of a basin of deposition or thin to zero thickness as the supply of sediment runs out.

Unconformities

All the evidence we have of the Earth's history comes from study of the rock. Where rock is not present, we have no means of grasping details of the Earth's history—other than to say that sediment was not deposited or that it eroded away subsequently. On the basis of his observations of rocks in the field, James Hutton (1726–1797), a Scot regarded widely as the founder of the science of geology, recognized that in a sequence of rock one often sees steeply inclined layers of rock underlying flat-lying or more gently inclined strata. Such a configuration marks an important event in the history of the Earth, a place where folding of the older rock took place before deposition of the younger rock (Figure 1-2). Hutton called the

(a)

(b)

Figure 1.2 **Examples of unconformities.** (a) Tilted sedimentary rocks and (b) flat-lying sedimentary rocks, overlain by much younger flat-lying sedimentary rocks; in each case there was a protracted interval of non deposition and erosion before the younger sediments were deposited. Images courtesy of A. Walton, University of Kansas.

surface separating the two an unconformity, which is now defined as a surface of erosion or nondeposition, usually the former, that separates younger strata from older rock. We now apply the term unconformity to any such surface, even if the rock above and below the unconformity are not folded at all. In his three-volume work *Theory of the Earth* Hutton developed his ideas of the long history of the Earth marked by numerous cycles and summarized succinctly his views on the great age of the Earth: "We find no vestige of a beginning—no prospect of an end."

Inclusions and cross-cutting relationships

You may not know how old your grandmother is, but you can be sure of two things: she is older than your mother, and your mother is older than you. Somewhere along the line in the development of geology this principle was applied to the rock, the idea that the sand grains that comprise a sandstone must pre-date the sandstone. Similarly, the chemicals that cement the sand grains together must be younger than the time of deposition of the sandy sediment. Finally, if one feature cuts across another, such as a fracture in the rock or a cave dissolved in limestone, it must be younger than the rock.

Faunal succession

William Smith (1769–1839) was a British surveyor and engineer involved in coal mining and in surveying and building canals. He observed that by taking note of the fossils in the rock and how one group of fossils succeeded another in layer after layer he could predict what layers of rock the canal builders and coal miners would encounter as they dug into the Earth. Because the characteristics of the rock have economic implications, Smith's discovery was of great practical importance. The reason for the succession of fossils, of course, is that layers of rock represent time, and organic evolution has brought about change in the fossils from layer to layer. Smith had no idea of evolution, having worked some 50 years before Charles Darwin presented his views on evolution, but he observed the result of evolution and put it to work.

Extinction

Thomas Jefferson adhered strictly to the idea of uniformity in nature. In 1799, while John Adams was Vice President, he published a description of *Megalonyx*, a large ground sloth (Figure 1-3), which he supposed to be a gigantic lion. Ground sloths are extinct, but Jefferson, as was true of most of his contemporaries, had no concept of extinction. He wrote, "Such is the economy of nature, that no instance can be produced, of her having permitted any one race of her animals to become extinct; of her having formed any link in her great work, so weak as to be broken." So confident in this principle was Jefferson that he instructed Lewis and Clark to be on the watch for living *Megalonyx* while on their expedition to open the American west.

In fact, three years before in 1796 Georges Cuvier presented the first irrefutable evidence of extinction among elephants, mammoths, and mastodons. Cuvier was a catastrophist and believed in worldwide extinction of nearly all animals, followed presumably by renewed creation of new forms. In spite of his catastrophic views, his recognition of extinction was an important step in the development of the use of fossils to compare the relative ages of rock units from widely separated areas.

Uniformitarianism and actualism

Charles Lyell (1797–1875), one of the pioneers of geology, was born the year James Hutton died. In his three-volume textbook, *Principles of Geology*, he developed his ideas on the uniformity of nature, a set of ideas that has since been termed *uniformitarianism*. The American paleontologist Stephen Jay Gould has shown that Lyell actually had two ideas in mind: related but logically distinct. The first of these

Figure 1.3 The giant ground sloth *Megalonyx*. Image by C. S. Prosser.

Gould labeled *substantive uniformitarianism*, the idea that rates of change have never in the past differed substantially from what they are at present. Lyell hoped to supplant the paroxysmal views of the catastrophists; the catastrophists envisaged the geological past as being marked by major catastrophic events rather than by the sort of slow change we see operating much of the time in the modern world. We now know that Lyell's view of rates of change as uniform and constant is simply wrong. Rates at which processes have operated have sometimes been radically different in the past. Glaciers have covered the northern half of North America: they swiftly advanced over North America and later they swiftly receded. In the Paleozoic Era mountains were folded up in the eastern part of the United States in at least three major episodes of tectonic activity that produced what are today called the Appalachian Mountains. From the catastrophists' perspective, it was hard to explain how such mountain belts might have been produced when the processes operating today seemed too slow and ineffectual to effect such vast changes. Substantive uniformitarianism has been tested and has been found to be false when applied in general: in this respect, the catastrophists were correct.

  The second aspect of uniformitariansim Gould labeled *methodological uniformitarianism*, the idea that the laws of nature are invariant in space and time. Lyell asserted this rule in the hope of establishing geology as a science in its own right; laws of nature could be discovered and invoked to explain geological phenomena: for example, the existence of a mountain range or a small river at the bottom of a deep river valley. This was in distinction to the catastrophist's who invoked supernatural intervention via catastrophes to explain such phenomena, for example, a catastrophic flood had produced the deep river valley. The idea that there are invariant natural laws is now a part of all science, including geology. However, this does not preclude the fact that major cataclysms can and do occur— the eruption of volcanoes, the occurrence of earthquakes of staggering intensity, or the in-crashing of a comet or meteorite. Such events, although catastrophic, are not catastrophism in the original sense because these catastrophes are not the result of supernatural miracles.

Facies

Facies is a Latin and French word that means aspect. The term was first applied in geology by Nicholas Steno, but the modern concept of facies of sedimentary rock was developed in 1838 by the Swiss geologist Armanz Gressly. To refer to the facies of a sedimentary rock unit is to refer to some distinctive aspect or feature of it. Perhaps it is sandy in one area and muddy in another, in which instance it could be said to have a sandy and a muddy facies. Perhaps one part of the rock unit was deposited near the shoreline (the nearshore facies) and another part was deposited farther out to sea (the offshore facies). Perhaps it has a fossiliferous facies and an unfossiliferous facies or a clam facies and a snail facies. The idea of facies is important in reconstructing the history of life because it brings home the idea that different environmental conditions exist in different places, often with profoundly different effects on the sediment and the organisms that live in and on them.

Geological Time

Understanding deep time—the immense age of the Earth—and subdividing geological time into manageable units that can be studied is one of the great intellectual achievements of Western Civilization. Paleontologists look at the ages of rocks and fossils in two ways: absolute time and relative time.

Absolute time

Absolute time is the term used when the age of a rock or fossil is expressed in years. Strictly speaking, it is not precisely absolute because absolute ages always involve some error, and even the best ages are expressed with an error term, such as 320 ± 3 Ma, which means that the best estimate of the age is 320 Ma but that the rock could be as old as 323 Ma or as young as 317 Ma.

The absolute age of certain rocks can be determined because of the phenomenon of radioactive decay, where one parent atom the rock contains transforms into a daughter atom at a constant, even rate (Figure 1-4). This graph implies the loss of a constant percentage of atoms through time. It can be converted to a semi-logarithmic plot by taking the logarithm of the values on the y-axis; applying such a conversion shows how the clock works in more detail (Figure 1-5). Here the straight line indicates a constant decay rate. At a certain amount of time half the atoms will be remaining; that time is called the half-life of the atom. If you know the decay rate of an atom, which can be determined in a lab, and you know how much of the parent and how much of the daughter atom is present in a rock (again, this can be determined in a lab) you can estimate the age of the rock. Some atoms, called radioisotopes, are better for determining the age of very old rocks. For instance,

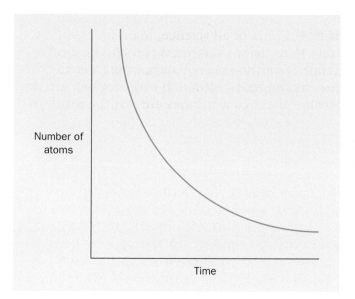

Figure 1.4 Radioactive decay. Diagram by M. Smith.

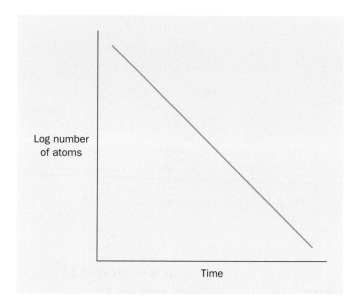

Figure 1.5 Semi-logarithmic plot of radioactive decay. Diagram by M. Smith.

potassium-40 decays to argon-40 with a half-life of 1.3 billion years: that is, after 1.3 billion years half of the potassium-40 originally in a rock will have decayed to argon-40. Because of its very long half-life, this would be an excellent radioisotope to use if you were trying to calculate the age of a very old rock. However, if you wanted to determine the age of a very young rock, or even a skeleton of a fossil human, it would not be a good isotope to use. This is because only a tiny amount of potassium-40 would have decayed out of the sample, and only a tiny amount of argon-40 would have been produced by the decay, such that these quantities would be very hard to measure accurately. For a particularly young volcanic rock, or a recent skeleton, one could use the radioistope carbon-14: it decays to nitrogen-14 with a half-life of 5,730 years. Such an isotope, while good for dating fairly young things, is not useful for determining the age of really old rocks, because all of the carbon-14 will have decayed away.

Radioactive dating methods have been successfully applied literally tens of thousands of times. One interesting application of these methods is when they are applied to determine the age of the oldest objects known on Earth. These oldest objects actually turn out to have originally came from outer space: they are meteorites and they typically are about 4.5 billion years old. These meteorites have a composition similar to that of the Earth, and they were the primordial objects in the solar system. They coalesced out of a swirling mass of dust and gas centered around a large object which was to become our Sun.

It is generally accepted that the Earth formed at around the same time as these meteorites. However, as of yet no rocks have been discovered on the Earth that are quite so old (although tiny crystals once part of rocks nearly this old may exist). Perhaps all of the rocks that were this old have subsequently been eroded away or they may still be present but are buried at great depths. It is more likely though that the Earth was mostly molten at this time and thus completely inhospitable to life as we know it. Still, the oldest Earth rocks are exceedingly old: they date from 3.9 billion years ago (Figure 1-6). These oldest Earth rocks are heavily metamorphosed and altered and thus could not contain any record of ancient life from quite so early in our planet's history, because any fossils or other organic matter present in the rocks would have been baked away. Still, there are sedimentary rocks (which are discussed in Chapter 2 and are the types of rocks that can potentially contain organic

Figure 1.6 **Very old rock**. A 3.8+ billion year old rock from Greenland. Image courtesy of Harald Furnes, University of Bergen, Norway.

remains) that are nearly as old as these rocks. These oldest sedimentary rocks, which date from around 3.8 billion years and hail from modern day Greenland, appear to contain evidence of organic activity. The implications of this will be explored more fully in Chapter 12, but it is worth recognizing that life had already evolved shortly after our Earth had solidified from its molten state.

Relative time

Most geological ages of rocks and fossils are given as relative ages. There are two principal ways of expressing relative ages. The simplest is the kind of relative ages we use in our everyday lives. You are younger than your parents. World War I occurred before World War II. The Victorian era came after the Elizabethan era. Paleontologists get this kind of information from such principles of stratigraphy as superposition, cross-cutting relationships, and inclusions. In a series of layered rock that has not been disturbed by subsequent geological activity, the oldest layer is on the bottom and the youngest is on the top.

Relative geological time is how we determine if a rock is older or younger than another rock when we lack an absolute age for the rocks. This determination can be accomplished through the use of fossils, recognizing the principle of faunal or biological succession which stipulates that life forms are unique to particular time intervals. Fossils principally occur in sedimentary rocks and these contain various facies (described earlier). Differences in the way a sedimentary rock looks, it feels, and yes, cvcn it tastes reflect differences in the type of environment it was deposited. Generally, the original environment will be under water, either seawater or freshwater, and conditions vary from place to place, or with the depth of the water, if there was a river nearby, etc. Because layers of rock tend to have a distinctive appearance, we can follow them for several miles. Following Steno's principle of superposition (also described earlier), an overlying layer of rock must be younger that the rock underlying it, at least in any one region. One might be tempted to think of any one layer of rock, for example, throughout the Grand Canyon, as representing a constant time line, but this is incorrect, especially if we were to follow a single layer over many miles as can be done in the Canyon. Why don't sedimentary rock layers represent constant time lines? Because any layer of rock represents a particular environment. Let's imagine the type of environment where some of the layers of rocks in the Grand Canyon were originally deposited: in an ocean somewhere near the shoreline at a beach (Figure 1-7).

At any time near this shoreline, depending on if one was looking nearer or further from the shore, there were actually several different types of environments on the seafloor where sediments were being deposited; each one of these places would eventually become one of the layers of rock seen in the Grand Canyon. Thus, to summarize, at any one time, several of the different rock layers we see today in places like the Grand Canyon were forming. Of course, the diagram shown is schematic, and the slope and spacing between the different rock types and the different sedimentary environments is exaggerated to make things clearer. However, the basic principle conveyed is still valid: sedimentary rock layers represent different environments, and at anyone time several environments and thus several future rock layers are being deposited.

In most marine settings like the one we've shown you, the level of the ocean tends to rise and fall through time. The result is that environments move back and forth, toward and away from the shoreline. Similar environments tend to end up being, much

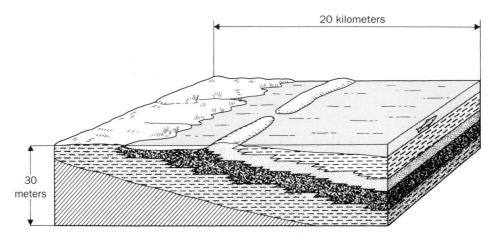

Figure 1.7 **Different environments correspond to different rock layers.** A modern beach environment showing that several different rock layers are being deposited at any one time. Diagram from R. S. Boardman, A. H. Cheetham, and A. J. Rowell (1987) *Fossil Invertebrates*, Blackwell Publishing; based on D. Eicher (1976) *Geologic Time*, 2nd edn, p. 43, Prentice Hall, Englewood Cliffs, NJ.

later, similar looking rocks. Thus, you cannot simply use the physical characteristics of a rock to say that this rock and that rock, because they look alike, were formed at the same time. There needs to be something to impart a direction or vector of change to such a cycling system in order to determine the relative age of a rock. This something is biology: the fossils provide a unidirectional vector or arrow of change within such cycles of rock because simply put life has evolved and a species can only evolve once. The idea that evolution is irreversible was codified as a law by the Belgian paleontologist Louis Dollo in the early part of the 20th century. His notion was not that there was some force or principle that prevented evolution from reversing but rather that organisms were so complex that it was extremely unlikely that the same type of organism could evolve more than once. The fact of evolution, discussed more fully in Chapter 4, and the fact that some animals and plants are preserved as fossils, provides us with that arrow we can use to tell relative time in the rock record.

Returning to the case of the Grand Canyon, to provide a concrete example, there are a set of trilobites that always occur in rocks of a certain age. They are found to cut across or occur in several of the individual rock layers in the Grand Canyon (Figure 1-8). They later are succeeded by a different set of trilobites that again cut across the rock units and define a later interval of time that can be recognized over great distances.

The Geological Time Scale

The more common and more useful way of dealing with relative ages of rock is to express ages in terms of the geological time scale, which is presented in Figure 1-9. Instead of saying, "This fossil is from 354 to 417 million years old," a paleontologist would say, "This fossil is Devonian." There are several reasons for this use of relative time. For one thing, the paleontologist rarely has direct knowledge of the absolute age of a fossil (unless it is a very young fossil and can be dated using carbon-14). Moreover, to express ages in terms of the geological time scale links the item to a great many other ages and events. To say, "This fossil is Devonian" is

(a)

(b)

Figure 1.8 Cambrian trilobites. (a) An Early Cambrian trilobite from Vermont, USA, in the collections of the American Natural History Museum. (b) Middle Cambrian trilobites from Antarctica, in the collections of the University of Kansas Museum of Invertebrate Paleontology. Images by B. S. Lieberman.

Time divisions			Began (Ma)
Cenozoic	Quaternary		
		Holocene	0.01
		Pleistocene	1.8
	Tertiary		
		Pliocene	5
		Miocene	24
		Oligocene	34
		Eocene	55
		Paleocene	65
Mesozoic	Cretaceous		144
	Jurassic		206
	Triassic		250
Paleozoic	Permian		290
	Carboniferous		354
	Devonian		417
	Silurian		443
	Ordovician		490
	Cambrian		543
Precambrian	Proterozoic Eon		2500
	Archean Eon		3800
	(age of Earth)		(4550)

Figure 1.9 Geological time scale. The major divisions are shown, and the numbers to the right approximate when the interval began. Diagram from R. Cowen (2005) *History of Life*, 4th Edn, Blackwell Publishing.

rather like saying, "This event occurred in Victorian times." Both statements link the fossil or event to a host of well-understood historical events that are different in nature, scope, and duration from those of other times in history.

The history of the development of the geological time scale has been laid out in a very interesting book by William B. N. Berry (see Additional Reading at the end of this chapter). The time scale has been developed over the past 180 years by geologists and paleontologists. The principles as to how to tell relative time in the geological record, when combined with the methods for determining absolute time, have allowed construction of a detailed geological time scale. In fact, the time scale was originally largely produced through the techniques of relative dating, but it was given expanded meaning with the addition of absolute dates.

Divisions within the geological time scale

The geological time scale is an encapsulated history of the Earth, and contains many intervals broken up by distinct boundaries which reflect major episodes of origination and extinction in the history of life. These episodes of origination and extinction can be thought of as the birth and death of various types of animals and plants. Thus, the time scale is not only a means of telling time in the geological record, but it actually represents an idealized history of life on Earth. If you worked for an advertising agency that was hired to create a one page ad summarizing life on Earth, to be distributed to a newly discovered alien civilization, you couldn't do any better than to show the geological time scale; if you wanted to be fancy about it you might also add some small images of the major animal and plant groups that were alive next to the various boundaries.

At any one place, it is not possible to find rocks that preserve the entire time scale; not even at places like the Grand Canyon where there is a lot of rock present and thus a lot of time preserved. However, because of the nature of the geological record, different parts of the time scale are present in different places, such that if we add up the rock from these places, globally, we can get a very complete picture of geological time.

It is worth noting that the geological time scale is hierarchically structured. For example, the time scale is broken up, at the largest scale, into three major eons. The Archean Eon and the Proterozoic Eon are often lumped informally into the Precambrian. The Archean Eon and the Proterozoic Eon combine a very large proportion of the history of the Earth (about 85%), but if we consider the geological time scale (Figure 1-9) there are far fewer boundaries during these time intervals than the succeeding, but far shorter Phanerozoic Eon. Archean is derived from the Greek word for ancient, bespeaking to the nature of these rocks, the oldest known from our planet; the Proterozoic succeeds the Archean, starting 2.5 billion years ago and ending 543 million years ago; the word translates in Greek, roughly, to early life. Fossils are quite rare in Archean rocks; these rocks typically have been intensively metamorphosed because they are so ancient; the fossils known from these rocks always represent the remains of microscopic bacteria. Proterozoic fossils are also scarce, though not as scarce as their Archean brethren, especially late in the Proterozoic. Again, just about all of the fossils known from the Proterozoic Eon consist of various microscopic life forms, principally bacteria, with a few important exceptions. We will discuss Archean and Proterozoic life more fully in Chapter 12, but in effect during these intervals there was not much going on evolutionarily in terms of large, multicellular life, though many evolutionary changes were transpiring among important and distinct bacterial lineages. In the absence of the origination, or extinction, of major fossil groups it is difficult to divide up geological time, and this is why these eons are not prominently subdivided.

The Phanerozoic, which began 543 million years ago, represents a prominent break from the Archean and Proterozoic. Its name is derived from the Greek words for visible life. This name encapsulates why there are many more divisions in the geological time scale during the third, but shortest, eon. A lot more is happening evolutionarily, especially when we consider the fossil remains of large, multicellular organisms. Abundant fossils of large organisms basically (with a few exceptions that will be considered in Chapter 12) do not appear in the record before 543 million years ago, at the start of the Phanerozoic (in fact at what is called the Cambrian Period, which we will discuss soon). During this eon there were many major origination and extinction events, first in animal life and later in plant life. For this reason the eon can be finely subdivided. For example, the Phanerozoic is further divided into three eras, the Paleozoic or time of ancient life, the Mesozoic or time of middle life, and the Cenozoic or time of new life. The boundaries separating the different eras represented times of major change in life on Earth. Between the Paleozoic and Mesozoic, maybe as many as 95 percent of all the species alive were wiped out. At the end of the Mesozoic the large terrestrial dinosaurs, along with many other species, went extinct; the mammals, the group we belong to, really started to become diverse in the Cenozoic. Within each of these eras there are also periods. Each period is also an interval of time defined by a characteristic set of animals, plants, or even microfossils.

The differences between the types of life forms present in any adjacent two periods is less than the difference between the life forms of adjacent eras, except in the case when period boundaries also correspond to era boundaries. For example, the boundary between the Permian and Triassic periods is equivalent to the boundary between the Paleozoic and Mesozoic. The boundary between the Cretaceous and Tertiary periods is equivalent to the boundary between the Mesozoic and Cenozoic eras.

The rocks that contain the fossils that were used to define the time interval representing any given geological period were originally defined by geologists in a relatively small region. For example, the rocks representing Cambrian time take their name from Cambria, the Latin name for Wales. There are rocks rich in Cambrian fossils in Wales, and this was a place where many early geological studies were conducted. Subsequently it was found that Cambrian fossils occur in rocks throughout the globe. There are now several places known to have a more complete record of Cambrian aged rocks then Wales.

It is very useful to learn the geological time scale because we will refer to the various intervals in it throughout the book. It is also very handy to learn because in a significant way this time scale corresponds to the history of life. Darwin is known for referring to the geological record as a book containing many chapters. Each one of the chapters or intervals in the geological time scale represents a time when a distinct set of life forms walked, crawled, or passively sat on the face of the Earth.

What is a Fossil?

A fossil is the remains or evidence of life from a previous geological time. That is a pretty loose definition, and it is intentionally so. What kinds of remains qualify as fossils? What kinds of evidence are sufficient? And how old does something have to be to be regarded as a fossil—that is, to be from a previous geological time?

Fine Tuning the Definition

In terms of our definition, there are two kinds of fossils. One comprises the actual remains of past life, typically the shells, bones, and teeth that occur in the rock. These are referred to usually simply as fossils, but they are more precisely termed body fossils. Paleontologists often refer to skeletal material as hard parts and to organic tissue as soft parts. They also refer to animals without mineralized skeletons as soft-bodied organisms, although in actuality the bodies of all organisms are soft. These terms, although jargon, are a useful way to refer to the skeletons and tissues of animals.

The tracks, trails, burrows, borings, nests, and coprolites (fossil excrement) of ancient organisms are called trace fossils, which are especially useful as a record of the behavior of ancient organisms, and we shall discuss these in greater detail in Chapter 2. Any remains of an organism and any sort of track or trail, even if the paleontologist cannot identify it or the plant or animal that made it, qualifies as a fossil if it was formed by a life form from the geological past.

Most fossils are very old, millions of years old at least and sometimes more than a billion years old, but technically any remains or evidence of life that is older than about 10,000 years is a fossil.

Far more important is the manner in which the object is studied. No matter how old an object is, if it is studied in the same manner as a very old fossil, as something that has come from rock or sediment, then it is often regarded as a fossil. Thus, because the frozen remains of the ice man found some years ago in the Alps are studied by the methods of biology and anthropology, most paleontologists would not regard the ice man as a fossil. On the other hand, a paleontologist interested in interpreting the interactions between organisms and their environments might study modern sea shells as if they were fossils, even if the organism that made them died only a few days ago. The term subfossil is sometimes used to refer to the remains of an organism that died recently if it is to be studied as if it were a fossil.

Kinds of Organisms Preserved as Fossils

The five principal kinds of body fossils are invertebrate fossils, vertebrate fossils, microfossils, fossil plants, and palynomorphs (fossil spores and pollen and thus both plant fossils and microfossils).

Invertebrate fossils

By far the greatest number of paleontologists are those who study invertebrate fossils, the remains of animals that do not have backbones (Figure 1-10). This is because these kinds of fossils are abundant and well preserved in many kinds of rocks; because they come from a wide variety of different kinds of organisms, many with long geological ranges; and because they are often preserved as whole organisms rather than as fragments. We shall deal with the many kinds of invertebrate fossils throughout the rest of this book.

Vertebrate fossils

The fossils of animals with backbones—the fishes, the amphibians, the various groups of reptiles, the birds, and the mammals—are termed vertebrate fossils (Figure 1-11). Their bones and teeth are hard and thus likely to be resistant to erosion, but they come apart readily so that most vertebrate fossils occur as isolated bones and teeth. In general, the more complete the fossil, the rarer its occurrence, but of course a vertebrate paleontologist gets a lot more information from a complete skeleton that has been preserved intact than from a few scattered bones and teeth.

Microfossils

Paleontologists use microscopes to study all kinds of fossils, even the bones of gigantic dinosaurs that may be more than a meter long and weigh many tens of kilograms. A microfossil is one that can be studied only with a microscope because it is so small. The most common microfossils are the remains of tiny, single-celled forms of algae, and many people are surprised to learn that even such tiny organisms sometimes secrete complicated mineralized skeletons that are often beautifully preserved (Figure 1-12). Those who study microfossils are called micropaleontologists, and many of them work for petroleum companies because, as we shall see, microfossils are useful for determining the ages of rocks, and they are likely to be the only kinds of fossils that are not ground up when oil wells are drilled.

Bacteria are preserved as fossils far more often than one would expect, perhaps because they are so abundant and hardy and because they alter their environments in such significant ways. The study of the interactions of bacteria with rocks and

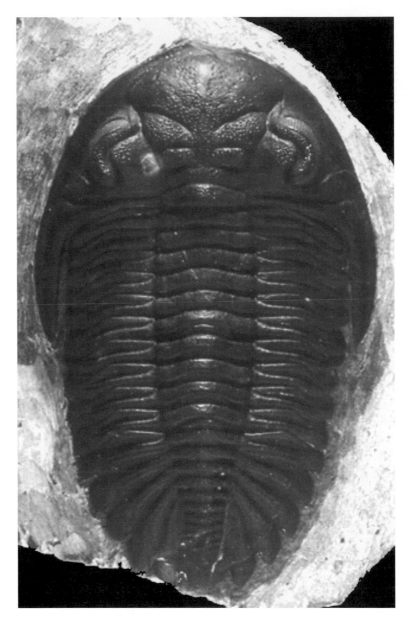

Figure 1.10 Invertebrate fossil. A Devonian trilobite from Morocco. Image by A. Modell.

Figure 1.11 Vertebrate fossil. A Cretaceous mosasaur. Drawing from R. Cowen (2005) *History of Life*, 4th Edn, Blackwell Publishing; after Merriam.

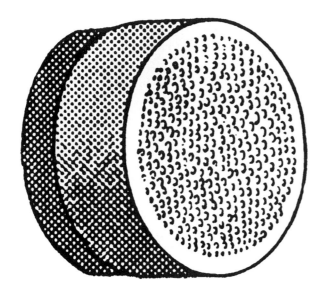

Figure 1.12 Microfossil. A diatom, diagram from R. S. Boardman, A. H. Cheetham, and A. J. Rowell (1987) *Fossil Invertebrates*, Blackwell Publishing.

sediment—termed geobiology—is one of the most promising and rapidly growing fields of geology.

Fossil plants

Most kinds of plants do not have any sort of mineralized skeleton, although some kinds of grass secrete tiny grains of opal that wreak havoc on the teeth of grazing animals. It is surprising, therefore, that the plants have such a good fossil record. In fact, the chances of any individual plant being fossilized are slim, but the abundance of plants in most environments ensures that when conditions are right plants will be fossilized (Figure 1-13). The sizes of tree trunks and the abundance of leaves also contribute to the abundance of the fossil record of plants. Large tree trunks can fossilize before they have time to rot; and if a great many leaves accumulate in an environment, they can actually alter the environment so as to enhance the likelihood of their fossilization. One surprising fact is that although vertebrate organisms and plants are interdependent in many ways, it is rather unusual to find both vertebrate fossils and fossil plants preserved together in the same kind of rock. The conditions necessary for their preservation are sufficiently different so that they are unlikely to occur together as fossils, even though we know that they generally lived together in the same places.

Palynomorphs

It is surprising that anything as tiny as a spore or pollen grain could be preserved in the rock for geologically long periods of time, especially since they have no mineralized covering of any sort. Nevertheless, fossil pollen grains are abundant enough (Figure 1-14) and resistant enough that a whole science has grown up around them, the science of palynology, and for this reason we distinguish fossil pollen grains from "fossil plants" even though they are of course parts of fossil plants. Different kinds of plants have different sorts of spores and pollen that can be distinguished by studying them with a microscope. They provide a lot of information about the evolutionary history of plants and, because plants are very picky about where they live, about ancient environments as well.

Figure 1.13 Fossil plant. Image from iStockPhoto.com © K. Jiang.

How do Fossils Form?

Paleontologists are often approached by people who have found fossils and would like to know more about them. One of the first questions they ask is, "Has it been fossilized?" Remember that a fossil is the remains or evidence of life from a previous geological time. All that has to happen, then, is that the remains have to persist for a long time: there is not a single process called fossilization. Anything that keeps the remains from being destroyed by geological processes is a type of fossilization.

Preservation of Original Material

All fossils have been modified to one degree or another from how they were when the organism was alive. It is just not possible to bury the remains of an organism deep in the Earth, perhaps for millions of years, without having some changes

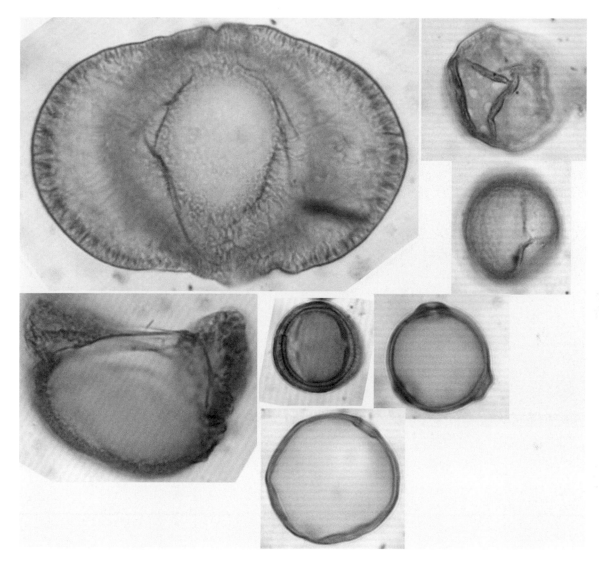

Figure 1.14 Fossil pollen. Pollen from the Miocene of the western United States. Image courtesy of O. Davis, University of Arizona.

occur. Some kinds of fossils, however, are much more stable than others and are remarkable for how little change they have undergone. Some kinds of fossils that have changed very little we shall discuss below in the section on special preservation. Here we deal with more common, less spectacular instances of preservation.

Mineralized skeletons of organisms—including both shells and bones—are more likely to be preserved with little alteration if they are chemically stable. The spicules of many sponges are siliceous, being made of the mineral opal—silicon dioxide with varying amounts of water included ($SiO_2.nH_2O$)—which is quite stable for long periods of time, even when buried deeply in the Earth. For this reason, sponge spicules are likely to be preserved with little change. The same is true of the skeletons of other kinds of organisms with siliceous skeletons, including the microfossil diatoms and radiolarians.

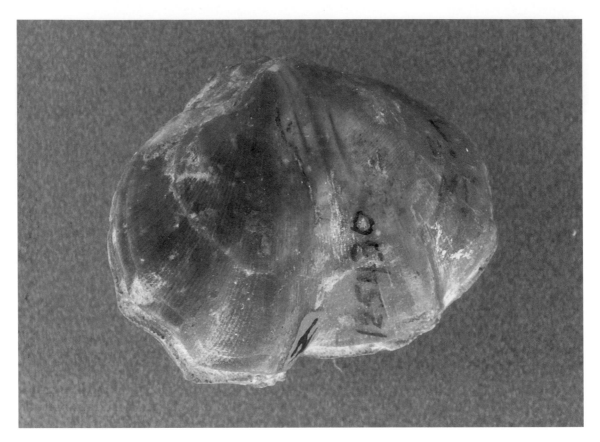

Figure 1.15 Fossil brachiopod. Fossil is from the Carboniferous of the central United States. Image by J. Counts; specimen in the collections of the University of Kansas Museum of Invertebrate Paleontology.

Among the most common Paleozoic fossils are the brachiopods (Figure 1-15), most of which secrete shells made of the mineral calcite—calcium carbonate ($CaCO_3$). Calcite is quite stable. The skeletons of brachiopods, moreover, are dense and contain very little magnesium (Mg), an element that is abundant in the calcite skeletons of many other kinds of organisms. Magnesium forms a very tiny ion that is easily removed from the calcite crystal lattice. As a result, it is not unusual to find fossil brachiopod shells that have almost the same chemical composition as that of living brachiopods. When a fossil shell hasn't been altered it tends to preserve many details.

Permineralization

Most skeletons are porous to one degree or another. Our bones are quite porous, the pores being occupied by soft tissue, and the shells of many kinds of invertebrates are also porous. When groundwater seeps through porous fossils after their burial, it typically deposits mineral matter into the pores, a process referred to as permineralization. The permineralizing material can be of the same composition as the skeleton, or it can be quite different.

Petrifaction

Petrifaction means literally to turn to stone. Its use implies that the substance being petrified must have started out without mineralized hard parts. That is, the organisms was probably soft bodied. The process of petrifaction involves typically the replacement of soft parts with mineral matter, usually silica in the form of microcrystalline quartz (SiO_2), calcite, or, more rarely, apatite—a calcium phosphate mineral with some other elements included, most notably fluorine [$Ca_5F(PO_4)_3$]. In some instances the preservation occurs on such a fine scale that individual cells are preserved, as occurs especially in some petrified wood.

Recrystallization

Recrystallization is a process of fossilization that interferes with good preservation. The most easily understood example is when shells made of one type of mineral recrystallize to a different type of mineral. This happens most frequently between two types of closely similar but not identical minerals, aragonite and calcite. Many invertebrate organisms secrete skeletons of the mineral aragonite; it has the same chemical composition as calcite ($CaCO_3$) but there are subtle differences in the molecular structure of these minerals. Most snails, clams, cephalopods, and corals from the Mesozoic and Cenozoic eras have skeletons made of aragonite. It turns out that the mineral aragonite is less stable than calcite, especially when it is removed from seawater. Further, molecular changes occur to aragonite (slowly perhaps but inevitably) when it is removed from seawater that cause it to be converted to calcite. That means that in the long term aragonite converts to calcite—which has the same chemical composition but a more stable crystal structure. As crystallization proceeds, much of the fine detail of the skeleton is lost, but the external form is likely to be preserved.

Casts and Molds

Sometimes a skeleton may dissolve before there is time for the replacement to occur. This can result in fossil-shaped openings in the rock that are called molds. When the opening is filled, the fossil is called a cast. A cast may have the same shape as the original fossil, but the composition is likely to be completely different and usually little of the fine detail is preserved.

Snails, which have aragonite shells, are especially likely to dissolve away. Sometimes the spiral shape of the inside of the snail shell is filled with calcite mud before the shell dissolves. The resultant shape, sometimes called a steinkern (a German word meaning stone kernel), is an internal mold—not a cast—because it is not a replica of the fossil but is a sort of negative, showing the shape of the interior of the shell.

Carbonization

Sometimes organisms are buried quickly before they have time to rot. As successive layers of sediment accumulate above them, they are buried deeper and deeper in the Earth. There, the heat of the Earth drives out all the volatile material, leaving behind a carbon film, often with excellent preservation of detail. Leaves are commonly preserved by carbonization. Animals are carbonized less frequently, but such fossils, when they occur, provide a unique glimpse into the ancient communities of the past, showing the nature of soft-bodied animals that have otherwise little likelihood of preservation. The most famous carbonized fossils are from the Burgess Shale of British Columbia, with which we shall deal in more detail in a later chapter.

Special Preservation

In the category *special preservation* we include instances of fossilization that preserve incredible detail, especially details about the soft parts of ancient organisms. Sometime, as we shall see, carbonization provides such detail, but here we have in mind kinds of preservation that take place under very unique circumstances.

Mummies

Mummies of Egyptian pharaohs preserved in the pyramids are not regarded usually as fossils, but the remains of ancient organisms are sometimes preserved in similar ways. Paleontologists think of mummies forming by having dried up before they have a chance to rot. Naturally, this sort of preservation is rare and can occur only under very dry conditions—in a desert or in a dry cave, perhaps. Mummies do not last very long, so no fossilized mummies are very old, and few of them qualify as true fossils. Climates change; caves collapse; bacteria invade and destroy the mummy. Nevertheless, when they occur they give a lot of information about the soft parts of ancient organisms.

Frozen mammoths

Freeze-drying is a special kind of mummification, but more spectacular are fossils that have frozen solid but not dried out. In 1900 some men hunting for fossil ivory from mammoth tusks in northern Siberia discovered a well-preserved mammoth embedded in the permafrost of a river bank (Figure 1-16). Such fossils, which are many thousands of years old and thus date from the Pleistocene Epoch, had been known for years, but this one was especially well preserved. Scientists visited the site, thawed the mammoth, cut it into manageable pieces, refroze it, and transported it to St Petersburg for study. In recent years scientists have proposed producing a living mammoth by collecting the DNA of frozen mammoths and combining it with elephant DNA. So far this project has not made much headway, but it lurks on the horizon as an interesting possibility.

Fossils in amber

Amber is fossil tree sap. Some kinds of trees, when damaged, ooze copious sap, an evolutionary defense mechanism that keeps insects from boring into the wood. Usually it is an insect that touches the sap and is entrapped, though sometimes lizards and even feathers can be trapped as well. More sap flows over it, and in time the bleb of sap may lose its volatile material, solidify, and become amber, with the fossil organism (or part of the organism) preserved inside it in exquisite detail. The oldest known amber is a tiny piece found in rock of the Midcontinent of North America that is nearly 300 million years old. Amber is more common in rock that is a few tens of millions of years old.

Figure 1.16 **Frozen mammoth**. This frozen carcass was discovered in Siberia in 1900. Image from R. Cowen (2005) *History of Life*, 4th Edn, Blackwell Publishing.

Phosphatization

Minerals rich in phosphate, especially calcium phosphate minerals, sometimes permeate the pore space in rock, forming phosphate nodules. In certain instances shells and even the soft tissues of organisms can be replaced. When this happens, the preservation can be remarkably good. Muscle fibers of fishes, larvae of invertebrates, and even individual bacteria can all be preserved by phosphatization.

Conclusions: Fossils as Curious Stones

The term *fossil* was originally applied to anything dug up from the earth. The word comes from the Latin verb *fodere*, meaning to dig. What we now regard as fossils—the remains or evidence of life from a previous geological time—were not yet recognized as being any different in nature from gemstones, ore, and other curiously shaped objects dug from the ground. One early attempt at a scientific explanation of rocks shaped like seashells—i.e., fossils—was that such shapes grew as a result of emanations from heaven—a *vis plastica* or plastic force that caused such things to form. Where the emanations impinged upon the water, marine animals grew. Emanations impinging on solid rock, however, caused nonliving shapes to grow in the rock. Of course we now realize that this early attempt was seriously inaccurate.

One of the things that is the most fun for a professional paleontologist is the opportunity to deal with members of the interested public who have unearthed curious stones. Sometimes the best and most important fossils found are discovered by so-called amateur paleontologists. They are called "amateur paleontologists"

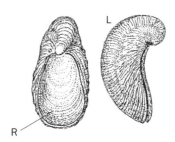

Figure 1.17 Devil's toenails?. Jurassic oyster fossils that were at one time mistakenly thought to be large toenails. Image from E. N. K. Clarkson (1998) *Invertebrate Palaeontology and Evolution*, 4th Edn, Blackwell Publishing; based on Zittel.

because they do not receive pay for their prospecting, yet their diligence and work ethic often exceeds those of the so-called professionals. Sometimes though, the finders of these objects attribute more interesting properties to such objects than their true nature merits. Rounded concretions in the rock have been interpreted as turtles or dinosaur eggs; stems of crinoids are mistaken for fossil sea serpents; in ancient times the heavy shells of fossil oysters were described as the devil's toenails (Figure 1-17). Once your textbook authors were asked to inspect an alleged fossil baby's head. Although the purveyor of this alleged fossil never showed up, it was likely to be simply a round rock with three holes—two eyes and a mouth—but one never knows just what curious stones might turn up.

Additional Reading

Berry, W. B. N. 1987. *Growth of a Prehistoric Time Scale*. Blackwell Scientific Publications, Boston, MA; 202 pp.

Cutler, A. 2003. *The Seashell on the Mountaintop*. Dutton, New York; 228 pp.

Gould, S. J. 1988. *Time's Arrow, Time's Cycle: Myth and Metaphor in the Discovery of Geological Time*. Harvard University Press, Cambridge, MA; 240 pp.

Repcheck, J. 2003. *The Man Who Found Time. James Hutton and the Discovery of the Earth's Antiquity*. Perseus Publishing, Cambridge, MA; 247 pp.

Rudwick, M. J. S. 1985. *The Meaning of Fossils*, 2nd edition. University of Chicago Press, Chicago, IL; 304 pp.

Rudwick, M. J. S. 2005. *Bursting the Limits of Time*. University of Chicago Press, Chicago, IL; 732 pp.

Rudwick, M. J. S. 2008. *Worlds before Adam*. University of Chicago Press, Chicago, IL; 800 pp.

Stanley, S. 2004. *Earth System History*. W. H. Freeman, San Francisco, CA. 567 pp.

Winchester, S. 2001. *The Map that Changed the World. William Smith and the Birth of Modern Geology*. HarperCollins Publishers, New York; 329 pp.

Chapter 2
The Nature of the Fossil Record

Outline

- Fossils in Sedimentary Rock
- Taphonomy
- Time Averaging
- Mode of Growth
- Colonial Organisms
- Trace Fossils
- Concluding Remarks
- Additional Reading

Fossils in Sedimentary Rock

In the previous chapter we discussed the means by which the remains of organisms become part of the fossil record. It should be clear that, even with so many different processes involved, the chances are slim that the remains of any one organism will be preserved. The net effect of this is that the fossil record is imperfect. Indeed, some kinds of organisms are often not preserved at all, especially those without mineralized skeletons. Further, the soft parts of most kinds of fossil organisms are not typically preserved, although we will deal with instances when this can happen in our discussion of the Burgess Shale in a later chapter. Among those organisms with hard parts, the likelihood of fossilization varies widely depending on several factors, including the kinds of minerals they are made of, the robustness of their skeleton, and the rapidity of burial. In this chapter we deal in greater detail with the imperfection of the fossil record. It turns out that the fossil record is biased in what is preserved, but it is this bias that makes interpretation of the fossil record both challenging and exciting.

Prehistoric Life: Evolution and the Fossil Record. 1st edition. By Bruce S. Lieberman and Roger Kaesler. Published 2010 by Blackwell Publishing.

The Nature of the Rock Record

One of the most important factors governing the quality of the fossil record is the nature of the rock in which the fossils are entombed. Most fossils are preserved in sedimentary rock, and sedimentary rock itself is replete with imperfections. Among other things, the sedimentary record is punctuated by unconformities, and as we shall see sedimentation itself is episodic so that in some environments rather long intervals of time can pass during which no sediment is deposited. If so, the plants and animals that lived at that time are unlikely to be buried at all and thus are unlikely to be preserved.

Kinds of Rock

Geologists classify rock according to mode of origin, texture, and mineralogical composition. The three kinds of rock according to mode of origin are igneous, metamorphic, and sedimentary rock. Igneous rock forms when liquid rock solidifies. An example is the igneous rock that forms from the cooling of lava erupted onto the surface of the Earth. Under these conditions, the lava cools quickly to form igneous rock with a fine-grained texture. Sometimes, magma remains buried deeply in the Earth where it is insulated from rapid cooling. As a result, its crystals have time to grow to large size as it cools gradually and solidifies, imparting a coarse-grained texture to this kind of igneous rock. Igneous rock, because it forms at initially extremely high temperatures, rarely if ever contains fossils. However, there are certain cases where a volcanic eruption releases a great quantity of ash that may blanket the environs for great distances around the volcano; this ash can trap and thereby preserve (sometimes beautifully if the ash is not too hot) fossils. This phenomenon is highlighted by the historical example of the eruption at Pompeii circa the year AD 79 where casts created by the subsequently solidified ash preserve human figures seeking shelter from the deadly ashfall.

Metamorphic rock forms when pre-existing igneous or sedimentary rock is put under conditions of great heat, pressure, or both, usually deep within the Earth. Different minerals have different ranges of temperature and pressure at which they are stable. As the conditions of temperature and pressure change, minerals may become unstable. As a result, the minerals or the chemical components of the minerals rearrange themselves to form new minerals that are stable under the new set of conditions. All these changes that take place to form metamorphic rock occur in the solid state and without any melting.

You can think of this process as occurring because the minerals change so as to relieve stress from pressure or temperature. Low-density minerals are transformed by great pressure into more dense minerals. Under conditions of high temperature minerals form that absorb heat as they develop from pre-existing minerals, again in the solid state and without any melting. Metamorphic rock that forms from pre-existing sedimentary rock—so-called metasedimentary rock—is more likely than igneous rock to be fossiliferous, but nevertheless a geologist could spend a lifetime studying metamorphic rock and never see a fossil.

Most fossils, of course, occur in sedimentary rocks, and to understand the fossil record you need a sound grasp of sedimentary rocks and how they form.

Table 2-1. Classification of sedimentary rock based on origin, texture, and composition

Siliciclastic (texture/grain size)	Rock type (origin)		
	Carbonate*	Evaporite*	Coal (extent of diagenesis)
Conglomerate (coarse) Sandstone Siltstone ↓ Shale and mudstone (fine)	Limestone Dolomite	Rock salt (halite) Anhydrite Gypsum	Peat (little) Lignite Bituminous ↓ Anthracite (great)

*Differentiated by mineral composition

A Classification of Sedimentary Rock

Sedimentary rocks can be classified into four broad categories according to their mode of formation and mineralogical composition: siliciclastic rock, carbonate rock, evaporites, and coal (Table 2-1). One of the quintessential aspects of nearly all sedimentary rocks is that they form under water.

Siliciclastic rock

A clastic rock is one that comprises primarily grains weathered and eroded from pre-existing rock. In a siliciclastic rock most of the mineral grains are quartz (SiO_2) or some other silicate mineral such as feldspar, some of the dark, dense, relatively unstable calcium- or magnesium-rich minerals, or some of the clay minerals. A classification of siliciclastic rock from coarsest to finest texture is conglomerate, sandstone, siltstone, and shale or mudstone (Figure 2-1; Table 2-1).

Siliciclastic rock is the end product of five sedimentary processes. Weathering, either chemical or physical, loosens mineral grains from pre-existing rock, breaks grains, and converts unstable minerals into stable ones. Erosion moves away the weathered grains. Transportation carries the grains to the area where they will be deposited ultimately. During transportation more weathering can occur, and grains may become rounded and sorted into different sizes. Deposition typically occurs when grains settle to the bottom of some column of water . Sometimes they settle into distinctive layers that can provide information about the environment in which they were deposited. Diagenesis comprises all those processes that change the sediment after it has been deposited. These include further weathering, cementation, dissolution of some unstable grains, compaction of fine-grained sediment to force out the water and form shale or mudstone, and a host of other processes. All five sedimentary processes also act on the remains of organisms and affect the fossil record and are relevant for our discussion of taphonomy, provided below; further, the processes that form fossils, which we discussed in the previous chapter, are all diagenetic processes.

The effect of the five sedimentary processes is to increase what is referred to as the maturity of the sediment. A mature sandstone or siltstone is one in which the grains are well rounded and well sorted for size and in which the grains are predominantly quartz, a very stable mineral. During the maturing process, the unstable minerals are converted by weathering to clay, which is fine grained and is carried away in suspension to be deposited in quieter water. Shale is also a mature siliciclastic rock when it is well sorted and consists only of chemically stable clay minerals, but it is much finer grained than sandstone or siltstone and is deposited only in quiet water.

(a)

(b)

Figure 2.1 **Different kinds of siliciclastic rocks**. (a) A conglomerate. (b) A sandstone (the thick unit) with a few thinner siltstone beds (towards the top). (c) A large cliff primarily made of sandstone (the thick and thin units) with some thinner interbedded siltstones. (d) Interbedded siltstones (the thicker units) and shales (the thinner units). (e) A shale (in the bottom half of the outcrop). Images from A. Walton, University of Kansas.

(c)

(d)

Figure 2.1 *continued*

(e)

Figure 2.1 *continued*

Carbonate rock

The two principal kinds of carbonate rock are limestone and dolomite (Figure 2-2). Limestone is made primarily of the minerals calcite and aragonite, both of which are calcium carbonate ($CaCO_3$). The rock type dolomite is made primarily of the mineral dolomite, similar to calcite but with a lot of magnesium in the crystal structure as well: $CaMg(CO_3)_2$. Most limestone forms from the accumulation of the shells of marine organisms. Sometimes the shells that comprise the limestone are the tiny skeletons of microfossils, in which case the limestone is very fine grained. The skeletons of large plants and animals may disintegrate into tiny calcite or aragonite needles, again producing a fine-grained limestone. In other instances large shells accumulate, but the large shells are likely to be broken up by wave action and subsequent compaction of the limestone. The spaces between the fossil fragments, called pores, are typically filled with calcium carbonate cement that is produced by processes of chemical change either without the aid of organisms or perhaps mediated by bacteria living in the tiny spaces.

Some limestone, especially freshwater limestone, and some dolomite are precipitated inorganically directly from the water. If so, the deposit is likely to be fine grained and rather homogeneous throughout with perhaps only a few fossils present. Some rock may start out as limestone, but then if groundwater seeps through the rock it can be converted to dolomite. Similarly, aragonite is converted to calcite by the action of groundwater. When this happens, it obliterates the fine detail of the fossils preserved in the rocks and sometimes removes the fossils altogether.

(a)

(b)

Figure 2.2 Limestones. Images from A. Walton, University of Kansas.

Figure 2.3 Halite crystals. Image from iStockPhoto.com © F. M. Catalin.

Evaporites

Seawater, and some freshwater, has a lot of mineral matter dissolved in it. When the water evaporates, the minerals precipitate from the water and form an evaporite deposit. Evaporites are typically composed of minerals called salts, including gypsum—hydrated calcium sulfate $Ca(H_2O)_2(SO_4)$, anhydrite ($CaSO_4$), and halite—and sodium chloride (table salt; $NaCl$), although a great many other kinds of evaporite minerals are known (Figure 2-3). Because the harsh environments in which they form are not conducive to teeming life, evaporites are not typically highly fossiliferous.

Coal

Coal is sometimes referred to as the rock that is not a rock (Figure 2-4). This is because the definition of the term rock specifies that a rock is made of minerals, which are substances that can occur only in inorganic nature. Coal is made of plant material that has accumulated in a swampy environment where it is protected from destruction by decay; it is not made of minerals. But if coal is not to be regarded as a rock, what could it be? There is a lesson here about humans and our urge to classify things—to tidy up by putting everything into pigeonholes—and nature's seeming urge to thwart us. One of the worst traps scientists can fall into is to expect their classifications to mirror exactly what nature has done, or for everything nature has done to fit into their scheme of classification. In 1924 the great geologist A. W. Grabau prepared a classification of sedimentary rock in which snow, composed

Figure 2.4 Coal. Image from iStockPhoto.com.

entirely of the mineral ice, was included as a sedimentary rock and labeled an atmolith. Strictly speaking, Grabau was right. Ice is a mineral, and snow, being composed of minerals, is a monomineralic rock. Little is gained, however, in the way of expanded knowledge or our ability to discuss geological phenomena by regarding snow as rock. Grabau's classification was simply an instance of trying to tidy up nature.

Coal is a substance that forms in several steps. The first step is the accumulation of peat. Peat is unconsolidated, semi-carbonized remains of plant material in which the structure of the plant material may still be present and the water content is very high, usually as great as 75 percent. The next step is the formation of lignite, a soft, brownish-black coal. As diagenesis proceeds more and more of the water is pressed out of the coal. Bituminous coal is the most common kind of coal, which is especially abundant in Carboniferous deposits. (The name of the Carboniferous Period means carbon or coal bearing.) Anthracite is the kind of coal containing plant material that is most altered from its original form. Anthracite forms only under conditions of high pressure, and based on the classification of rocks already introduced it is often regarded as metamorphic coal. Anthracite is highly prized for the heat it produces, but it is much less common than other types of coal.

Taphonomy

We begin our investigation into the preservation of fossils and the bias in the fossil record by considering taphonomy. The term comes from two Greek roots: *taphos*, meaning grave, and *nomos* meaning laws, knowledge, or study. Strictly speaking, taphonomy is the rules of the grave.

A Definition of Taphonomy

Taphonomy is best described as comprising all the things that can happen to the remains of an organism as it moves from the biosphere into the lithosphere, that is, from the time it is alive to the time when it becomes a part of a rock and is perhaps discovered ultimately by a paleontologist. As used originally the term dealt primarily with the information lost during the processes of fossilization. Taphonomy was seen as an impediment to our interpreting the fossil record. There is much more to it, however, than this rather negative view. Since taphonomy comprises the things that happened to the remains of an organism, it follows that careful study of taphonomy can provide a lot of information about the environments in which the ancient organisms lived, died, and were preserved. Specifically, the shapes and original compositions of most organisms (their shells, bones, leaves, etc.) are controlled by the genetics of the organisms that formed them. The forces of nature—taphonomic forces—may change the shape and composition of the remains of organisms. Because we know the original nature of the fossil material and understand the taphonomic forces that operate today, we can use our knowledge to help determine what was going on in the ancient environment. We deal more with this idea in a later section of this chapter.

Components of Taphonomy

Taphonomy comprises three kinds of events that change the remains of ancient organisms. These three processes sometimes overlap so that their boundaries are not as clear as we may wish, but they include all the things that can happen in the sedimentary environment to the remains of an ancient organism. They do not include metamorphism.

Necrolysis

As is true of taphonomy itself, necrolysis comes from two Greek roots. The first, *nekros*, means a dead body; *lyo* means to break up. Necrolysis is all the things that happen to the remains of an organism while it is dying. It includes damage done during predation, including drilling or crushing by predators, chewing, and, if the organism is consumed, perhaps its complete destruction during digestion. Most organisms that are eaten by predators leave behind no trace. Consider your own diet, for example. Where are the soup and salad you may have had yesterday for lunch? How likely is either to make it into the fossil record? Although we should say that of course there are cases where fossil dung, called coprolite is preserved, and coprolites can provide important information about an animal's diet (but alas it is not always possible to figure out which type of animal did the eating and which type of plant or animal was eaten).

By studying necrolysis we can sometimes determine the cause of a fossil's death. Perhaps it was drilled by a predatory snail (Figure 2-5) or attacked by a shell-crushing fish. Crabs sometimes attack snails, peeling their shells away much as we would peal an orange to get to the edible part. The organisms may have been buried alive by a sudden sedimentary event, perhaps due to a storm that moved layers of sediment into its habitat. Maybe nothing happened. Some organisms die of old age or disease, and when this happens no amount of study is likely to reveal the cause of death. It is unfortunately true that in most instances determining the specific cause

Figure 2.5 **Predation**. A clam drilled by a predatory snail. Image from J. Hendricks, San José State University.

of death is impossible: a great many different causes of death can produce the same pattern in the fossil record.

Biostratinomy

Once an organism is dead and the predator that killed it has finished with it, the remains—shells, bones, and soft parts—continue to be a part of the sedimentary environment where they are acted upon by all the sedimentary processes except diagenesis, which we shall consider later as a separate process. Biostratinomy combines three roots: *bios*, Greek for life; *stratum* meaning bed or layer in Latin; and again *nomos*. It includes all the things that can happen to the remains of an organism once it is dead and before its ultimate burial, that is, while it is behaving as a sedimentary particle and is in the process of being added to the sedimentary record. In short, the remains of an organism can be attacked by scavengers, weathered, eroded, transported, and deposited.

Biostratinomy includes the decay of soft parts by bacteria and their destruction by such scavengers as nematode worms and crabs. Shells lying about on the surface are often colonized by boring sponges, which are not predators but use the shell as a site for attachment and for protection, boring their way into the empty shell in the process. [To say that all sponges are boring in one way or another is an old paleontologists' pun. Try to resist saying it!]

Other shells or shell fragments may be covered by marine algae that secrete many layers of the mineral calcite, sometimes entirely coating the shell. Wave and current action roll the shells around, causing them to become abraded. In the process, fragile shells may be broken, small or immature shells may be destroyed or carried away to be deposited in a different area, and shells may be moved into a more stable position, for example turned with their convex sides up or stacked like shingles so that the current can no longer move them.

Study of biostratinomy can tell us a lot about the ancient sedimentary environment in which an organism lived. It provides clues as to the amount of wave and current energy, the rate of sedimentation, and even the abundance of scavengers. A great deal of research has been done in this area of taphonomy, but paleontologists have only begun to scratch the surface in their understanding of the interactions of organisms, the remains of organisms, and the sediments in which they lived.

Diagenesis

Diagenesis is one of the five sedimentary processes. Strictly speaking, the term refers to all the changes that occur to a sediment or fossil after it is first deposited, and as it resides in the geological record. These are apart (in Greek, *dia*) from the origin of the organism. Here we treat diagenesis as a group of taphonomic processes separate from biostratinomy because it involves things that happen to the remains of an organism after burial.

Diagenesis can involve either physical or chemical changes. When mud is deposited, it may contain as much as 75 percent water. As the mud is compacted into shale or mudstone, the water is squeezed out, causing the sediment to lose much of its volume. As this dewatering continues, shells deposited with the mud may be crushed or distorted by compaction (Figure 2-6). As groundwater seeps through the sediment, it converts aragonite to calcite, it could also replace calcium carbonate shells with silica (SiO_2, the same composition as quartz), fill the space between shells where the animal lived with sediment or cement, and perhaps dissolve the shells altogether. The common means of forming fossils that we discussed in the previous chapter all occur during diagenesis: permineralization, petrifaction, recrystallization, formation of casts and molds, and carbonization.

Fuzzy boundaries

We mentioned earlier that the boundaries between the three taphonomic processes are not always distinct. This is true in large part because bones, shells, leaves, etc., may be buried, acted upon by diagenetic processes, and then brought again to the surface by wave and current action or by the action of burrowing organisms. When this happens, biostratinomic processes take over again, even though diagenesis may have begun, and the processes of burial, diagenesis, exhumation, biostratinomy, and reburial can be repeated time and time again.

Taphonomy—Implications and Parting Words

With this background, let us consider a few examples of the information that is lost due to taphonomy and the information that we can gain from studying it.

Figure 2.6 A distorted trilobite. A Cambrian trilobite from India. Image from N. Hughes, University of California, Riverside.

Taphonomy as information loss

Taphonomy has been described as a veil that conceals a great deal about ancient environments and the organisms that lived in them. In this sense it results in loss of information because some organisms have been removed by taphonomic processes and are, thus, not available for study. Soft-bodied organisms, especially those that have been attacked by predators, are unlikely to show up in the fossil record, no matter how abundant they may have been when alive. Similarly, aragonitic shells may be dissolved, and other kinds of shells may be destroyed during compaction of fine-grained sediment.

Taphonomy has also been regarded as an overprint. In this sense it is analogous to using the same piece of paper to print two different pages of a term paper. Overprinting results when fossilized remains of organisms are preserved together, even though the organisms themselves may have lived in different places or at different times and never had any possibility of interacting in life. This is much like what would happen if telephone books from two different cities in different area codes were to be merged. The resulting book would be twice as large, but it would not contain twice as much useful information. Chaos would result if someone tried to use it literally to contact people in the two cities.

Taphonomy as information gain

Besides the negative view of taphonomy as information loss, there is a positive view that is equally valid because taphonomy can provide a lot of useful information about biostratinomic processes and diagenesis and, to some extent, necrolysis as well. This is true in large part because we know what the shapes of ancient organisms were when they were alive, and we understand what minerals different major groups of organisms secrete as they build their mineralized skeletons. Thus, through careful comparison of the shapes of modern organisms or very well-preserved fossils, a paleontologist can tell if a fossil shell of the same species or a closely similar one has been transported before ultimate burial. The transported shell is likely to have had its spines broken off and perhaps various nodes, corners, and projections worn down.

In the marine environment many kinds of organisms, called epibionts, live attached to hard surfaces. Other kinds of organisms, such endobionts as the boring sponges we discussed earlier, penetrate hard surfaces. A natural source of such hard surfaces, of course, is the shells of other organisms, whether alive or dead (Figure 2-7). If a fossil clam shell is found that has no other organisms attached to it, a reasonable hypothesis is that it was an infaunal clam, that is, one that burrowed into the sediment and lived where no epibionts could survive; another reasonable hypothesis is that the clam could have been free of epibionts in life and upon death was buried very rapidly. If the outer surface of a fossil clam is covered with epibionts, a reasonable interpretation is that the clam lived on the surface in an area with a lot of nutrients in the water on which the epibionts could feed. Sometimes, however, the inside of a shell has been colonized by epibionts. This can happen only after death when the inside of the shell is exposed, and it suggests that the clam died and lay on the surface of the sediment for a while before burial, during which time the larvae of the epibionts found the clam and colonized it. Endobionts wreak havoc on shells. If a shell is exposed on the seafloor for a long interval of time, it can be shot through with the borings of endobionts and finally reduced to an unrecognizable shell fragment.

Time Averaging

Time averaging is the process whereby organisms that did not live at precisely the same time come to reside together in the same fossil deposit such that it is impossible to distinguish their ages. They may lie on the same bed and appear to be

Figure 2.7 **Encrusting epibionts**. A sponge, the light patch in the center of the image, growing over the dead skeleton of another organism. Image from J. Watanabe, Hopkins Marine Station, Stanford University.

of precisely the same age, yet they are not. This phenomenon is exceptionally common in the fossil record. It is something nature does to the fossil record that can make the job of the paleontologist more difficult.

Exhumed Clams from Sonora

One of the most vivid examples of time averaging comes from the work of the paleontologist Karl Flessa and his students. In their study of clams from modern environments along the Sonoran coast in Mexico, they used radiocarbon dating to determine the ages of clam shells on the beach. Radiocarbon dating, sometimes referred to as carbon-14 dating, is a means of determining the time in years since the clam died. Flessa and his students were not surprised to find clams on the beach that had died very recently. They were surprised, however, when some of the clams on beaches that were equally well preserved, and looked indistinguishable from the recently dead clams, had in fact sometimes been dead for more than 3,000 years. What happened, no doubt, is that the clams died, were buried, and then were brought again to the surface either by the activities of burrowing organisms (termed bioturbation) or by storm waves that removed the sand and mud and left the coarse shells behind.

Because of the limits of radiocarbon dating described in Chapter 1, Flessa's technique can be applied only to relatively young fossils. Still, if Flessa's results are generally applicable, and there is no reason to suppose that they are anomalous, then they have important implications for the work of paleontologists. They suggest that two fossils found in the same place in the same layer of rock may have been separated in life by thousands of years.

Snapshots and Time Exposures

What does time averaging mean for the study of fossils? Well, for one thing, it implies that there are some kinds of biological processes that are often hard to study in the fossil record. For instance, a biologist studying living organisms in the ocean can establish easily whether two animals are alive at the same time and thus determine whether they were potentially interacting with one another ecologically. For instance, one of the animals might typically prey on the other one, or it might prevent it from occupying the region during certain times of the day. Because of time averaging, it is not always possible to consider these issues with fossil deposits. Imagine a limestone bed packed with fossils that is only a few centimeters thick. If time averaging was prominent in this bed, it could make it hard for a paleontologist to determine which of the individual organisms were alive at the same time. This in turn could make it hard to determine which organism was preying on which one and what other types of ecological interactions might have been occurring between various animals.

It is appropriate, therefore, to compare the work of a biologist and that of a paleontologist by using an analogy from photography. The biologist works with the living fauna, which is like working with a snapshot. A snapshot tells you precisely what was going on at some time for a fraction of a second. It does not tell you what happened in the past. [Consider your driver's license photograph. For one fleeting moment you actually looked like that, and the camera recorded it.]

The paleontologist, on the other hand, when he or she deals with time-averaged samples, in effect works with a time exposure. Details are smeared out. However, sometimes this can be quite advantageous. For example, the paleontologist can get an idea of change through time and a sense of things that went on at different times in the past in a way that a biologist usually cannot.

The net effect is that there are a lot of things a biologist can learn from study of the modern fauna that a paleontologist cannot learn about the past. On the other hand, the paleontologist can deal with long-term, historical change that is beyond the reach of a biologist, although the paleontologist cannot consider his or her samples with the same precision as the biologist.

Mode of Growth

An important aspect of the fossil record, one that influences the kind of information available from fossils and the nature of the fossil record, is the mode of growth of the organism. Different major groups of organisms have radically different modes of growth. Some grow by adding layers to their shells; others grow by addition of skeletal components or by molting; and some modify existing parts. Some organisms are solitary, as is true of humans; others live in colonies. However, we mentioned earlier that nature cannot always be fit into the neat pigeonholes that correspond to the way we humans like to group things. Thus, it should not surprise you to learn that some kinds of organisms employ a combination of modes of growth.

Accretion

One of the simplest and most common modes of growth is by accretion. Organisms that grow in this manner do so by adding layers to existing skeletal components, much as a tree adds a growth ring every year. Animals that grow by accretion secrete a tiny skeleton early in life, and as the animal grows its shell keeps up by the addition of very thin layers. Some kinds of organisms add a layer every day, and others do so less frequently or at irregular intervals.

Corals, brachiopods, and molluscs (snails, clams, and cephalopods) grow by accretion. Corals are among those kinds of organisms that add a single, very thin layer each day. Corals that live in areas with seasonal temperature changes also have annual cycles so that daily layers secreted at some times of the year are more closely packed together than those formed at other times of the year. The paleontologist John Wells, by counting the number of daily layers in an annual cycle, was able to determine that in the Devonian Period there were more than 400 days in a year instead of 365 as we have today. This is the result of the slowing of the Earth's rate of rotation due to drag imposed by the Moon's gravitational attraction for the Earth, something that physicists predicted but were not able to test directly: as paleontologists we find it intriguing (and kind of cool!) that knowing about the mode of growth of a 400 million year old fossil coral helped scientists directly study the interactions between the Earth and the Moon over hundreds of millions of years.

Clams secrete thin layers, too, and some kinds of clam shells are marked by the presence of major growth ridges. The significance of these ridges varies. Some are annual growth lines; some are monthly growth lines; some are the result of the clam's being stressed by storms or other environmental perturbations; and some have unknown causes. To interpret them correctly and not be led astray, you have to know your clams pretty well.

Growth by accretion provides two important advantages to an organism. The organism is always protected by a shell, and its development (termed its ontogeny) is never disrupted. Growth goes on unimpeded and typically uninterrupted. A principal disadvantage is that this mode of growth provides the organism with very limited opportunity for making a radical change in shape as it grows. Adults are, thus, likely to be shaped very much like immature forms. Perhaps of greater importance, an organism that grows by accretion must carry around with it all the shell that it has ever secreted. A snail, for example, has to carry its twisted, spiral shell, which it continues to occupy, even though carrying extra skeletal minerals seems as if it would waste a lot of energy. Most snails have no mechanism for getting rid of shell material that was secreted earlier in life.

Addition

An altogether different mode of growth is growth by addition of skeletal elements. This is the mode that is characteristic of most echinoderms: the starfishes, echinoids (sea urchins), and crinoids (Figure 2-8). The stem of a crinoid is a stalk-like appendage that allows the crinoid to feed high in the water instead of being restricted to feeding at the sediment–water interface. The stem comprises a number of discs stacked like poker chips. As the crinoid ages, it adds new discs to the top of

(a)

(b)

Figure 2.8 Crinoids. (a) Living and (b) fossil crinoids. Images from T. Baumiller, University of Michigan.

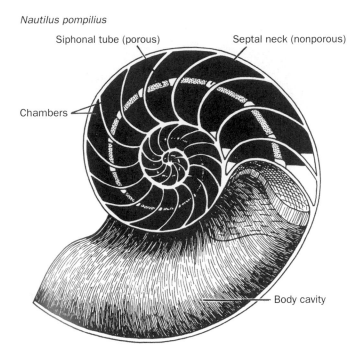

Nautilus pompilius

Siphonal tube (porous)

Septal neck (nonporous)

Chambers

Body cavity

Figure 2.9 **Nautiloid**. A cross-section through the shell of a modern *Nautilus*. Diagram from E. N. K. Clarkson (1998) *Invertebrate Palaeontology and Evolution*, 4th Edn, Blackwell Publishing; modified from Denton and Gilpin-Brown.

the stem just below the bud it supports, which is called a calyx. It is the combination of the stalk and the calyx that gives the crinoids their common name: the sea lilies. As a crinoid or other kind of echinoderm grows, it has to modify the shapes of its skeletal components so that they all fit together, a process we will discuss later under the heading **Modification**.

As we mentioned above, some kinds of animals employ several different modes of growth. Consider the ammonoids and nautiloids (Figure 2-9), which are molluscs that are closely related to squids. They add chambers to their skeletons from time to time and they do this by accretion. However, the formation of a chamber is a discrete event that can be regarded also as a form of growth by addition. Another good example comes from the arthropods, which includes the insects, crabs, lobsters, centipedes, and trilobites (to name only a few); they are in many respects characterized by an entirely different mode of growth (see **Molting** below). Nevertheless, as they grow many kinds of arthropods add segments to their body at each growth stage, which is also a way of growing by addition.

The advantages and disadvantages of growth by addition are the same as those for growth by accretion. The animal is protected at all times by a skeleton, and its growth can proceed unimpeded, although in a stepwise manner. There seems to be no mechanism for getting rid of unused skeletal elements, however, so the echinoderm has to carry around parts of its skeleton that may be no longer functional.

Molting

Molting is an important mode of growth common to arthropods, the most diverse animal phylum, and also several closely related phyla. In arthropods molting results in growth in a series of discrete step-like jumps. First they shed the outer hard layer

Figure 2.10 Arthropod. A soft-shelled crab that has recently molted. Image from iStockPhoto.com © J. Gecewicz.

of the external skeleton, revealing a softer skeleton underneath. When the hard exoskeleton is gone, the animal increases rapidly in size, typically nearly doubling its volume; then the exoskeleton hardens. The animal is locked into the new larger size by its hard shell and then no growth takes place until the next molting.

Have you ever eaten soft-shelled crabs? A soft-shelled crab is not a different kind of crab (Figure 2-10). It is a crab that has molted and does not yet have a hardened exoskeleton. At this stage of its growth it is quite vulnerable to such predators as fishes, other arthropods, and hungry people. Moreover, without a rigid skeleton, its muscles do not work very well, preventing it from escaping predators. These might be seen as disadvantages to molting; however, there are advantages as well. One advantage is that molting allows an organism to radically change its morphology from one growth stage to another: consider the changes in a butterfly from caterpillar to winged adult.

Modification

How do humans grow? A newborn baby has an essentially complete skeleton. The bones have a lot of growing and hardening ahead of them, but they are essentially all in place. The same is true of all vertebrate organisms—fishes, amphibians, reptiles, birds, and mammals. Within some species of fishes, the number of vertebrae is greater among individuals that live in cold water than among warm-water forms. Nevertheless, once the body plan has been laid down very early in life, the number of bones does not change.

We all know that vertebrate organisms grow during their lifetimes. As the bodies of vertebrate animals grow, their bones change continually both in size and in shape so that their parts fit together at all times. This is called growth by modification. Besides the vertebrates, this mode of growth occurs also among the echinoderms, although they grow primarily by addition—crinoids, starfishes, and echinoids (sea urchins). During the life of an echinoderm, calcite plates are added continually. Once added they change in size and shape as the individual grows, a necessity so that the animal is covered completely with a protective external skeleton.

Growth by modification is possible only among organisms with mineralized skeletons that are permeated by living material. Thus, our bones and the plates of echinoderms are porous and, in life, occupied by soft tissue that secretes the mineralized skeleton and that dissolves and reprecipitates it to change the shape. Secreting a mineralized skeleton is metabolically very costly: it takes a lot of energy to do. Growth by modification provides great economy of material, however, so that the organism does not have to carry around unused skeletal structures. It also facilitates the rapid repair of broken bones and plates and provides the organism with a skeleton that is functional at all times. On the down side, the vertebrates especially are prevented by their mode of growth from making radical changes in skeletal morphology because all their bones are present from the start.

Colonial Organisms

In our discussion of modes of growth we have been focusing on individual organisms, like a crab or a nautiloid or a vertebrate. However, not all of the organisms we see in the fossil record are exactly analogous to these. Instead, some organisms are colonial and composed of an integrated "super-organism" consisting of several more or less semi-independent "individuals." A sponge, for example, is essentially a colony that comprises a number of cells. A colony of corals consists of what are in many ways effectively individual organisms rather than cells, but these individuals are all linked together by soft tissue (Figure 2-11). Bryozoans are tiny animals that live in colonies in which each individual is on its own and there is no sharing of soft parts.

Colonial life presents a number of advantages and disadvantages. On the one hand, living in a colony enables organisms to evolve large size—as a colony—while not having to cope with the many physiological problems that stem from being large individuals. (Some of the evolutionary advantages and disadvantages of large size are discussed in greater detail in a later chapter.) A colony can be attacked by a predator, for example, that destroys most of the semi-autonomous individuals in the colony, but the few individuals that remain can continue life as normal and repopulate the colony. On the other hand, the small size of individuals in the colony makes these individuals more vulnerable to chance events.

Trace Fossils

Up to now we have dealt primarily with fossils that are the preserved parts of long-dead organisms. Recall, however, that the definition of the term fossil specifies

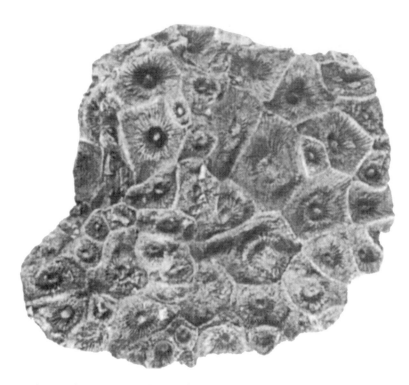

Figure 2.11 **Colonial animal.** A Carboniferous fossil coral. Image from E. N. K. Clarkson (1998) *Invertebrate Palaeontology and Evolution*, 4th Edn, Blackwell Publishing; by J. Jameson.

that a fossil is any remains or evidence of prehistoric life. In this connection, it is useful to differentiate fossils into two types. Body fossils are the actual bodies or body parts of long-dead organisms. Trace fossils are trackways and related forms that provide evidence of the behavior of ancient organisms. A body fossil may be preserved in exquisite detail, but without other information it may be hard to tell how it lived, how it behaved, and how it interacted with its environment. A trace fossil, on the other hand, provides a record of behavior, but determining exactly what kind of organism made the trace is usually not possible. Indeed, many kinds of soft-bodied organisms have no fossil record other than the traces they have left behind.

Trace fossils come in many forms and include tracks and trails (Figure 2-12), burrows, borings, and coprolites (fossilized excrement). Trace fossils record events that took place while the trace maker was alive. A hole in a fossil clam shell drilled by a boring, predatory snail, for example, was drilled by a living snail, usually into the shell of a living clam (Figure 2-5). If the drilling snail is successful, the event results in a good meal for the snail and the death of the clam. Similarly, coprolites left in the fossil record were formed by animals that inhabited the area where the coprolites occur.

Trace Fossils and American History

Among the most spectacular trace makers is a boring clam named *Teredo*. It is called the shipworm, although this clam is no worm, and its boring activities may have made life a little too exciting for some unfortunate folks. *Teredo* inhabits tropical seas and bores into wood, including the wood of sailing ships. One reason so many

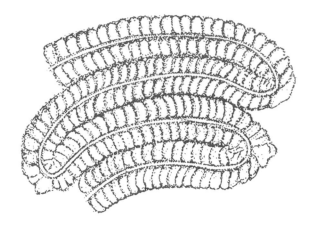

Figure 2.12 Trace fossil. A fossil trackway produced by an animal crawling over the seafloor. Diagram from E. N. K. Clarkson (1998) *Invertebrate Palaeontology and Evolution*, 4th Edn, Blackwell Publishing; redrawn from *Treatise on Invertebrate Paleontology*, University of Kansas Press and Geological Society of America, Lawrence, Kansas.

Spanish galleons laden with gold from the New World sank on their way back to Spain is that the bottoms dropped out of the *Teredo*-infested wooden ships. Such freshwater harbors as New Orleans were favored by the seafaring crowd because shipworms can survive only in seawater. They die when exposed to freshwater for short periods of time. Thus, a good maritime tactic was to moor a galleon in a freshwater harbor for a long enough time to kill all the shipworms that had infested the hull. To Spanish sailors out of sight of land in the Gulf of Mexico or Caribbean Sea, these little clams turned out to be not as boring as they might seem at first glance.

Concluding Remarks

Many factors determine how and why an organism becomes a fossil. We have tried to outline some of these factors in this chapter, while also concentrating on the nature of the fossil record itself. To summarize, although imperfect, and indeed any source of scientific data is imperfect, the fossil record is our one true chronicle of the history of life. As such, it preserves a rich and varied archive containing abundant information on the modes, habits, and evolution of prehistoric life.

Additional Reading

Behrensmeyer, A. K., Damuth, J. D., DiMichele, W. A., Potts, R., and Wing, S. L. 1992. *Terrestrial Ecosystems Through Time*. University of Chicago Press, Chicago, IL; 588 pp.

Benton, M. J., and Harper, D. A. T. 2009. *Introduction to Paleobiology and the Fossil Record*. Wiley-Blackwell, Oxford; 592 pp.

Briggs, D. E. G., and Crowther, P. R. 2001. *Palaeobiology II*. Blackwell Science, Oxford; 600 pp.

Bromley, R. G. 1990. *Trace Fossils*. Unwin Hyman, Boston, MA; 280 pp.

Cowen, R. 2005. *History of Life*, 4th edition. Wiley-Blackwell, Oxford; 336 pp.

Gayrard-Valy, Y. 1994. *Fossils. Evidence of Vanished Worlds*. Harry N. Abrams, New York; 191 pp.

Martin, R. E. 1999. *Taphonomy. A Process Approach*. Cambridge University Press, Cambridge; 508 pp.

Rudwick, M. J. S. 1985. *The Meaning of Fossils*, 2nd edition. University of Chicago Press, Chicago, IL; 304 pp.

Chapter 3
Organizing the Fossil Record

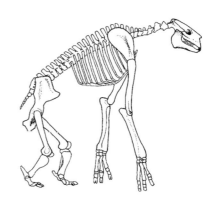

Outline

- History of Ideas on Biological Classification
- Applying Linnaeus' Hierarchy
- What is a Species and How Does a Paleontologist Identify Them?
- Conclusions: the Difference Between Inanimate Atoms and Living Things
- Additional Reading

History of Ideas on Biological Classification

We have already mentioned the tendency of humans to classify things and how some classifiers try to shoehorn too much of nature into a classification, which can be unproductive. For example, Grabau's classification of sedimentary rock included snow. He was correct: snow fits the definition of a sedimentary rock. Such a classification, however, contributes little to our understanding of snow or of sedimentary rocks and the environments in which they occur. Other examples of nonproductive classifications exist. However, classification schemes when successful and appropriate can be extremely important because they allow scientists to accurately organize knowledge and better understand nature. In this chapter we will focus on two important aspects of the fossil record that have been successfully organized by classification schemes. These classification schemes have contributed fundamentally to our understanding of the patterns and processes that govern the history of life on this planet. The first important aspect of organizing the fossil

Prehistoric Life: Evolution and the Fossil Record. 1st edition. By Bruce S. Lieberman and Roger Kaesler. Published 2010 by Blackwell Publishing.

record involves how we classify living organisms; the second involves dividing the geological record into different temporal units. Along the way we will consider the differences between living and nonliving things.

First, consider living organisms. People have sought to classify living organisms almost since the beginning of intellectual history. As shall be discussed in Chapter 4, ultimately the information from classification provided important evidence that evolution had happened.

Classification of living things is important because it provides a framework that can be used to discuss organisms. If the classification is a useful one, it helps us explain why a kind of organism is like it is, that is, to determine the pathway it followed in its evolutionary history. A useful classification can even let us predict where to look for characteristics of the organism that are of interest.

Using a Classification for Prediction

Let us look at an example of the importance of classification for prediction. The biochemical taxol, which occurs naturally in the bark of the California yew tree, has been discovered to be useful in the treatment of breast cancer. When the medical effectiveness of this biochemical was discovered, people began harvesting the bark of yew trees, the removal of which, of course, caused yews to die. The concern was raised immediately that this species of yew would soon be extinct. Where else in the plant kingdom could we search for a supply of taxol? The yew belongs to the genus *Taxus*, and most species of these are green bushes or trees with bright red berries. Perhaps other green bushes and trees with red berries would have taxol. There are in fact many green bushes and trees with red berries in the plant kingdom. One example is *Ilex*, the holly with which many people decorate their houses for the winter holidays. *Ilex*, however, is unlikely to provide a source of taxol because holly is a flowering plant. That is, it belongs to a group called angiosperms that includes many of the plants humans are familiar with from garden vegetables to cactus to trees. However, evergreens, called conifers, are not angiosperms and belong to a different group of plants called gymnosperms. *Taxus* is a conifer and is thus more closely related to pine and fir trees than to *Ilex*. That does not mean that we should expect to find taxol in all pine and fir trees, but there are species of conifers that are quite closely related to the California yew, including species of the genus *Yew*. These are the plants where scientists should search for taxol. Here is a case where classification can be used to make important predictions potentially vital for human health and medicine. Although taxol has now been synthesized in the laboratory, thus saving a great many yew trees from an untimely demise and providing an endless supply of the biochemical, this type of "bioprospecting" will likely be one of the primary ways humanity discovers new medicines in the future.

The Founding Fathers of Classification

Aristotle (384 BCE to 322 BCE)

The classification of organisms, like so many intellectual achievements upon which we rely in our day-to-day lives, began with the 5th century BCE Greeks. A noteworthy achievement in the classification of organisms was made by Aristotle (Figure 3-1), who arranged organisms on the basis of their similarity into large

Figure 3.1 **Aristotle**. The famous ancient Greek philosopher and the founder of biological classification. Image from iStockPhoto.com © S. Negavonovic.

groups, each of which he called a genus (plural, genera), a term we still use today. Aristotle's genera were very broad compared to those we recognize today. For instance, he regarded birds as comprising one genus and fishes as another, whereas today we recognize that each of these groups consists of a very large number of genera. Aristotle developed what is call the *scala naturae* or ladder of life, in which he arranged organisms linearly according to their perceived degree of perfection from very primitive to advanced (Table 3-1). We now know such a view is inaccurate, but it is noteworthy as perhaps the earliest preserved example of an attempt to classify nature.

John Ray (1628–1705)

The English biologist John Ray was interested primarily in plants, and he collected and described hundreds species of plants during his lifetime. Working with co-authors he also made important contributions to ornithology, the study of birds. Ray's work was done at a time when people's understanding of the vast diversity of life was beginning to develop as a result both of growing interest in natural history and increasing contact with the organisms on continents outside Europe. John Ray (and other early scientists) began to recognize that within Aristotle's categories in the ladder of life were a great many kinds of organisms that were most similar to each

Table 3-1. Aristotle's *scala naturae* or ladder of life, in which everything in nature was arranged according to its perceived degree of perfection. Aristotle placed humans at the top of the scale. Today we know this is not an accurate view. For example, Cetacea are mammals and should be grouped with these mammals; the evolutionary affinities of the Malacia lie with the Ostracoderma (they are all molluscs) and further Malacia cannot be said to be above Ostracoderma, Malacostraca, Entoma, or any of the other listed groups for that matter

Humans
  Mammals (quadripeds that give birth to live offspring)
    Birds
      Reptiles and amphibians (egg-laying quadripeds)
        Cetacea (porpoises and whales)
          Fishes
            Malacia (octupuses and squids)
              Malacostraca (crabs, crayfish, and similar organisms)
                Ostracoderma (molluscs, including snails and clams)
              Entoma (such arthropods as insects and spiders)
                Zoophyta (such simple invertebrates as sponges and jellyfishes)
              Higher plants
            Lower plants (including algae)
          Inanimate matter

other and different from everything else—what we call species today. Ray recognized that members of species are united by their ability to reproduce with each other, an idea that forms the foundation of our modern understanding of species. Ray recognized also that some species are more similar to other species and could be lumped together into what we call today genera. As a result of Ray's work, scientists classifying nature began to arrange groups of organisms according to their physical similarities.

Carolus Linnaeus (1707–1778)

The great Swedish naturalist Carl von Linné (Figure 3-2), whose name was Latinized to Carolus Linnaeus, made several major breakthroughs in biological classification. First, he developed a method of describing species that made clear precisely the manner in which species differed from closely similar species. Second, he popularized binomial nomenclature, in which species are given two names, a generic name followed by a species name, such as *Homo sapiens*, our own species; *Equus caballus*, the modern horse; or *Canis familiaris*, the familiar dog we keep as household pets. Third, he developed a method of classifying organisms in a hierarchy.

    Let us deal with his third contribution first. A hierarchy is a system of classification in which objects are grouped at various levels called categories, each of which is subordinated to the next above it. For example, in a company workers are subordinated to a manager and he or she is subordinated to a boss. According to television shows like the *Soprano's*, in a mafia family there are associates who are subordinates of the soldiers; these in turn are subordinates of capos who are subordinates of the underbosses; they are all subordinates of the boss or "don." Finally, in a university setting, professors are subordinates of the department chairs; chairs are subordinates of the assistant deans; these in turn answer to the dean;

Figure 3.2 Linneaus. The famous early naturalist, as portrayed on a Swedish postal stamp. Image from iStockPhoto.com © M. Caven.

Table 3-2. The linnaean hierarchy and two examples from the modern world

Kingdom	kingdom Animalia	kingdom Animalia
Phylum	phylum Chordata	phylum Chordata
Class	class Mammalia	class Mammalia
Order	order Perissodactyla	order Carnivora
Family	family Equidae	family Canidae
Tribe		
Genus	genus *Equus*	genus *Canis*
Species	species *Equus caballus*	species *Canis familiaris*

deans are subordinated to a vice chancellor or vice president; vice chancellors are subordinated to the chancellor, and he or she may answer to some sort of board. Another example of a hierarchical classification, also drawn from the university setting, has departments as the lowest level in the hierarchy, which are a part of colleges or schools, which are components of the university as a whole.

Of course Linneaus' key insight involved nature. The modern version of Linnaeus's hierarchy is shown in Table 3-2, where each level in the hierarchy is referred to as a category, and the group of animals or plants that are grouped into a category are called a taxon (plural, taxa). In addition to the main headings, additional layers can be incorporated into the hierarchy. The category "Families," for example, may be grouped into superfamilies or subdivided into subfamilies. Orders are sometimes divided into suborders and again into infraorders, which are

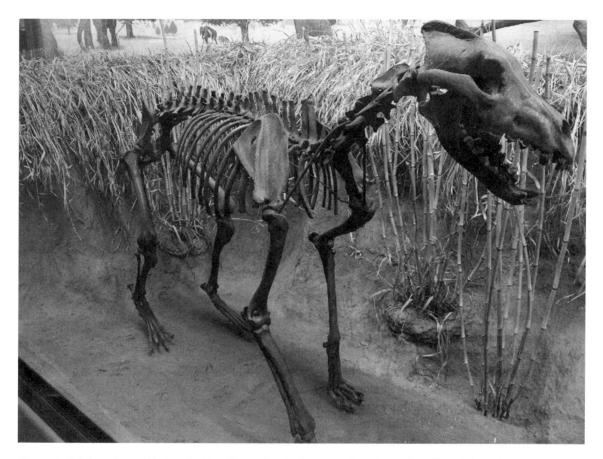

Figure 3.3 **The Dire wolf.** *Canis dirus* from the Pleistocene La Brea Tar Pits of Los Angeles, California. A close, extinct and larger relative of the modern wolf. Specimen in the University of Kansas Natural History Museum. Image by B. Scherting.

subordinated to suborders. In some groups of organisms, such as those shown in Table 3-2, the category tribe is not widely used.

Linnaeus's hierarchy, which is still in use today by nearly all biologists and paleontologists, is a classification most akin to the second example from the university setting we presented, where subordinate groups are members of the groups above them. The taxon *Canis*, for example, the genus of dog-like animals, contains its member species: *Canis lupus* (the wolf), *Canis latrans* (the coyote, literally the laughing dog), *Canis familiaris* (the familiar or domesticated dog), a few other, more obscure species of dog-like animals, and extinct species, including *Canis dirus* (the Dire wolf) (Figure 3-3). Consider our example from the tv show the Soprano's and how it is different: "capos" are not at the same time also "soldiers."

Linnaeus's second contribution mentioned above was his popularization of the use of binomial nomenclature. Species of organisms are referred to by two names, the first a generic name, in our case *Homo*, and the second a trivial name, for us *sapiens*. Our genus *Homo* has several species, four of which are *Homo sapiens* (thinking man—that's us, at least excluding certain talk show hosts), *Homo neanderthalensis* (Neanderthal man), *Homo erectus* (upright or erect man), and *Homo habilis* (handy man—alas referring to neither of the authors of this text, but at the time of its discovery this species was thought to be the first tool maker). It is not proper to refer to any of these organisms by the trivial name alone, *sapiens*, *neanderthalensis*, *erectus*, or *habilis*. The name of the genus is unique. There can be only one genus of animals and one genus of plants with the same name. The same trivial name, however, can be applied to species in any genus. This is a good provision given that there are several million species. If duplication of trivial names were not allowed, we would have long ago run out of names.

Linnaeus himself named many of the common species of organisms with which we are familiar, including a great many species of plants. Originally genera and species were given names with Latin or Greek meanings. *Homo*, for example, is Latin for "man." (Note, biologists in Linnaeus' day were mostly men and probably all sexists; today we of course endorse the view that "man" in this context means "human" and includes both "men" and "women." It is also interesting to note that the word "*Homo*" in Greek has a different meaning: same or similar.) We have already seen the meanings of four of the trivial names of species of *Homo*. The word *canis* means dog in Latin, and *rattus* means rat, both of which are the generic names of the two groups of species that comprise those genera. Later names were applied that recognized or honored individuals, usually biologists or paleontologists. Two genera of fossil fusulinids, a group of microfossils, *Dunbarula* and *Dunbarinella*, were named to honor Professor Carl O. Dunbar, a specialist of these organisms. *Darwinula stephensoni*, an ostracode species (a type of crustacean arthropod), was named to honor Charles Darwin, and its trivial name means *of Stephenson*. A trivial name made of a person's name usually but not always implies that the person had something to do with the finding or naming of the species. Sometimes genera or species are named for the locality where they were first found or where they occur commonly. *Kansanella* is another genus of fusulinid that is a prominent fossil in the upper Paleozoic rocks of Kansas. *Henryhowella tenmilecreekensis* is a species of fossil ostracode in the genus that honors the great micropaleontologist Henry Howe and that was first found in an outcrop near a body of water called Tenmile Creek.

Two complex sets of rules exist, codified in books that are guaranteed to cure insomnia, that govern the naming of species and other taxa and their arrangement into the Linnaean hierarchy. These are the *International Code of Zoological Nomenclature* (ICZN), which deals with animals and animal-like organisms in other kingdoms, and the *International Code of Botanical Nomenclature* (ICBN), which covers plants, algae, and fungi. The science that deals with these matters is referred to as taxonomy or systematics, and most biologists and paleontologists who specialize on the study of a group of organisms are intimately familiar with the provisions of these codes.

Jean Baptiste Pierre Antoine de Monet, Chevalier de Lamarck (1744–1829)

Linnaeus was a skilled botanist, but he made rather a mess of classifying animals. For instance, he classified some organisms we now recognize to be animals as plants. Many of the problems with his classification of animals were cleared up by Lamarck (Figure 3-4), a French zoologist whose ideas on evolution we deal with in the next chapter. At the time Lamarck worked and until the mid-19th century, much of the groundwork of biology and paleontology was being laid by French scientists, and many of the genera and species they named are still regarded as valid concepts. Lamarck is criticized for his ideas on evolution, but he is clearly one of the greatest biologists who has ever lived.

Figure 3.4 Lamarck. The famous French biologist. Diagram from J. Pizzetta.

Applying Linnaeus' Hierarchy

We know today, of course, that the reason a hierarchical biological classification scheme (like Linnaeus') works so well is because of evolution, a topic we shall introduce in detail in the next chapter. For now, recognize the fact that species in a genus share a more recent common ancestral species with each other than with species in another genus. Similarly, genera in a family share a more recent common ancestor with each other than with genera in another family, and so on. The Linnaean hierarchy is a way of grouping organisms into more and more inclusive groups, each one, if classified correctly, sharing a more recent common ancestry with the members of its respective group than with other such groups of organisms. It so happens that the Linnaean system is not the only hierarchical classification system used by paleontologists: some use other systems, but one thing all these systems have in common is that they allow organismic relationships to be expressed in a hierarchy showing shared patterns of common descent. It is also important to realize that groupings of organisms into genera, families, and higher taxa can change as more research is done. For example, for much of the 20th century it was thought that whales were most closely related to an extinct group of carnivorous mammals; now, thanks to information from DNA sequences, it appears they may be more closely related to the modern hippo.

One implication of the hierarchical nature of life is that evolution comprises an array of branching sequences. In the kingdom Animalia, for example, there are some 37 phyla (singular, phylum) depending on whose classification you use. One of those phyla, the phylum Mollusca, has three principal living classes, all of which are well represented in the fossil record: the Gastropoda (snails), Bivalvia (clams,

(a)

(b)

(c)

Figure 3.5 Molluscs. (a) A fossil snail or gastropod; (b) a fossil clam or bivalve (both are in the University of Kansas Museum of Invertebrate Paleontology, images by J. Counts); (c) the modern *Nautilus*, a cephalopod (diagram from E. N. K. Clarkson (1998) *Invertebrate Palaeontology and Evolution*, 4th Edn, Blackwell Publishing).

mussels, oysters, and scallops), and Cephalopoda (squids, octopuses, nautiloids, and the extinct ammonoids) (Figure 3-5). All the phyla are divided into classes, all the classes are divided into orders, and so on down the sequence of Linnaeus' hierarchy as is shown in Table 3-2.

Linnaeus' Hierarchy at Work

To give you an idea of how the Linnaean hierarchy can be applied to major groups of organisms, let us consider briefly three examples: the perissodactyls, the artiodactyls, and the hominids.

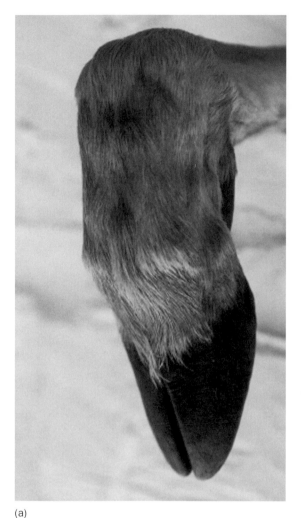
(a)

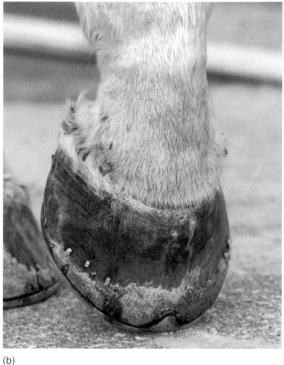
(b)

Figure 3.6 Ungulate feet. (a) An artiodactyl (deer) hoof, with two toes (© L. Wiberg), and (b) a perissodactyl (horse) hoof, with a single toe. Images from iStockPhoto.com.

Artiodactyls: even toed mammals with a cloven hoof

The ungulates are an informal grouping of mammals that includes such four-legged herbivores as horses, rhinoceroses, pigs, cattle, deer, and camels, some of which are not especially closely related evolutionary. The artiodactyls comprise the even-toed ungulates, that is, those with two toes and typically a cloven hoof such as what is found in cows and deer. Horses, by contrast, have only a single toe on each foot (Figure 3-6; Table 3-3) and are not artiodactyls but instead perissodactyls, a group we shall discuss shortly. There are six main groups or families of living artiodactyls and at least one major group that is now extinct.

The artiodactyls have been remarkably successful in their evolution. They include 79 genera in the modern world that are distributed among six larger families: the cow-like Bovidae; the deer-like Cervidae; the forest deer, called Tragulidae; the camels and their kin, the Camelidae; hippopotamuses and their relatives of which there are only two extant species, broadly organized into the Anthracotheridae; and

Table 3-3. Some general trends in the evolution of the horses

Size increases
Legs and feet become proportionately longer
Number of toes decreases, and middle toe becomes prominent
Back becomes straighter and stiffer
Cheek teeth become increasingly complicated and higher crowned
The face in front of the eye becomes increasingly elongated
The brain becomes proportionately large and complex

Source: Modified from Colbert, E. H. 1955. *Evolution of the Vertebrates*. John Wiley & Sons, New York.

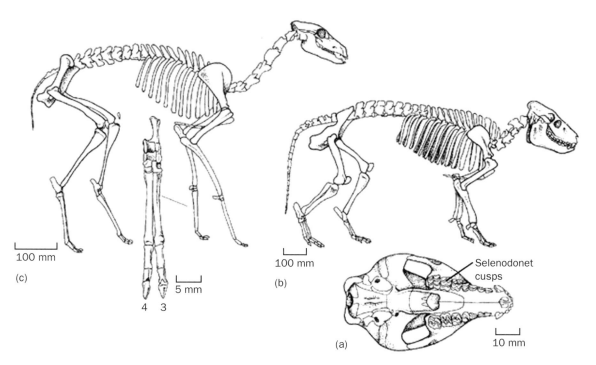

Figure 3.7 Oreodonts. Some representative skeletons of this extinct family of artiodactyls. Diagrams from M. J. Benton (2004) *Vertebrate Palaeontology*, Blackwell Publishing.

the pigs and their kin, the Suidae. The Oreodontidae are an extinct group of pig-like artiodactyls (Figure 3-7) that evolved in the Oligocene and went extinct in the late Miocene.

Perissodactyls: odd toed herbivorous mammals

The perissodactyls are odd-toed ungulates. That is, modern forms have either three toes or one toe on each foot, but never only two as is true of the artiodactyls. They include three living and nine extinct families. *Equus caballus*, the modern horse, is a well known perissodactyl. However, when compared with the artiodactyls the modern perissodactyls are rather species poor: today they are represented by only six genera; they had, however, a much more impressive evolutionary history earlier in the Cenozoic Era, especially during the Eocene, which could have been called the

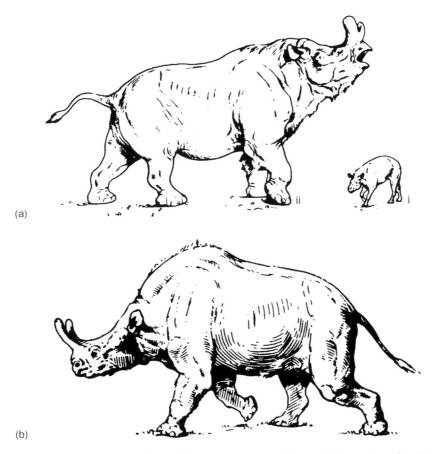

(a)

(b)

Figure 3.8 Titanotheres. Reconstructions of some representatives of this extinct family of perissodactyls. Diagrams from R. Cowen (2005) *History of Life*, 4th Edn, Blackwell Publishing; based on Osborn.

heyday ("hayday" for you horse lovers) of the perissodactyls. It was during its prolific early evolutionary history that the largest land mammal that ever lived, *Indricotherium*, a perissodactyl, evolved. This beast was 25 feet long and nearly 18 feet tall.

The extant perissodactyl families are the Equidae, horses, asses, and zebras; the Tapiridae, the tapirs, a rare group today; and also the Rhinocerotidae, the rhinoceroses, which sadly today are nearing extinction because of the depredations of poachers. Two important extinct families are the Titanotheriidae and the Chalicotheridae. The titanotheres first evolved in the Eocene from a small animal similar to the ancestor of the horses, but later they became truly gigantic creatures reaching 2.5 meters high at the shoulders (Figure 3-8). They went extinct in the middle Oligocene.

The Chalicotheriidae are most unusual herbivores because instead of hoofs they had heavy claws (Figure 3-9). Their distinctive nature and anatomy would have represented a challenge to the famous paleontologist Baron Georges Cuvier's pronouncement that merely by examining a small part of an animal he could deduce what the animal looked like and whether it was a herbivore or a carnivore; if shown the claw of a chalicothere, without being aware of the rest of its anatomy,

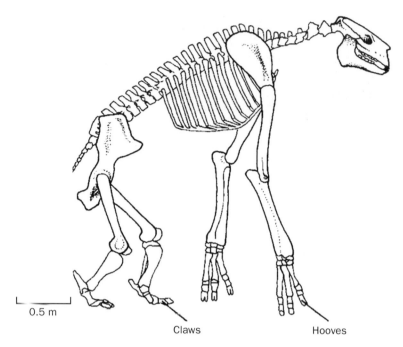

0.5 m

Claws   Hooves

Figure 3.9 Chalicothere. A skeleton of this unusual, extinct family of perissodactyls. Diagram from M. J. Benton (2004) *Vertebrate Palaeontology*, Blackwell Publishing; after Zapfe.

Cuvier might have pronounced the animal a carnivore. Little could he have guessed that they used their claws not to capture prey but either to dig roots or drag down tree limbs for browsing. We might say that although the chalicothere's evolutionary heritage could allow it to evolve claws, it did not allow it to evolve the other traits of a carnivore including sharp teeth.

Hominids: upright evolution

As will be discussed in the next chapter, evolution involves a branching, tree-like pattern with various diverging species appearing through time; indeed, this tree-like pattern produces the hierarchy of relationship we have been describing in this chapter. Thus, it is perhaps ironic that the origins of our own group, the hominids, involved largely leaving life in the trees behind and becoming more attuned to a mobile lifestyle on the African grasslands.

Darwin states in the last line of his classic 1871 book *Descent of Man* (if published today it would have been called *Descent of Humans*) that "Man still bears in his bodily frame the indelible stamp of his lowly origin." Our only dispute with him is his use of the term "lowly origin". There are no animals more emotive, more human like, than our closest evolutionary relatives, the primates in general and the chimpanzees and gorillas in particular; they can use tools, learn sign language, and associate in complex societies. Genetically and evolutionarily we are quite close to chimpanzees (Figure 3-10), having more than 98 percent identity of our DNA sequences. Divergence estimates for the separation of the last common ancestor of chimps and humans indicate these two lineages began to diverge about six to seven million years ago.

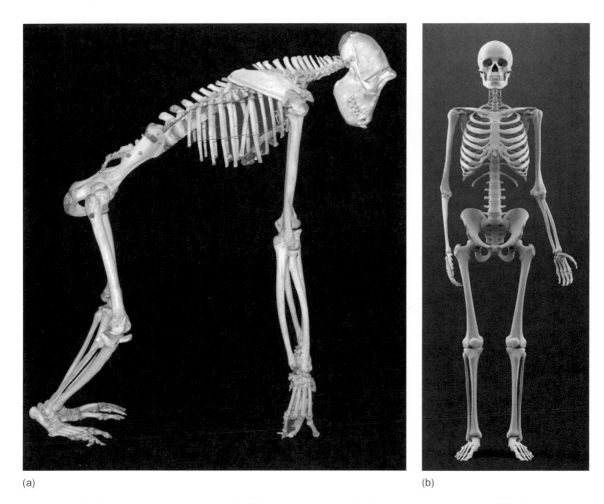

(a)   (b)

Figure 3.10 Two primate skeletons. (a) Chimp, image by B. Lieberman, courtesy of D. Freyer, University of Kansas. (b) Human, image from iStockPhoto.com © H. Aydin.

Darwin also made one of the understatements of the century in his *Descent of Man* when he presciently recognized the notion "that man is descended from some lowly organized form, will, I regret to think, be highly distasteful to many." Still, distasteful or not, the fact remains that the narrow evolutionary distance separating chimps and humans is completely filled in: there is a family (in the Linnaean sense) of primates, now almost wholly extinct, called the Hominidae or hominids, and we are the last surviving representative of the group. The hominids comprise our own species, *Homo sapiens*, and several other (more than 20) now extinct species that evolved (and in all but one case went extinct) in the last six million years or so (Figure 3-11). Included in this tree are such genera as *Sahelanthropus*, *Ardipithecus*, *Kenyanthropus*, and the familiar genus *Homo*; *Homo* includes our own species and at least five other species including the well known *Homo habilis*, *H. erectus*, and *H. neanderthalensis*.

The first hominid species to evolve were very much chimp-like, having a brain close to the size of a chimp and significantly smaller than the brain of a typical human. They do differ though in at least one critical respect from chimps in that their posture was much more upright, and in this respect they much more closely

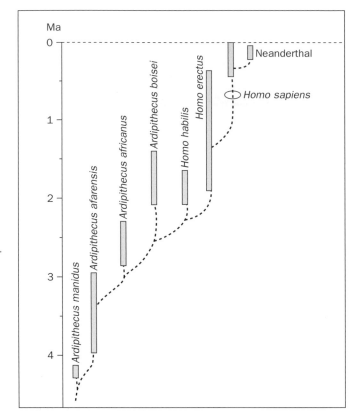

Figure 3.11 Hominid evolutionary tree. This tree relates some of the known species of hominids. Diagram from M. J. Benton and D. A. T. Harper (2009) *Introduction to Paleobiology and the Fossil Record*, Blackwell Publishing.

resemble humans; what may be surprising to some is that the distinctive trait that first separated hominids from their chimp like ancestors was not a large brain but rather an upright posture. Maybe our parents really were onto something when they emphasized maintaining a good posture?

In any event, early hominids spent much less time in the trees and much more time moving about on the ground with a bipedal or two legged walking gait than previous primates. As we move forward through time from six or seven million years ago towards the present the species that occur in the fossil record overall became more and more human-like, less and less chimp-like, and their relative brain size dramatically increased. There is a clear evolutionary trajectory preserved through time documenting this transition.

This is not to say that there was only one line of hominid evolution. We might perceive things that way because of our relatively large brains and the other obvious differences that separate us from chimps; there were also other what might be called "side branches" within the hominid evolutionary tree, including *Paranthropus boisei* and *P. robusti*. (At times these two species are instead referred to as *Ardipithecus* or *Australopithecus*.) These species lived in Africa about two million years ago and had large skulls and broad teeth that would have been very useful for crushing and chewing grains, nuts, or even bones; their brains were relatively small and not much bigger than a chimpanzee's (Figure 3-12).

It seems that almost every few years another new species of fossil hominid is found in Africa, the evolutionary cradle of hominids and also our own species; these finds continually push paleontological knowledge closer and closer to that point when the last common ancestor of chimps and hominids first diverged

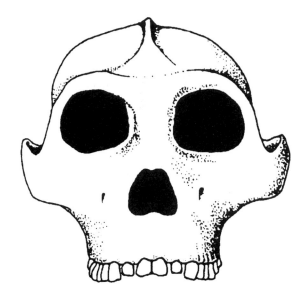

Figure 3.12 *Paranthropus boisei*.
A skull of this extinct hominid.
Diagram from R. Cowen (2005)
History of Life, 4th Edn, Blackwell
Publishing; after Howell.

evolutionarily. One of the oldest hominid fossil species is called *Sahelanthropus tchadensis*; the name translates as Sahel man from Chad, the region and country within Africa whence it was found. It is known from several roughly 6–7 million year old fossils, including a nearly complete skull. Finds like these and other fossil species give us a fairly good idea of what that earliest ancestral hominid looked like. Other important fossils include the well known *Australopithecus afarenis*, translating as "Southern Ape from the Afar region of Ethiopia", which includes the famous fossil known as "Lucy" (Figure 3-13), and *A. africanus*, "Southern Ape from Africa". These species first appeared respectively 3.5 million and three million years ago. Each possessed a mix of what we might call chimp and human characters including relatively small brains when compared to humans, and they probably stood fully upright; they were also relatively small in stature. The two species do differ in subtle ways. Ultimately the aforementioned *Paranthropus boisei* and *P. robusti* share close evolutionary affinity with these lineages, though based on their teeth *A. afarensis* and *A. africanus* ate a different (and shall we say mushier) diet.

  Australopithecus is also the closest relative of the genus *Homo*. Within this genus *A. africanus* is likely the most similar species to the last common ancestor of *Australopithecus* and *Homo*. The average brain size in even the earliest species of *Homo* is relatively much larger than those of other hominids; their skulls show a flatter face, less prominent ridges above the eyes, and bear smaller teeth. *Homo habilis*, translating as "The Handy Man" because fossils of these are associated with early tools including choppers and hand axes, was described by Louis and Mary Leakey and represents one of the oldest species in the genus; it first appeared nearly two million years ago. Evolving closely on its heels is *H. erectus* (Figure 3-14): the name translates as upright man; this species had an even larger brain than *H. habilis* and concomitantly its larger brain may explain the more sophisticated tools this species produced. It was also the first hominid species to make the journey out of Africa, spreading throughout parts of Asia, Europe, and possibly even journeying to Australia.

  One interesting aspect of hominid evolution, in addition to the prominent evolutionary trajectory we have been describing, is that in terms of hominid diversity the modern world is unique, or nearly so. It is perhaps the only time,

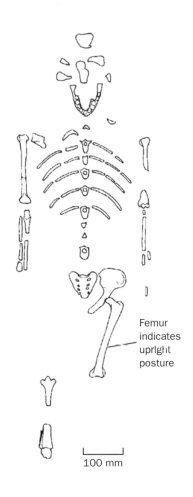

Figure 3.13 The fossil called Lucy. The 3.5 million year old hominid fossil from Ethiopia, *Australopithecus afarensis*. Diagram from R. Cowen (2005) *History of Life*, 4th Edn, Blackwell Publishing; based on Lewin.

Femur indicates upright posture

100 mm

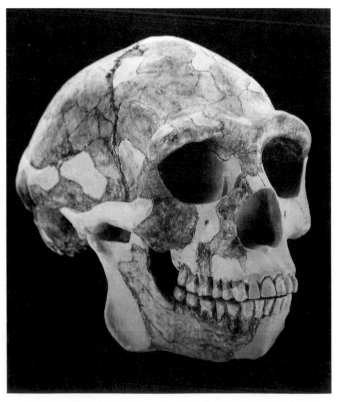

Figure 3.14 *Homo erectus*. A skull of this hominid. Image from I. Tattersall, American Museum of Natural History.

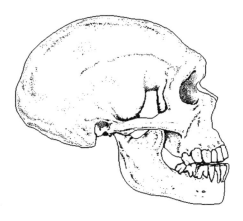

Figure 3.15 *Homo neanderthalensis*. A skull of a neanderthal. Diagram from R. Cowen (2005) *History of Life*, 4th Edn, Blackwell Publishing; redrawn, idealized, and simplified from Trinkaus.

except for six or seven million years ago when the family originated, that only a single species of hominid walked the planet. For example, around 1.5 million years ago, as many as five hominid species may have coexisted. They may not have been actually living together, but they certainly did live near one another at the same time; thus presumably there was at least some form of contact among the different species. Even as recently as 50,000 years ago there may have been three hominid species extant, the aforementioned *H. erectus*, our own species *H. sapiens*, and our intriguing closest evolutionary cousin *H. neanderthalensis* (Figure 3-15).

The neanderthals first appeared around 300,000 years ago and are known from various sites including Europe, Asia, and the Middle East. Their brains were quite large, as large as our own in fact, while their bones indicate they were far more physically powerful than we were, or are. They produced complex tools, though in variety these tools were not as diverse as those made by early humans; like humans they also buried their dead. Neanderthals are known to have survived until as recently as 25,000–30,000 years ago in parts of Europe. Their large brains and massive bodies begs the question of why we are the ones that lived, whereas they did not. The truth is we likely will never know although William Golding, the author of the famous book *Lord of the Flies*, described an interesting imagined view of what might have happened in his classic book *The Inheritors*. (In effect he posited that we killed them off.) In spite of the fact that we may never be able to answer this important question, because Neanderthal bones are so young and often well preserved, it is possible to learn other intimate details about them. In particular, at times it is possible to isolate and sequence parts of their DNA. Recent studies of Neanderthal DNA sequences (this type of research is referred to as the study of ancient DNA and is discussed in Chapter 7) suggest that although their DNA is very similar to human DNA, showing far greater than 99 percent similarity, there is no evidence that the two species ever interbred. There is, however, excellent evidence that Neanderthals lived in the some of the same caves in Europe that housed modern humans, albeit at different times.

Neanderthals appear to be our closest relatives, although we are not descended directly from them; instead we share a unique common ancestor with them likely to the exclusion of all other known hominids. Our own species appeared more than 400,000 years ago in Africa, and these fossils are sometimes referred to as archaic *Homo sapiens* because of subtle differences in their skulls that separate them from what are called anatomically modern humans or *Homo sapiens sapiens*. The earliest anatomically modern humans are represented by roughly 100,000 year old fossils in

Africa; they had entered the Middle East shortly thereafter, and arrived in Europe perhaps 40,000–50,000 years ago; Australia was populated roughly at the same time Europe was, while humans entered North and South America tens of thousands of years later.

What is a Species and How Does a Paleontologist Identify Them?

Biologists sometimes define a species as an actually or potentially reproducing community of organisms, and this definition is sometimes called the "Biological species concept." Such a definition is useful but not always practical even with modern organisms because although it is easy enough to collect distinctive looking organisms rarely do we get the chance to observe populations mating in the wild; even more rarely do we get the chance to bring them back to the laboratory and manage to get them to survive long enough to reproduce in a controlled setting. These problems with the Biological species concept are even more severe with paleontological remains; when it comes to fossils it is almost always impossible to determine which organisms were interbreeding (the case of humans and Neanderthal's represents one of the only exceptions to this).

For this reason, another definition is often used that recognizes a species as the smallest distinct group of organisms that shares a unique set of anatomical features in common. Most of the more than one million species that have been described on Earth were described largely without information about mating patterns or genetics. Instead they were defined on the basis of their possession of one or more unique anatomical features that allowed them to be distinguished from all other species. This concept works amazingly well and is on the whole the one that paleontologists use because they typically lack the ability to study which populations were mating but do have the ability to study the anatomy of organisms.

There exist good scientific reasons as to why we expect there to be broad consonance between species defined in this manner and those defined using the Biological species concepts. Partly these concepts define similar species because the individuals that most resemble one another are those that are reproducing; further, because they are reproducing with one another, their offspring will also tend to closely resemble each other. It is true that there are times when these different species concepts may indicate different patterns. For example, some organisms look identical but do not reproduce with one another: these are called cryptic species. A well known case of cryptic speciation involves the cichlid fishes. There exist more than 1,000 species of these fish in the large lakes in eastern Africa. In one of these large African lakes the skeletons of many species of cichlids are effectively identical but their skin color patterns are different; further, these coloration patterns are used by the different sexes to recognize the appropriate mates: fish of like colors will only breed with like colors and thus represent biological species; it is the female that chooses which fish will mate with her. If discovered in a fossil deposit the fish would all be called a single species because their skeletons are identical and one might argue that therefore a paleontologist would have an incomplete picture of the diversity preserved in the lake. Something interesting, however, is happening in the lake which suggests that this may not be the problem it first seems to be. Because of the negative effects of the growing human population in the region the flow of pollutants into the waters of these African rift lakes has increased, making the waters cloudier. As the waters get cloudier it is harder for the various cichlids to accurately see and distinguish different color morphs from one other, and what is happening is that "species" that once were not interbreeding now are. The result is that the many different so-called species that had started on the pathway of evolutionary divergence are now homogenizing back into one, or a few, distinct species.

Although the effects of humans in this case are unfortunate, they point out the often ephemeral nature of these types of species; in the long term, cryptic species likely will not persist in the same region and will not present a major problem for a paleontologist. Instead, one of two things will happen, either the various "species" will eventually interbreed, swamping out any evolutionary differentiation that has accumulated, or, being so similar, they will come into direct competition with one another for the same resources and most of the so-called "species" will be driven to extinction. In summation, because these types of cryptic species are ephemeral a paleontologists' inability to study and recognize them does not represent a major challenge to our understanding of the major features in the history of life. These after all involve the rich spectrum of anatomically different species that have evolved through time.

It is also worth mentioning here that sometimes there are organisms that look fairly different that can and do reproduce with one another, producing fertile offspring; these are sometimes called polymorphic species, i.e., a species that has several different forms. The objection might be raised that a paleontologist using a species concept based on anatomy would treat these different looking organisms as different species when in fact they belong to the same Biological species. However, again it is doubtful that this represents a serious challenge to a paleontologist's ability to study life preserved in the fossil record. This is because if organisms that look different are interbreeding with one another over time we would surely expect the following pattern: their offspring would resemble a single form representing an average of the different anatomical types. Again, as with cryptic species, polymorphic species tend to be ephemeral, and anatomically and Biologically defined species will be largely congruent.

Conclusions: the Difference Between Inanimate Atoms and Living Things

After our discussion of biological classification it is worth briefly revisiting the geological time scale (Figure 1-9). It is hierarchically structured, just like the Linnaean taxonomic hierarchy that we described in this chapter. Further, in Chapter 1 we described the difference between relative and absolute geological time. Recall that a relative clock allows you to place geological events and the rocks representing those events in the order in which they occurred without reference to actual time or dates measured in years. An absolute clock, by contrast, provides a specific age for a rock but not every type of rock, especially those that contain fossils, can be analyzed with techniques used to determine its absolute age.

One topic that is worth considering in greater detail is why the technique of radiometric dating is so effective for telling absolute geological time. The reason why the radiometric dating method works so well is that each radioactive atom does not really behave as an individual. What we mean by this is that one atom of potassium-40 behaves the same as any other; it also behaves the same in England as in the United States. This similarity in fact implies that they are very simple objects. The general equation for radioactive decay is like the equation for a random process. In effect you can think of each radioactive atom like a coin that gets flipped every minute. Imagine there's a one in 1,000 chance that the atom will decay: that chance is always the same.

Let's imagine for a second what it would entail if that were not the case: if the radioactive decay line we showed you earlier in Chapter 1 was not linear, it would imply that each atom has a history. The fact that atoms are truly random means that they have no history and that decay is always constant. The randomness means that it can be predicted. This might not seem intuitive but randomness implies predictability. Predictability and randomness are why in the long term it is impossible to make a good living in Las Vegas playing at the Craps tables (without cheating, and based on viewing the Martin Scorsese movie *Casino* we would strongly discourage that). The odds are always the same and they always favor the house. So recognize that a collection of millions of atoms of potassium is something that is very simple. That simplicity, its regularity, is what allows geologists to tell time.

The famous paleontologist and evolutionary biologist Stephen Jay Gould pointed out how different radioactive atoms are from more complex objects like populations of humans or the set of all the species on the African grassland. Humans and the group of species on the African grassland are very large collections of complex objects that may interact in a variety of different ways. The complexity, and indeed the individuality of each of the humans in a population, or the different species in the African grassland, means that we cannot necessarily predict what will happen to them through time.

We will illustrate by showing a graph very similar to the radioactive decay curve we showed in Chapter 1 (Figure 1-5) but now instead of just tracking the decay of atoms through time imagine that what is also being graphed is the life expectancy of individual organisms in a population. Such graphs are often used in the field of ecology, which is the study of different organisms, and it is called a survivorship curve. This type of graph shows the number of individuals that are born and considers how long they live. To put it in the same form as the radioactive decay curve we'll look at the logarithm of the number of surviving organisms. Further, let's consider the fate of 1,000,000 humans and 1,000,000 fish from the point when all are born to a point when all expire (Figure 3-16). These curves are concave or

Figure 3.16 Survivorship curves. For fish, atoms, and people. Diagram by M. Smith.

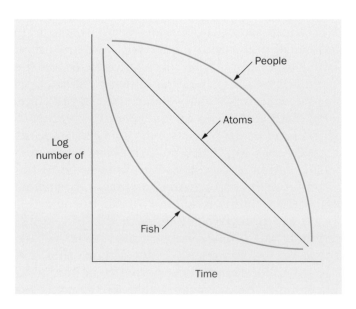

convex because these organisms have individuality and history. For example, in the United States, Canada, and most European countries infant mortality is relatively low (actually, infant mortality is higher in the United States than in these other countries which can be attributed to a variety of sociological causes). As humans age the probability of dying increases, but not at an even rate. By contrast, in fish the situation is reversed. Early on in their lives fish are very susceptible to predation because they are small, such that only a few fish survive to an advanced age. Further, very old fish, being larger, are much less likely to be eaten as prey. Thus, once fish reach an advanced age their probability of mortality is relatively low. The ultimate message here is that it matters where a person or a fish is born, who that person is, and even whether you are a fish or a human. The survivorship curves in complex biological things show individuality. If the survivorship curve of human populations fit a straight line, that is, matched a radioactive decay curve, it would mean that at every minute of our lives there was an equal chance of dying; in humans that is not the case because we are complex individual objects. This is what makes the study of evolution and the history of life so exciting and so interesting: complex objects do not behave randomly and thus are unpredictable.

Additional Reading

Blunt, W. 2001. *Linnaeus. The Compleat Naturalist.* Princeton University Press, Princeton, NJ; 264 pp.

Brooks, D., and McLennan, D. 2002. *The Nature of Diversity.* University of Chicago Press, Chicago, IL; 668 pp.

Doyle, P., Bennett, M. R., and Baxter, A. N. 2001. *The Key to Earth History: An Introduction to Stratigraphy.* John Wiley & Sons, Chichester; 304 pp.

Mayr, E. 1982. *The Growth of Biological Thought.* Belknap Press, Cambridge, MA; 974 pp.

Prothero, D. 2003. *Bringing Fossils to Life*, 2nd edition. McGraw Hill, New York; 512 pp.

Tattersall, I. and Schwartz, J. 2001. *Extinct Humans.* Westview, New York; 256 pp.

Chapter 4
Introduction to Evolution

Outline

- Introduction
- A Biological Definition of Evolution
- The History of Evolutionary Thought
- Science and Religion
- Darwin and Wallace: Never Ask a Stranger to Present Your Paper at a Meeting You Cannot Attend
- Natural Selection
- Conclusions: Why was Natural Selection Not Endorsed at Once by Many Scientists?
- Additional Reading

Introduction

What is evolution? This is a topic of great relevance to the geological and biological sciences for evolution has left a profound stamp on our biosphere, the atmosphere, and the fossil record. This chapter focuses on the meaning of evolution, the history of the idea, why evolution should be treated as a fact as well resolved as the fact that the Earth revolves around the Sun, the fallacies of creationism, and other related topics. Evolution will also be the focus of several other chapters in the book. Here, though, the stage is set for what evolution is all about and why it matters so fundamentally to the geological and biological sciences.

Prehistoric Life: Evolution and the Fossil Record. 1st edition. By Bruce S. Lieberman and Roger Kaesler. Published 2010 by Blackwell Publishing.

What is Evolution?

Evolution is a term that deserves careful consideration, because it turns out that the meaning of this term has evolved through time, and evolution means different things depending on the context being used. It is important to focus first on popular examples of phenomena that resemble evolution. In essence, evolution is indeed something we are all familiar with, and it is not a divisive or even a naturally controversial topic.

Evolution and Popular Culture

In an important sense, baseball teams evolve. For instance, the Atlanta Braves baseball team used to be the Milwaukee Braves, and before that they were the Boston Braves. In this colloquial sense, evolution of ball teams means that they show a history during which they also experience change. What other things can be said to have evolved using this definition of evolution? Well, a historical collection of Raggedy Ann dolls will also show that these objects have evolved.

   Other things familiar to popular culture evolve too, such as rap music (Figure 4-1). The origins of rap music probably date to Gil Scott-Heron's "The Revolution Will not be Televised," which appeared in the late 1960s. This music is based strongly in the rhythms and musical stylings of jazz. The next key innovation documented in our tree is The Sugar Hill Gang's "Rapper's Delight," which is musically much more centered and focused on the beat: not surprising considering it was produced during the disco era. Grand Master Flash and the Furious Five are viewed as the next major stylistic innovation in rap music, and they occupy the next branch of the evolutionary tree. They were the first to introduce to a large audience a distinct brand of socially conscious rap music that also had very complex and sophisticated

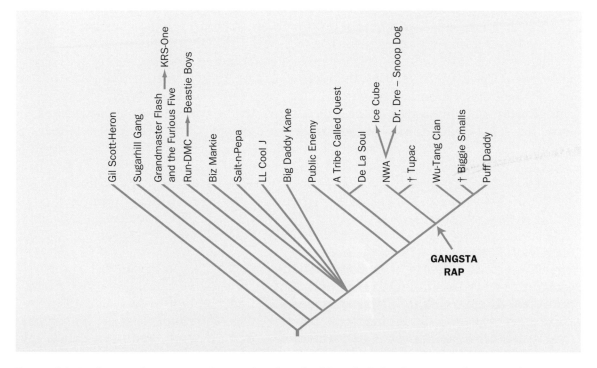

Figure 4.1 **Evolution of rap music**. A tree showing the historical development of rap music, emphasizing several major artists; daggers are next to deceased artists, which are akin to extinct lineages. Diagram by F. Abe, University of Kansas.

lyrics; they also featured a prominent role for the DJ for the first time. KRS-One and his Boogie Down Productions, because of his socially conscious lyrics, are derived from the tradition of Grand Master Flash and the Furious Five. Run D-MC occupies the next transition because they added, among other things, rock sounds and guitars to their music, and the Beastie Boys are derived from this tradition.

Rap music then underwent an explosion of new artists in the mid- to late 1980s, only some of which are shown here. This is when the sampling of pre-existing musical tracks began in earnest. Public Enemy changed things by adding a lot of different sounds (noise) to their samples and they also focused much more on political issues; we see them as another important stylistic change. Around the same time the music of De La Soul and A Tribe Called Qwest, which has been referred to as hippie rap, also appeared. Toward the top of the tree a large branch appears comprising Gangsta Rap. West Coast rap is towards the left-hand side of the tree, and East Coast is towards the right-hand side. Such groups as NWA spawned other careers including Ice Cube and Dr. Dre, which thus in a round about way led to Snoop Doggy Dog. In the history of life there are extinct lineages that are no longer with us today, such as the trilobites; the tree of rap also contains extinct lineages: the artists Tupac Shakur and Biggie Smalls. Certainly other examples of evolution, defined as a history with change, could be contrived, but these all represent good examples of a certain type of evolution.

What Evolution Meant in the 19th Century

Finally, in another important respect people are evolving. Indeed, the term evolution was first used in the 18th century to describe the changes that any individual organism went through as it developed from a fertilized egg to an adult. For instance, the musician Ozzy Osbourne underwent a series of evolutionary changes throughout his lifetime. These were not as profound as the changes that he went through during gestation inside his mother's womb, but they were profound nonetheless.

It was this definition of evolution, changes of individual organisms, that was still prevalent in the 19th century, and it explains why Charles Darwin, in his book *On the Origin of Species* (1859), used the term sparingly. In fact, the term appears only once in this book. "There is grandeur in this view of life, with its several powers, having been originally breathed into a few forms or into one; and that, whilst this planet has gone cycling on according to the fixed law of gravity, from so simple a beginning endless forms most beautiful and most wonderful have been, and are being, evolved" (Darwin, 1859, p. 490). This, the last sentence of Darwin's *On the Origin of Species*, is probably as close to poetry and literature as one can find in science, and we hope that it shows how science is not just about fact finding and tedious drudgery but at its best is about creativity, insight, and even art.

A Biological Definition of Evolution

Now, thus far we have used the term evolution in several contexts, and these are all appropriate uses of that term. But when we talk about biology or life, evolution takes on an entirely different meaning. It is the proposition that all life is descended from a single common ancestor and branched from that point. That is, it entails the idea of common ancestry and descent. Different groups share a common ancestor. The more closely they are related the more recently they shared a common ancestor. One of the first evolutionary trees ever published is shown in Figure 4-2;

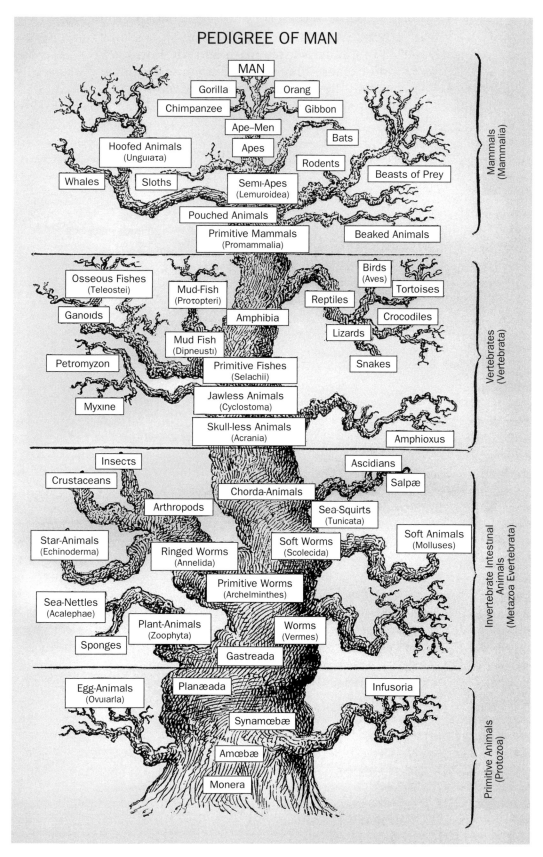

Figure 4.2 Early evolutionary tree. Ernst Haeckel's 19th century evolutionary pedigree or tree of man (humans), showing his take on the relationships of many major living groups. Diagram by E. Haeckel in *The Evolution of Man* (1879).

it comes from the work of Ernst Haeckel, a famous 19th century scientist, and although anachronistic for its title, *Pedigree of Man*, and its positioning of humans at the apex of life, views not accepted by modern-day evolutionists, it is illustrative of this principle that evolution is about common descent.

Furthermore, common descent implies a period of time when lineages shared history and subsequently diverged. For instance, chimpanzees and humans share a common ancestor that lived more than 5 million years ago. Humans and jaguars (Haeckel's *Beasts of Prey*) share a common ancestor that probably lived more than 60 million years ago. Humans and snails (Haeckel's *Soft Animals*) share a common ancestor that lived perhaps 550 million years ago. This is why chimps look more like humans than they look like jaguars, and chimps, jaguars, and humans look more like one another than any of them looks like a snail.

Implicit in this notion of evolution by common descent is the idea of a nested hierarchy or a tree relating groups of organisms. The more closely related the organisms are, the closer they sit on the evolutionary tree of life. Closely shared evolutionary relationship implies a common history or a period of time when groups were part of the same branch on the tree or the same evolving lineage. A key point about evolutionary biology is that there are different levels of shared common history in the tree of life, starting out with a single kind of organism. The definition of evolution does not concern itself with the mechanisms that may have been responsible for the actual pattern of the common descent on this tree. Rather, it deals only with the idea that there is a shared pattern of that descent.

A clear analogy exists in the study of the genealogy of our close relatives. When lineages in families or genealogies of humans are traced, this is tracing back patterns of reproduction. In evolution, genealogies are also uncovered, that is who is related to whom and the evolutionary relationships of different groups of organisms are deciphered. For instance, we may want to know the relationships of the frogs, mammals, crocodiles, and birds. Here is where the difference comes in between the biological definition of evolution and the term used above when discussing the evolution of rap. In a tree depicting evolutionary relationships among species or organisms the only criterion that we use is the history and the pattern of descent of lineages. By contrast, in the tree of rap, we arranged artists partly by stylistic considerations and also partly based on their order of appearance. One could debate whether or not KRS-One belonged with Grand Master Flash and the Furious Five or with Public Enemy or even on some other branch, depending on how much we factor in different stylistic considerations. In biology there is only one true evolutionary tree of life which scientists known as systematists reconstruct through their research. (Systematics is discussed in greater detail in Chapter 7).

The pattern of shared history implies a period of time during which the ancestors of organisms reproduced with each other. When branches in this tree of life separate it means that the organisms on the separate branches are no longer able to reproduce with each other.

The History of Evolutionary Thought

It is useful to clarify how ideas on the biological definition of evolution developed over the centuries. Too often discussion of the topic starts and ends with Charles Darwin and the publication of *On the Origin of Species* in 1859, but to do so gives an inaccurate picture of the history of evolutionary thought. Darwin is a great intellectual figure, but no idea of any consequence is developed in a vacuum (Figure 4-3).

Figure 4.3 **Charles Darwin**. A statue of the iconic man from the Natural History Museum, London. Image from iStockPhoto.com © A. Holwerda.

Stretching back many millennia is a long tradition in western thought of thinking about evolution. For instance, the Greek philosopher, Anaximander in 550 BCE wrote that humans were descended from fish. There were a few other early Greek philosophers who articulated evolutionary views. There was not much written about evolution from then on, however, at least little that is recorded, until about the end of the 17th century when there began a flurry of scientific activity that lasted until the last half of the 18th century. This included scientists and philosophers like Leibniz, who, along with Newton, invented the calculus, and the great French scientist Buffon.

In their works they wrote in places about how different species had been transformed into other species. It is unclear whether they understood evolution in the strict sense, that is, in the sense of common descent, because they sometimes wrote conflicting ideas. At times they seemed to assert that different species were descended from one another. Other times they did not, but suggested rather that all species had been created separately. In still other places they suggested that still living species were continually transforming into other species that were also still living. That is, some frogs were becoming turtles while other frogs were happy just to be frogs. Meanwhile some of the turtles would be evolving into alligators while other turtles were contented turtles. What this means is that the state of turtleness was evolving several different times over and over again.

This makes little sense if common descent has occurred, because it implies that species that had diverged already were evolving into other species and there was some shared pattern of history. It would be as if the Atlanta Braves transformed into the New York Yankees, while the Kansas City Royals suddenly moved to Atlanta and called themselves the Braves. The new Atlanta Braves would not be the same as the original Atlanta Braves. With musical groups, it would be as if one of those rap artists we mentioned, for example, Ice Cube, decided to form a new group with KRS-One, and one member of the Beastie Boys decided to call himself Ice Cube. This Ice Cube would not be the same as the original Ice Cube.

Part of the increased tendency towards evolutionism had to do with the enlightenment, the development of the age of reason, and the growth of science. It was no longer necessary to invoke supernatural or miraculous causes to explain natural phenomena. For example, the nature of lightning was discovered; it was found that the Earth rotates around the Sun; and scientists began to understand that gravity causes objects to fall. If such a wide variety of patterns in the physical world could be explained by scientific reasoning, patterns in the biological world could be similarly explained.

The First True Biological Definition of Evolution

The first person who assuredly described something akin to biological evolution and published on it was the great French biologist Lamarck (1744–1829) (Figure 3-4) who in 1809 published his ideas on evolution in a book entitled *Zoological Philosophy*. Lamarck believed that in animals there was a graded series of perfection, and animals were continually striving to be more perfect so that they were moving up a chain of being by transforming from one species to another. In this manner, Lamarck believed that simple species that started out on the bottom of the chain, like worms, transformed to more complex ones, like humans. The external environment, such as the temperature or the types of plants around, influenced how a species strove toward perfection by causing it to use or not use some of its parts. Thus, his was a belief that organisms were becoming more fit to their environment, and he emphasized the influence of the environment on evolution. The writings of Lamarck on evolution preceded those of Darwin by 50 years.

In the intervening period between the publication of Lamarck's book and Darwin's *On the Origin of Species*, many other people suggested the idea that life might have evolved, and it is clear that interest in the idea of evolution by common descent was building in Europe in the early 19th century. Although some people did not accept Lamarck's explanation for the mechanism of evolution, they were convinced by Lamarck and others that evolution had occurred, such that there was a groundswell of evolutionary thought leading up to the work of Charles Darwin.

Darwin's *Voyage of the Beagle*

In this milieu of increasing interest in evolution, Darwin in 1832 took a sea voyage around the world aboard the HMS *Beagle*. This five-year voyage influenced him profoundly, and he later claimed that it was the source of all of his ideas about evolution. On this trip Darwin saw several phenomena that were of relevance to evolution, but three of them had the greatest significance. These involve one pattern of change in life forms through time and two patterns of change in life forms across geographical space. They all involve observations he made in and around South America (Figure 4-4).

Pattern one

First, since Darwin was both a geologist and a biologist, he studied both fossil and living animals. In South America, he found some fossils in the Miocene rocks (now known to be about 10 million years old, although Darwin had no idea precisely how old they were) and determined that they looked somewhat like animals living in modern South America, specifically the modern llamas (Figure 4-5). So let us think of this pattern that Darwin found as an interesting pattern of change that occurred through time.

Pattern two

Second, Darwin found two species of large flightless birds, called rheas, in South America that look much like the ostrich of Africa (Figure 4-6). The two species of rhea look very similar to one another, although they are not identical. He identified an interesting pattern in the distribution of these rheas. The species did not occur in the same region. One species is found only north of a large river in southern South America. Another is found only south of this river.

Pattern three

Darwin's third observation is in some ways related to the second: it involves birds and their distribution. Off the coast of Ecuador lie the Galapagos, a chain of volcanic islands. Darwin observed that several species of birds, particularly mockingbirds, live in the Galapagos. All the species are similar, yet slightly different from each other, and each species is restricted to its own island. (Despite popular belief, the finches of the Galapagos, often termed Darwin's finches, were not used originally as an example of evolution.) The same is true of the large Galapagos tortoises. Different types, in fact now recognized to be subspecies, occur on each island.

Interpreting the patterns

Patterns two and three can be thought of as interesting geographical patterns of change that Darwin recognized, that is, patterns of change across space. The

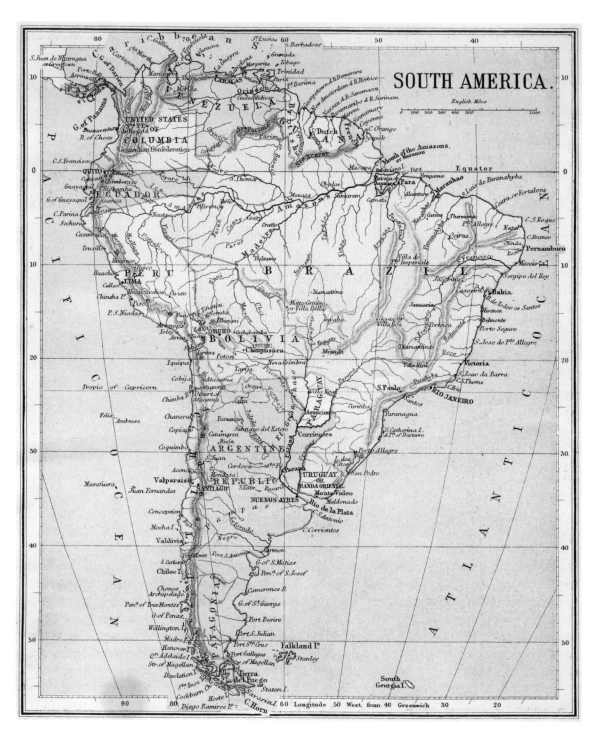

Figure 4.4 **South America.** A Victorian era map. Image from iStockPhoto.com © L. Steward.

Figure 4.5 **Llama**. From Ecuador. Image by B. S. Lieberman.

patterns involve similar species that were separated by geographical barriers. For the Rhea the barrier is a large river, and although these flightless rheas can swim fairly well, they do not do so well enough to cross this river consistently.

In the Galapagos the birds and tortoises on the different islands were separated by short stretches of ocean, a few miles or so across. Although, all of these birds in the Galapagos can fly and cross the short stretches of ocean, it is interesting that they do not do so, demonstrating that at some level all living beings are lazy.

Now, the reason we know that Darwin observed these things is because he published a popular book shortly after returning to England in 1839 entitled "*Journal of Researches into the Geology and Natural History of the Various Countries Visited by the HMS Beagle under the Command of Captain Fitzroy, Royal Navy, from 1832 to 1836.*" This not very catchy title was eventually abbreviated to simply "*Voyage of the Beagle.*"

Darwin's book described many of his observations during this trip, but Darwin really did not synthesize any of them, and at least when he wrote the book he still did not accept evolution by common descent and instead believed that all the species had been created independently. Shortly after he finished his book, however, he opened what is referred to by scholars in the history of science as his first notebook. These notebooks were a set of private diaries in which he started to

Figure 4.6 Rhea. Darwin's rhea, from Argentina. Image from iStockPhoto.com © G. Coles.

lay forth evidence for the idea that life had evolved. The chief reason he began to accept descent with modification was that he put together the information from South America. The ideas that species can change both across geographical space, as did the rheas and the birds and tortoises from the Galapagos, and also through time, as did the modern llamas and the fossil llama-like animals that we mentioned. Darwin's idea was that if you could study how things across geographical space change in the present you could develop a theory of evolution through time. The pattern of variation in two dimensions could be converted to a pattern of change through the third dimension, time.

There was another reason why Darwin became convinced that evolution by common descent had occurred. Evolution by common descent means that all of life is descended from a single common ancestor. The more closely related evolutionarily two species are, the more recently they shared a common ancestor. Think back to our discussion of Haeckel's tree of life and the way evolution happens. Branches split off the main trunk. Then twigs on that branch separate

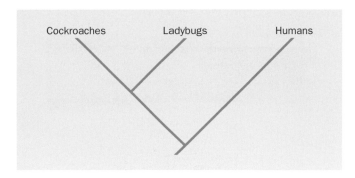

Figure 4.7 Cladogram. The hierarchical relationships among three living species. Diagram by F. Abe, University of Kansas.

later until you get the leaves at the very tips of the tree. This pattern of branching produces what is called a hierarchy. A hierarchy is a set of ever expanding relationships, like the Russian dolls that come one inside the other. If you look closely at all of life, you can see that there is a hierarchy that can be expressed as an evolutionary tree or cladogram (Figure 4-7). A cockroach and a ladybug look a lot more like one another than they look like a human.

The idea that there is such a hierarchy was accepted in Darwin's day, although most scientists then did not think it implied evolution. In spite of this opposition to evolution, scientists tried to classify life nonetheless. They expressed their classification of life in the familiar set of taxonomic categorics developed by the 18th century Swedish natural historian Carolus Linnaeus whom we mentioned in Chapter 3 along with the Linnaean hierarchy: Phylum, Class, Order, Family, Genus, Species. To reiterate, in this hierarchy, species that are closely related are in the same genus, genera that are closely related are in the same family, etc.

The Darwinian Notebooks

The patterns in South America that Darwin recognized took on special significance as he jotted down these and other observations in his notebooks. As discussed in Chapter 1, the study of geology and biology is about putting together different patterns we see in the real world and explaining these patterns using scientific processes. Darwin, in his notebooks, focused not only on the pattern of evolution and the evidence that evolution had occurred, but also he offered a mechanism to explain the pattern of evolution by common descent. Darwin called this mechanism natural selection: the idea that there was a struggle for existence among animals or plants within a species. We will discuss natural selection later in this chapter and break it down into its component parts, but for now understand that as Darwin developed these ideas he shared them with only his closest friends.

Darwin was a wealthy man of privilege. He was friendly with other rich and famous scientists, and these were the ones first exposed to his ideas about evolution and natural selection. One of these friends was Charles Lyell, the uniformitarian geologist introduced in Chapter 1. Darwin engaged in passionate debates with Lyell and other scientists about whether evolution occurred and what caused it. Initially, however, he kept his ideas within his close circle of friends and was hesitant to publish on his natural selection theory.

Figure 4.8 Alfred Russel Wallace.
The famous evolutionary biologist.
Image from M. Ridley.

Alfred Russel Wallace

Into this historical setting there stepped a man whose role in the history of biology is enigmatic and often under appreciated, Alfred Russel Wallace (1823–1913; Figure 4-8). Wallace was a poor man of limited education who had to struggle just to get by. He earned his living by voyaging to places like South America and the Far East and collecting many species of animals and plants including butterflies and birds from tropical rain forests. Wallace sent these back to England for sale to interested collectors and natural history museums.

Like Darwin, Wallace was a very astute observer. Wallace knew a lot about paleontology, and he also had immersed himself in the study of modern plants and animals. Alone, he put together two facts. First, the groups of species that occur in the fossil record tend to change through time. Second, if you look across geographical space, you tend to find very similar species living nearby one another, yet separated by geographical barriers. In a paper published in 1855, Wallace set forth the idea that species change through time. Also, he noted that as one crosses geographical barriers one tends to find similar but not identical species. That is, he recognized that species also change across geographical space. In short, he had come up with the idea that there was evolution by common descent, four years before Darwin published the idea. In response to Wallace's publication, Lyell urged Darwin to publish his own ideas in a scientific outlet, yet Darwin refused. His reticence is baffling. Perhaps he wanted to avoid arousing controversy, and feared offending many people's religious sensibilities. Although at this stage of his life Darwin was no longer religious, his wife still was. We shall return to the saga of Wallace and Darwin, but now let us stop for a moment to consider science and religion.

Science and Religion

In the 19th century many felt that religious beliefs and evolution were incompatible. Today, although many religious people see no conflict between accepting evolution and their religious faith, some fundamentalist religions deny the existence of evolution because it is in conflict with their doctrine. Yet interestingly, these same religions often do not deny other aspects of the scientific approach.

This is paradoxical because in reality the same scientific principles and approaches underlie and explain how a ball falls when released from a person's hand, how a microwave oven heats an object placed inside it, how cancer drugs are designed, and how life has evolved. The pattern of evolution is indisputable. This is not to say, however, that scientists do not still debate the processes that govern evolutionary change or the exact sequence of all evolutionary transitions. They do! The work of science is not finished. Just as physicists debate the origins of the universe or the various forces that have produced the structure and shape of our universe and just as biologists and biochemists debate how to design and identify the best potential cancer treatments, biologists and paleontologists debate the mechanisms of evolution and the pathways it has followed in forming the vast diversity of life.

Is Evolution a Fact or a Theory?

We shall treat the idea that evolution has happened as a fact as well resolved as the fact that the Earth revolves around the Sun. This claim might be jarring to some because we frequently hear evolution referred to in the news as a theory or, sometimes, as "just a theory." However, the word theory has very different meanings in science and everyday language. In everyday language, theory means a hunch or a guess. In science the word theory means an idea or groups of ideas that is supported by overwhelming evidence, that has never been refuted, and that answers a lot of questions about the way the world works.

Technically and in a scientific sense, the idea that the Earth revolves around the Sun is a theory: no person has sat out in space watching the Earth spin around the Sun. The evidence, however, that the Earth revolves around the Sun is overwhelming, and cannot be explained in any other conceivable way so that effectively, and at least in popular jargon, we can treat this idea as a fact.

Similarly, no scientist was present during the major evolutionary transitions that have occurred over the last 3.5 billion or so years. The overwhelming evidence, however, is that these evolutionary changes have happened and that evolutionary change continues to happen. We can treat evolution as a fact. There are various theories, however, to explain why evolutionary change occurs and why some lineages have survived, whereas others are now extinct. Darwin's ideas on natural selection are one such theory that we discuss more fully below.

What is the Evidence that Evolution has Occurred?

The evidence that evolution has occurred falls into four primary categories: small-scale, direct evidence of change documented in the laboratory; small-scale, direct evidence of change documented in the wild; large-scale, direct evidence of change documented in the fossil record; and large-scale, indirect evidence of change.

Small-scale, direct evidence from the laboratory
First, there is the evidence that evolution on the small scale has occurred in the laboratory. This is direct evidence, and there is much of it. Indeed, thousands of

pages of scientific journals are filled with a large litany of this type of evidence. For example, scientists have witnessed populations of fruit flies evolve through time in the laboratory. Strains of bacteria also evolve in lab populations through time.

Small-scale, direct evidence from nature

There is another type of direct evidence that comes from the documentation of evolutionary change known to occur at the small scale and in the wild. One famous example involves populations of moths in England and how they have changed over the last 100 years or so. During the industrial revolution, pollution levels rose in England. In parts of the country, levels of soot increased to such an extent that the trunks of trees became darker. This effected the moths in an interesting way. These moths spent part of their lives on tree trunks, on which they frequently land. The primary predators of these moths are birds, who identify their prey by sight. Against the soot darkened trunks of the tree, light colored moths, which originally dominated, were more easily visible to the predatory birds and thus were eaten such that the proportion of dark-colored moths in the population increased. An important result of the Clean Air Act of 1956, and subsequent updates to this act in 1968 and 1993, was a decrease in pollution levels in England over the last part of the 20th century. As pollution levels have declined in England, the amount of soot on tree trunks has declined concomitantly. Now, it is the darker-colored moths that are more visible against the tree background, and these are preyed upon more intensely, while the lighter colored moths are camouflaged better and harder for predators to see. Because of this, scientists have witnessed a decline in the number of darker-colored moths and an increase in the number of lighter-colored moths.

Another example of this type of direct evidence of small-scale evolutionary changes observed in the wild comes from studies of industrial areas. These tend to be contaminated by pollutants, and in these areas biologists have seen an increase in the tolerance of plants to these pollutants over time. A final, and unfortunate example worth mentioning involves the evolution of resistance to antibiotic drugs by bacteria. Through time bacteria that cause infections in humans evolve resistance to the antibiotics we employ to kill them. In particular, those that have resistance to a particular antibiotic are far more likely to survive and produce more bacteria. This is why scientists need to continually develop new antibiotics and why our society must be careful to use, and not abuse, those available.

Large-scale, direct evidence from the fossil record

In addition to this small-scale evolutionary change, there is also direct evidence at the large scale, visible in the fossil record, that records evolutionary transitions. An example was the shift from reptiles to dinosaurs to birds, which are, in effect, small, flying dinosaurs, that took place over many tens of millions of years (Figure 4-9).

In the sequence shown in Figure 4-9, the bird *Archaeopteryx* is transitional between dinosaurs and modern birds. Like modern birds, *Archaeopteryx* had feathers. It also had a well-developed breast bone, to which the flight muscles were attached, and relatively hollow bones, which are lighter and thereby make flying easier. Unlike modern birds, however, *Archaeopteryx* has a long bony tail and also teeth set into a bony jaw, both traits that we associate with large, nonflying dinosaurs. Moreover, in some of the dinosaur groups most closely related to *Archaeopteryx* and birds, the bones also became relatively hollowed, and,

(a)

(b)

Figure 4.9 Large-scale, direct evidence of evolution. (a) The Galapagos tortoise, image by B. S. Lieberman. (b) The dinosaur *Allosaurus*, image from iStockPhoto.com.

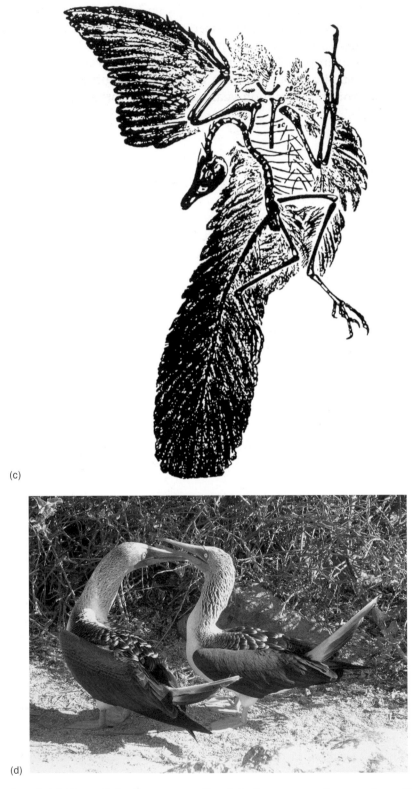

(c)

(d)

Figure 4.9 *continued*. (c) One of the famous fossils of *Archaeopteryx*, diagram from R. Cowen (2005) *History of Life*, 4th Edn, Blackwell Publishing. (d) Birds from the Galapagos, image by B. S. Lieberman.

furthermore, there was a structure in the wrist that is very similar to that of *Archaeopteryx*. There is even evidence, based on some recent fossil finds, that some of these nonflying dinosaurs may have evolved feathers.

Another excellent example of such direct evidence recovered from the fossil record is the numerous links of species in human evolution joining chimpanzee-like forms, dating from roughly 5 million years ago, to anatomically modern humans (see Figures 3-11–3-15).

Large-scale, indirect evidence

In addition to this overwhelming and persuasive direct evidence there is finally the indirect evidence. This indirect evidence reflects the theme of oddities and imperfections that record life's history. The noted paleontologist and evolutionary biologist Stephen Jay Gould was among the first to point out the significance of this indirect type of evidence. If all organisms were perfectly adapted to their environments because they were perfectly designed by a creator, then there would not be the pervasive evidence that life has a history.

Before we discuss actual examples of this kind of evidence of evolution from the history of life, we want to bring up an everyday example that may be more familiar to you and to which you can perhaps more easily relate. This example shows that things familiar to us through our popular culture have evolved and preserve indirect evidence of this history of change, thereby cementing the notion that parallel examples from life's rich pageant indicate that evolution has occurred.

One beautiful example of this involves our calendar system and the months of the year, specifically September, October, November, and December. Now if we look at the linguistic or etymological roots of these words from the Latin, we find that *sept-* is the root for seven; *oct-* is the root for eight; *nov-* is the root for nine; and *dec-* is the root for ten. In our current calendar, however, these are the ninth, tenth, eleventh, and twelfth months. What does this mean? It implies that our calendar has changed through time or evolved from one in which September to December were the seventh to tenth months to our present system in which they are the ninth to twelfth months; yet, during this evolution the new calendar did not become perfectly adapted and instead retained a record of its earlier history.

In the *Origin of Species* there are numerous examples of life's quirky little oddities. For example, the animals that live in the Galapagos, the islands we mentioned in reference to Darwin's *Voyage of the Beagle*, are most similar to animals found in western South America in the region of Ecuador (Figure 4-10). This happens to be the part of South America that is closest to the Galapagos. The two, however, have very different climates. If every organism were created to be adapted perfectly to its environment, why would very similar types of animals be found in very different environments? Furthermore, why would the organisms on the Galapagos resemble most closely the species of Ecuador—those that are the closest to them in geographical space? This makes sense only if life has descended from a common ancestor so that the species in the Galapagos share a common ancestor with species that live in present-day Ecuador.

Another beautiful example of such indirect evidence comes from the work of Stephen Jay Gould and involves the panda (Figure 4-11). Panda's are related evolutionarily to bears. In contrast to other bears, however, they are not carnivores but eat only bamboo. Unlike humans, bears do not have an opposable or grasping thumb. Instead, the thumb is joined to the other fingers. Bears have vicious claws,

Figure 4.10 Galapagos marine iguana. This lizard, with its distinctive marine lifestyle, is presumed to be related to iguanas from South America. Image by B. S. Lieberman.

but it would be difficult for a bear, being thumbless, to install a memory chip into a personal computer or perform other tasks that require great manual dexterity. The structure of the bears' hand is a feature that has characterized bears for millions of years. It is complex and thus probably difficult to undo evolutionarily. The way the panda eats bamboo is first to strip the outer woody layer from the bamboo to get at the juicy shoot inside. This is not easy to do without precision grasping. It turns out that pandas have a thumb-like projection from the hand that allows them to grasp and strip the bamboo. In short, it functions as an opposable thumb. If you study the skeleton of a panda's hand, however, you will see that it has the five digits with a structure like that of all other bears. The so-called thumb is actually a false sixth digit. It functions as a thumb, but it is a projection off the wrist bone and is, thus, not really an extra digit: it has no flexibility and lacks muscle. The wrist bone was all the panda had available to adapt in order to strip bamboo so that it could eat. It is not a perfect adaptation; a true, opposable thumb would have been much more efficient. If all species were created to be perfectly adapted to their environment, the panda should have been endowed with an opposable thumb. That it is not so endowed, because it would have been very difficult or impossible to evolve such a thumb based on the structure of the bear's hand that had characterized the bear lineage for tens of millions of years, means that we can trace the pathway of history in the panda's origin. This demonstrates that evolution has occurred.

Figure 4.11 A giant panda. Showing the hand of this herbivorous bear, featuring five digits and a "thumb." Image from iStockPhoto.com © E. Isselee.

Darwin and Wallace: Never Ask a Stranger to Present Your Paper at a Meeting You Cannot Attend

Now back to our narrative about Darwin and Wallace. What happened next is a fascinating story. Wallace, racked with a high fever in the tropics of southeastern Asia, came up with the concept of natural selection: the idea that all organisms were engaged in a struggle for existence, and the survivors were those that went on to populate the world. Those that tended to survive and give rise to future generations were more fit or better adapted to the environment than those that failed to make it. Thus, the Earth's flora and fauna changed through time.

After Wallace recovered from his fever, he wrote a paper on this topic and who should he send it to, but Charles Darwin? He did not know Darwin, but he had heard of him and was indeed very familiar with his name and his work because of Darwin's book *The Voyage of the Beagle*. Wallace asked Darwin to present the paper for him at a prestigious English scientific meeting. The year was 1858. (In those days many scientific papers were first presented at a meeting and then published in a volume describing the proceedings of the meeting.)

Darwin, upon receiving the manuscript, probably went into a state of shock. Here was someone who had come up with what Darwin thought was his idea alone. He realized that all his years of work and all his careful thought could go down the drain because he had not published. Then, Darwin feverishly went to work writing. He first took Wallace's paper and added to it a long section of his own. He added his name to the paper as a co-author; then he included a letter signed by two famous scientists of the day stating that Darwin had been thinking of these ideas long before Wallace. Then the paper was read at the meeting and published, while Wallace was still in Asia. The rest is, as they say, history. A year later, Darwin's book was published with the title *On the Origin of Species by Means of Natural Selection or the Preservation of Favored Races in the Struggle for Existence*, the book that is referred to typically as Darwin's *Origin of the Species*. Darwin received almost all the credit for the idea of natural selection described in the book, and he also received the credit for convincing people that evolution by common descent had occurred. Indeed, in 1889, seven years after Darwin's death, Wallace himself published a four-hundred paged book entitled *Darwinism*.

The moral is, never ask somebody you do not know to read something to an audience when you cannot attend the meeting. The interesting thing is that Wallace and Darwin became friends. Wallace was very magnanimous; that is, he did not feel as if he needed to get most of the credit for the idea. It was more important that the idea gained exposure. Of course, perhaps we might also wonder if Wallace was a bit naïve, but he probably appreciated the fact that Darwin gave him any credit at all for his ideas. In those days in England, if you were poor you had little prospect for success. Darwin's money and prestige made it easier for him to obtain most of the credit at the expense of Wallace, but also, we think, Darwin's social position made the idea more palatable to the general public. The general public and also the scientific community could understand how a wealthy, well-known gentleman scientist could come up with a great idea, but would sadly probably not be able to understand how some poor, relative unknown could come up such with a great idea.

It is sad, but in some respects things still can work something like this in science. Now, of course, wealth or social status are not such important factors, but the perceived prestige or power in the scientific community matters. Prestigious and powerful scientists find it much easier to get their ideas accepted than do young, unknown scientists.

In the end, it might be hypocritical to fault Darwin. How many of us, in the face of seeing 18 years of work go down the tubes, when given the opportunity to obtain credit for an idea, would not have done what he did? In the end, it is best to view Darwin as we would any other great thinker: he was human and therefore flawed. Those who came before him had created an atmosphere in which ideas on evolution could be accepted. Darwin was smart, and in the right place at the right time, which certainly beats being in the wrong place at the wrong time.

Natural Selection

A key point about Charles Darwin is that his name is not synonymous with the concept of evolution. He is a tremendously important figure in the history of biology and geology and a very important figure in the growth of ideas about evolution, but he is not the only figure. Questioning or challenging aspects of his ideas (or indeed those of other evolutionary biologists) does not call into question the idea that evolution has occurred.

Meaning of Natural Selection

Remember that there is a distinction between the idea that evolution occurs and the mechanism that causes evolution to occur. Darwin was certainly not the first to develop the idea that evolution occurs, not by a long shot. He was the first, however, along with Alfred Russell Wallace, to develop a reasonable mechanism to explain how evolution occurred. These two developed the idea that evolution occurred by natural selection. Now you probably grasp intuitively what natural selection means from our brief mention of it, but we would like to go into greater depth as to what it means.

Superfecundity

Natural selection is a statistical rule that is derived from the recognition of four important facts. The first is called the principle of superfecundity. This is a fancy term describing the idea that all organisms produce more offspring than can possibly survive. An excellent example of this is the cod fish. A female cod fish produces as many as a million eggs, but only a small proportion of these survive to produce eggs of their own. Imagine if they all had 1,000,000 offspring that lived and these offspring in turn had 1,000,000 offspring that lived. Very shortly we would be up to our ears in codfish.

Variation

The second crucial fact in the formulation of natural selection is that there is variation. Offspring do not look exactly like their parents, especially among species that reproduce sexually. Further, within populations there is variation.

Inheritance

The third crucial fact is that much of this variation is inherited. That is, offspring resemble their parents.

Survival and reproduction

The final fact is that sometimes these differences may make it more likely for the organisms to survive and eventually reproduce successfully. Perhaps some of those codfish differ in ways, perhaps not visible to our eyes, that make them more likely to survive or attract mates.

Nothing from a Vacuum

Darwin came to appreciate facts two and three as part of his interaction with animal breeders in England, especially, those who bred pigeons. Sometimes a new pigeon with a distinct type of color marking on the wing would hatch. If the pigeon breeders fancied it, they would try to breed it to get a large number of similar pigeons. In fact, animal breeders have been applying these principles for thousands of years: selecting for favorable traits by allowing those with the desired traits to breed.

It has been suggested that Darwin came to appreciate the first fact from reading the work of Thomas Malthus. Malthus is a famous 18th century figure who reasoned that populations of humans can increase very rapidly or geometrically

(2, 4, 8, 16, 32, . . .) like the codfish, whereas the food supply can increase only algebraically (2, 4, 6, 8, 10, . . .), which is much more slowly. Thus, the world's human population will perpetually be in a state of near famine, unless there is some control of population size—what amounts now to birth control. As we move through the early part of the 21st century, we can see how prescient Malthus was. The ramifications of massive human population size will be the greatest challenge our children and grandchildren will face, as we shall see in later chapters.

So from these facts came the idea that natural selection was the means by which evolution occurred. Only some organisms will survive out of the many that are born. As in acting: many are called, but few are chosen. Organisms vary; that is, they differ in some of their characteristics. Some of these characteristics may make an organism more or less fit or adapted, either to its physical environment or to compete with other organisms. On average, those that are more fit survive and go on to give rise to the offspring that make up the next generation. Those that do not survive or do not reproduce do not give rise to subsequent generations and are said to be selected against.

This does not mean necessarily, however, that organisms are actually getting better or smarter or faster. Some traits might be beneficial in one local environment but not in another. For instance, imagine a rodent that spends all of its life living in narrow tunnels. It might not confer much survival advantage to have a large skull with a big brain in such an environment. In fact, such a trait might be selected against, because the rodent's large head would get stuck in the tunnels, leading to a bad end. (By contrast, in other types of rodents, perhaps those that mostly live above ground and bury or hide nuts for future use, it might be beneficial to have a larger brain to remember where they buried all those nuts!)

This is the mechanism for evolution that Darwin laid out in his private notebooks, that Wallace laid out in his 1858 paper that he sent to Darwin, and that Darwin laid out in his addendum to that paper and in the *Origin of Species* in 1859.

Conclusions: Why was Natural Selection Not Endorsed at Once by Many Scientists?

Although natural selection is very much accepted now as a mechanism of evolutionary change, this was not true until the 1930s. This can be attributed to several factors. First, in Darwin's day biologists did not have an understanding of how the mechanisms of heredity worked, that is, how genetic information is passed from parent to offspring. Darwin believed that it occurred by the complete blending of the traits of both the mother and the father. If inheritance did occur in this manner, then there would be no discrete traits of organisms that could be selected for and inherited. Instead, traits would continually be watered down, making concerted evolutionary change due to natural selection difficult.

In the late 1860s the Austrian monk Gregor Mendel discovered the principles of genetic heredity that we accept today. His paper was published in a very obscure magazine where it was not seen by many scientists. Although he had mailed a copy of his article to Darwin, the pages were left uncut and Darwin never read it. Mendel's paper was quickly forgotten, and the principles of genetic heredity were not rediscovered until the early 1900s in laboratories at Columbia University through experiments on fruit flies.

Another possible reason why natural selection was not accepted originally relates to some interesting issues that still plague the lay public's understanding of the term evolution. We said earlier that in Darwin's day evolution was not a term that meant that all animals and plants are descended from a single common ancestor. Rather, it was a term that was originally from the science of embryology and was used to describe the changes that an organism went through during its lifetime. Evolution also had another meaning, which was even more significant in Darwin's day and also still today. Evolution has often been taken as being equivalent to progress. In this sense, evolution could reflect a belief that human civilization had progressed through time, which may be true, or it could reflect a belief that progress had occurred throughout the history of life. Now, we are not asserting that this is what Darwin believed or even what we believe, but rather it is what many scientists in the 19th and early 20th century believed and also what many people continue to believe even to the present day. Such a view is highly consonant with Lamarck's views on evolution. This has often been expressed in the iconography of our own human lineage (Figure 4-12), where it is assumed that we have evolved through a progressive series of less and less stoop-shouldered descendants, until we stand fully upright in the modern world.

This view of evolution and progress is still so pervasive throughout our culture that it is readily adapted by cartoonists, showing a deep-seated bias. That is, people may believe that this is the way evolution happens (even though it does not) for social and psychological reasons that continue to support our own perceived superiority (Figure 4-13).

Thus, there is this idea that the use of the word evolution implies something about progress. It was true in Darwin's day, and it remains generally true today.

Figure 4.12 **Evolution as progressive change.** A diagram illustrating the manner in which many in the public assume evolutionary transitions happen: a linear, progressive pattern. Diagram from iStockPhoto.com © C. Alcantara.

Figure 4.13 Evolution in cartoons. A caricature of linear, progressive evolution, showing key stages in aging from toddler to high school graduate. Diagram from iStockPhoto.com © D. Purcell.

Yet Darwin used the word progress in his writings only sparingly. Why was that so, and what does it have to do with the initially limited support for his ideas on evolution? Part of the paradox of Darwin was that he was wealthy and from a socially conservative background, yet in some ways he was a committed liberal. He raised strong opposition to slavery, for instance, whereas others from his social class may have favored it. Darwin also had a distaste for the way most people of his day, particularly nonscientists, viewed the biological world. Many religious figures and educated lay people in his day believed in what is termed the argument of design, the idea that all organisms and all of the world's ecosystems are completely perfect and harmonious. They argued that this was evidence that the Earth and its life had been designed by a deity. The more radical side of Darwin, along with the data he gathered and the observations he made, caused him to completely reject this notion of perfection in design. First, he believed that the world was replete with poorly designed organisms that were simply good enough to get by. He also observed lots of misery in the world and felt that life had not been designed to be perfect and good. Furthermore, Darwin believed that whatever good design existed was due to natural selection and to organisms being engaged in a struggle for existence, with only those well-adapted kinds of organisms winning out and giving rise to future generations. So, Darwin failed to believe in this principle of perfection and felt that whatever good design did occur was the result of natural biological mechanisms. In Chapter 5 we will consider in greater detail the idea that the history of life is about progress.

Additional Reading

Brooks, D., and McLennan, D. 2002. *The Nature of Diversity: An Evolutionary Voyage of Discovery*. University of Chicago Press, Chicago, IL; 668 pp.

Brooks, J. 1984. *Just Before the Origin*. Columbia University Press, New York; 304 pp.

Browne, J. 1996. *Charles Darwin: Voyaging*. Princeton University Press, Princeton, NJ; 622 pp.

Browne, J. 2003. *Charles Darwin: The Power of Place*. Princeton University Press. Princeton. 600 pp.

Eldredge, N. 2000. *The Triumph of Evolution and the Failure of Creationism*. W. H. Freeman, New York; 223 pp.

Eldredge, N. 2005. *Darwin: Discovering the Tree of Life*. W.W. Norton, New York; 288 pp.

Gould, S. J. 1992. *Ever Since Darwin*. W.W. Norton, New York; 288 pp.

Gould, S. J. 1992. *The Panda's Thumb*. W.W. Norton, New York; 352 pp.

Lieberman, B. S. 2000. *Paleobiogeography*. Springer, New York; 222 pp.

Mayr, S. 1985. *The Growth of Biological Thought: Diversity, Evolution, and Inheritance*. Belknap Press of Harvard University Press, Cambridge, MA; 992 pp.

Prothero, D. R. 2007. *Evolution: What the Fossils Say and Why it Matters*. Columbia University Press, New York; 408 pp.

Raby, P. 2002. *Alfred Russel Wallace: A Life*. Princeton University Press, Princeton, NJ; 368 pp.

Scott, E. C. 2005. *Evolution vs. Creationism: An Introduction*. University of California Press, Berkeley, CA; 298 pp.

Ward, P. 2001. *Future Evolution*. W. H. Freeman, San Francisco; 192 pp.

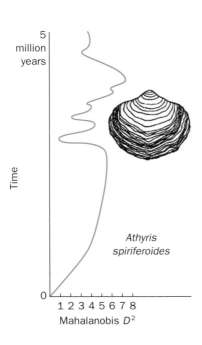

Chapter 5

Macroevolution, Progress, and the History of Life

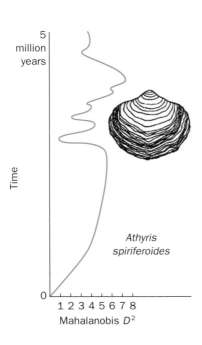

Athyris
spiriferoides

Time

5 million years

0

1 2 3 4 5 6 7 8
Mahalanobis D^2

Outline

- Introduction
- Competition and Macroevolution
- Does Evolution Happen Gradually or Episodically?
- Natural Selection Operating Above and Below the Level of the Individual Organism
- Progress and the History of Life
- Conclusions: Patterns and Processes of Increasing Complexity
- Additional Reading

Introduction

This chapter will focus on three fundamental questions.

1 Is it competition that drives major evolutionary changes and causes the extinction of major groups of life?
2 Does evolution occur gradually or in spurts of change interspersed between long periods of stability?
3 Does evolutionary change favor the production of more progressive and more complex lineages?

These questions are fundamental topics in macroevolution, the area of paleontology and evolutionary biology that focuses on the birth, death, and persistence of species and groups of species.

Prehistoric Life: Evolution and the Fossil Record. 1st edition. By Bruce S. Lieberman and Roger Kaesler. Published 2010 by Blackwell Publishing.

Regarding competition, it has been shown that competition causes natural selection and can favor particular characters in certain organisms. Thus, competition can lead to the evolution and maintenance of adaptations. But, to what extent, at the grand scale among groups of species, has the history of life been structured by competition. Specifically, are all organisms and groups perpetually engaged in a struggle for existence driven by natural selection? Answering questions like this is relevant because it helps us understand which evolutionary processes are most important. At one level the answer to the question is yes, but by the same token that does not necessarily mean that when we study the evolution of individual species or groups of species in the fossil record that we can see the pervasive stamp of competition. As part of the consideration of competition, and in light of our discussion of natural selection in Chapter 4, evidence will be presented that not only organisms but also genes and species can be selected. This represents the recognition that evolution is a process that operates both above and below the level of individual organisms.

Similarly, there has been considerable debate about the tempo of evolution and whether evolutionary change occurs gradually, as Darwin argued, or episodically and in sudden pulses. As we shall show, the answer to this question partly depends on the scale being considered, but the view that evolutionary change always occurs gradually is inaccurate.

The final focal point of this chapter considers whether more progressive and complex lineages are evolutionarily favored. In some respects this issue is complicated, because concepts like progress and complexity are difficult to define. Still, scientific results suggest that on the whole possibly more progressive and certainly more complex organisms have appeared on the planet since life first evolved. As we shall explain, however, this fact does not mean necessarily that there is an active evolutionary force that favors more progressive or complex organisms.

Competition and Macroevolution

The idea that there is a struggle for existence that drives evolutionary change via the mechanism of competition is explicit in the writings of Darwin and Wallace. Actually, the idea has an even longer intellectual heritage than that. Perhaps the first person to suggest it was the French botanist Augustin de Candolle, who in 1820 said that all the plants and animals are at war with one another. The ready acceptance of this idea by 19th century Victorian England, embodied in Darwin's *On the Origin of Species*, at least partly has to reflect cultural biases. In Darwin's day the economist Adam Smith's ideas on economics, presented in his book *The Wealth of Nations*, which was published in 1776, were accepted currency. The idea posited there and accepted by Victorian society was that it was bad to regulate the economy. Instead, Smith believed that if trade were free, unrestricted, and competitive, it would benefit the economy and the nation. In a book about the history of life, not economics, it is not of primary importance whether Smith's ideas were correct. We wish merely to assert that the culture in which scientists live influences their ideas. Smith's ideas are expressed in Darwin's writings: Darwin argued that competition was a highly important evolutionary force. In fact, Darwin was correct in many ways, as today it is well known that within populations of living species organisms frequently compete for resources and mates.

Paleontologists have also sought to find evidence for the importance of competition in the fossil record. They have specifically focused on whether through

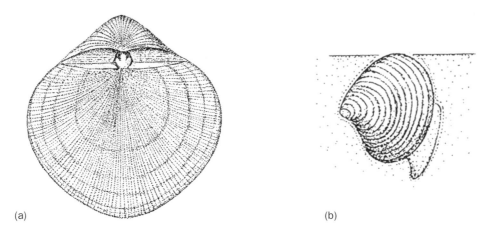

(a)   (b)

Figure 5.1 **Brachiopod and clam**. (a) A Silurian brachiopod and (b) a modern clam in life position. Diagrams from E. N. K. Clarkson (1998) *Invertebrate Palaeontology and Evolution*, 4th Edn, Blackwell Publishing, (b) redrawn from Stanley.

time the success of one lineage has driven another lineage to extinction; it turns out that they have only rarely been able to discern such a pattern.

Case Studies of the Role of Competition in Macroevolution

One of the classic cases where competition has been invoked is to explain the evolutionary history of two major groups of animal phyla: the brachiopods and the clams (Figure 5-1). An examination of the Paleozoic fossil record reveals that brachiopods were once very diverse, comprising thousands of species, and also very common and abundant. If you could walk along an early Paleozoic beach it would be littered with brachiopod shells; you might find some clam shells, but they certainly would not be as abundant, nor would they be represented by as many diverse types of species. By contrast, if you walk along a modern beach, you are apt to find clam shells everywhere, whereas one would be very hard pressed to find brachiopod shells on most beaches. The brachiopods are now only moderately abundant and diverse in such regions as New Zealand and Antarctica, and then only in deeper water. A traditional explanation for this pattern is that the clams competitively displaced the brachiopods from their preferred environments because they were more fit and better adapted. Physiological studies have shown that clams are more fit at performing some tasks needed to sustain them than brachiopods, and they are also much more capable of activity and movement. Indeed, Martin Rudwick, a well-known and highly respected paleontologist and a specialist of brachiopods, argued that "the decline of brachiopods was due to their constitutional inaptitude to compete successfully with the Mollusca under changing conditions."

Competition has been invoked to explain the differing trajectories of diversity and abundance in these groups based on two arguments. First, it has been recognized that brachiopods and clams superficially look alike. They are animals that live enclosed between two calcareous shells; they both live in the ocean and often near the seashore. (Although the clams and brachiopods share superficial similarities, they are only distantly related. The characters they share are believed to be what are called convergences that evolved independently. See Chapter 7 for an extensive

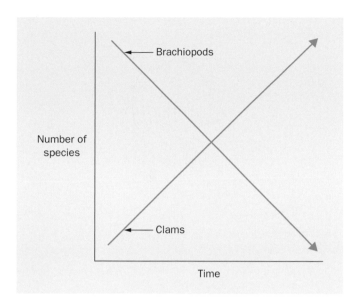

Figure 5.2 Negative diversity correlation. The supposed diversity history of clams and brachiopods throughout the Phanerozoic. Diagram by M. Smith.

discussion of how scientists determine the evolutionary relationships of organisms and how they sort out which characters are valid evidence that organisms are related and which characters that appear similar evolved independently.) The second fundamental tenet of this argument is based on the idea that diversity appears to rise in one group, the clams, while it falls in the other, the brachiopods, such that the diversity of the two groups can be said to be negatively correlated (Figure 5-2).

Scientists, however, need to be very careful when they invoke these two types of argument for several reasons. The first argument only superficially considers the similarities and shared habitats of these two very distinct types of organisms, so we cannot be sure that they are actually in direct competition with one another for food and other resources.

Furthermore, with regard to the second argument, there are lots of false or spurious, negative (and positive) correlations that can occur just by chance. In other words, correlation does not equal causation. It may imply it, but without additional information it does not demonstrate it. As examples of so-called spurious positive correlations consider the following example. The price of gasoline in 2008 and the distance between the Earth and Halley's comet are both positively correlated. The price of gasoline has generally been going up (at least as this book was being written) while the distance between the Earth and Halley's comet has also been increasing. Is there any causality to this positive correlation? Assuredly not. Another amusing, but spurious positive correlation, is that the number of teachers and alcohol consumption are both increasing in the US population. Is there any causality here? Educators everywhere certainly hope that there is not. More likely they are both correlated with a general rise in population. The lesson: always be careful whenever there is the tendency to compare two things that may be changing in the same direction. The correlation can be spurious.

It turns out that when the pattern of diversity in clams and brachiopods was considered in greater detail, it was found that even the pattern of a simple negative correlation was not realized. The relevance of the actual recovered pattern, and its significance, was pointed out in a landmark study by the paleontologist Stephen Jay Gould and the statistician Calloway. They tabulated the different number of types

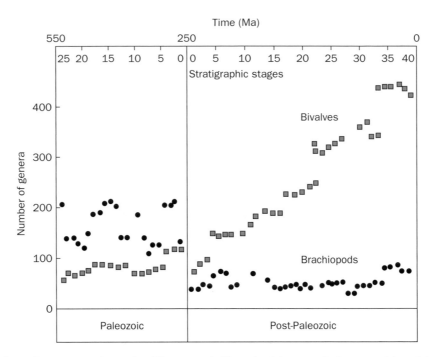

Figure 5.3 **Ships that pass in the night.** The actual diversity history of clams and brachiopods throughout the Phanerozoic. Diagram from M. J. Benton and D. A. T. Harper (2009) *Introduction to Paleobiology and the Fossil Record*, Blackwell Publishing; based on Gould and Calloway.

of brachiopods and clams in the fossil record through time, and they found that the diversity of these two groups does not seem to be strongly associated throughout their evolutionary histories (Figure 5-3). Neither group shows much change in its general trajectory or pathway of diversity either before or after the end of the Permian: brachiopod diversity is flat; clam diversity slowly climbs, though the rate of this climb does increase somewhat, but not until after the start of the Triassic when brachiopod diversity had already fallen. More importantly, the relative position of the trajectory of the groups changes primarily because something dramatic happened to the brachiopods at the end of the Permian period. This event was the Permo-Triassic mass extinction event, the most severe extinction event in the history of life. (The nature of mass extinctions in general and the Permo-Triassic mass extinction in particular will be considered in Chapter 6.) The brachiopods were hit harder at the Permo-Triassic boundary than the clams. The diversity of both groups declined at this event, but the clams declined much less precipitously; by the earliest Triassic clams had a greater diversity, which they never relinquished. Thus, it would be inaccurate to argue that these groups were interacting competitively, at least in such a way so as to influence their macroevolutionary patterns. The two groups responded differently to a short but profound event, and this is in large part why clams are abundant today while brachiopods are relatively rare.

Other paleontological studies have also generally upheld the notion that one group of animals typically does not drive another group to extinction. For example, paleontologist Mike Benton has investigated the role that competition plays in effecting macroevolutionary patterns in the group of vertebrate organisms called tetrapods. These include amphibians, reptiles (thus dinosaurs and birds), and also our own group, the mammals. After sampling fossil groups containing thousands of species of tetrapods, Benton found only limited evidence that one group of tetrapods may have driven another group to extinction. This does not mean that competition never occurs or is unimportant, but it does mean that we need to be cautious with assumptions about the relative role that competition plays in the history of life at this scale.

One reason that competition may not have always been important in the fossil record is that the history of life records several major mass extinctions. During these times of mass extinction the basic competitive rules no longer applied because the environmental changes that occurred were so profound. As a result, organisms went extinct not because they were being outcompeted but rather because they could not survive the major changes (which may have been quite rapid as well) that occurred. More examples of this will be presented in Chapter 6, when mass extinctions are discussed in greater detail.

Does Evolution Happen Gradually or Episodically?

The traditional view held by various scientists including Darwin was that evolutionary change happens gradually in a manner termed phyletic gradualism. Such gradual evolution was assumed to involve one species slowly transforming into another species by what is referred to as anagenesis (Figure 5-4). One of the reasons Darwin supported gradual evolution was that if evolution occurs gradually he felt it would be impossible to invoke some supernatural miracle to explain how new species originate. His belief was codified in the Latin term *"Natura non facit saltum,"* which translates as "Nature does not take jumps."

Thomas Henry Huxley was one of Darwin's greatest supporters in the 19th century. He was known as Darwin's Bulldog because he crossed England giving lectures praising Darwin's ideas. He supported many of Darwin's ideas, but Huxley disagreed with Darwin on the principle of *Natura non facit saltum*. Huxley was convinced that the evidence that evolution had occurred was overwhelming, but he

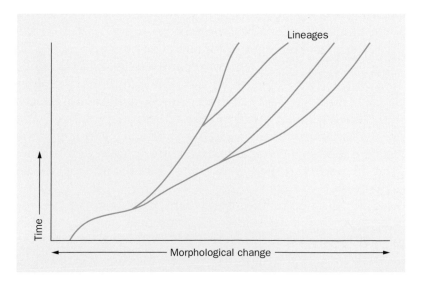

Figure 5.4 Phyletic gradualism. The way evolution is often assumed to occur, via anagenesis, involving the gradual transformation of one species into another. Diagram from R. S. Boardman, A. H. Cheetham, and A. J. Rowell (1987) *Fossil Invertebrates*, Blackwell Publishing.

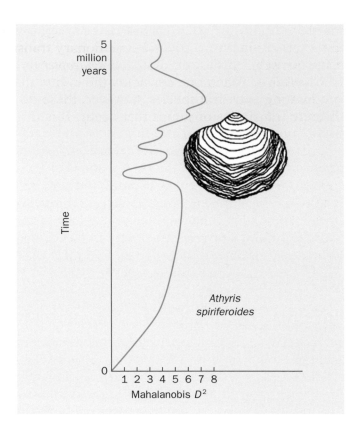

Figure 5.5 Stasis. Most species show little change throughout their evolutionary history in the fossil record. A representative pattern from the Devonian brachiopod shown is illustrated with time on the vertical axis and morphology on the horizontal axis. Diagram from Eldredge, N., Thompson, J., Brakefield, P., et al. (2005) The dynamics of evolutionary stasis. *Paleobiology* 31: 133–145. The Paleontological Society.

felt that the fossil record provided ample evidence that evolution did sometimes occur in jumps. Darwin was actually well aware of this characteristic of the fossil record. To get around it, he argued in the *Origin of Species* that the fossil record was far too imperfect to preserve the gradual sequences of evolutionary change he predicted. Darwin posited that change between species might look relatively rapid, but this is only because the intervening rock layers are missing. If they existed, Darwin believed it would be possible to trace a complete chain of intergrading species through time.

It is now known that although the fossil record is not a perfect chronicle of the history of life, it is fairly accurate and complete in several respects (Chapter 2). One basic pattern of evolution in the fossil record was first discussed in detail in a landmark paper by the noted paleontologists and evolutionary biologists Stephen Jay Gould and Niles Eldredge. As they documented, and as subsequent studies have continued to reiterate, the typical picture of what happens to individual species throughout their evolutionary history is stability: often species are stable for millions of years. They start out with a given anatomy or morphology, and they do not change much if at all, and then they go extinct. Eldredge and Gould termed this stability stasis (Figure 5-5).

Eldredge and Gould argued further that when change happens, it seems to occur in discrete jumps. Now the actual size of the jump need not be very large; Eldredge and Gould were not invoking sudden transitions between things as different as cows and whales in a single speciation event. They were not even invoking more moderate transitions such as those that might be required to change something like a gorilla into something like a chimpanzee. Instead, they were talking about much smaller-scale anatomical changes and transitions. Thus, in one respect, as discussed already, the scale being considered determines whether evolutionary transitions are treated as being gradual or sudden. At the grand scale of the history of life, there are a set of effectively gradual evolutionary transitions that are preserved like those between turtles, dinosaurs, *Archaeopteryx*, and modern birds mentioned in Chapter 4. When one considers the events that produce new species from pre-existing ancestral species, however, these do involve discontinuities: relatively discrete transitions or jumps that occur. It is this pattern of evolution that Eldredge and Gould first brought to the fore.

In their paper, Eldredge and Gould presented examples from groups as different as trilobites and snails to show that the way transitions between ancestral and descendant species occur is in small but discrete and relatively sudden transitions. The transitions appear in the fossil record to have been relatively rapid, particularly when compared to the duration of individual species, which seems to be on the order of millions of years. It is these small-scale jumps or punctuations, bracketed by the long periods of stasis or equilibrium, that Eldredge and Gould described using the term punctuated equilibria.

Eldredge's research concentrated on a group of common and well-known trilobites from the Devonian Period of New York State that belong to the genus *Phacops* (Figure 5-6). Eldredge studied the evolutionary history of four species of trilobites in the genus *Phacops* known from these rocks. Trilobites had compound eyes with many lenses, in some respects like the modern housefly, and the species of *Phacops* differed from one another primarily in the number of rows of lenses in the eye (Figure 5-7).

The four species Eldredge studied occur in rocks that were deposited over a period of approximately six million years. These trilobites are extremely common in

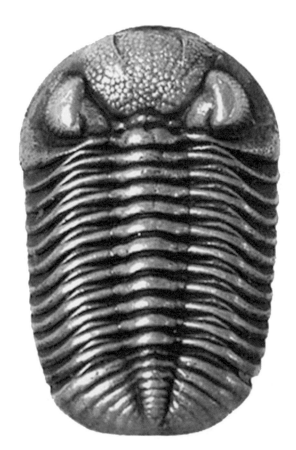

Figure 5.6 *Phacops*. One of the species of Devonian trilobite studied by Niles Eldredge that served as the hallmark example of punctuated equilibria. Image from N. Eldredge, American Museum of Natural History.

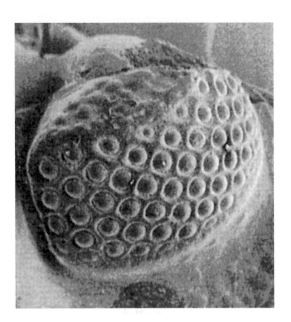

Figure 5.7 Trilobite eye. A close up of the eye of *Phacops*. Image from E. N. K. Clarkson (1998) *Invertebrate Palaeontology and Evolution*, 4th Edn, Blackwell Publishing.

these rocks and known from literally tens of thousands of specimens. Eldredge found that the oldest species started out with 18 rows of lenses in its eyes and then persisted unchanged for two million years before it went extinct. Sometime during its existence, however, a small population on the periphery of the species territory (in what is now New York State, which was under seawater at the time) became isolated and lost one of the rows of lenses in its eye (leaving only 17 rows of lenses). This species also persisted for a very long time, four million years, without changing. Twice more populations of this species became isolated on the margins of the species range and then speciated: a species with 16 rows in each eye evolved; and a species with 15 rows in each eye evolved. Each of these species was also stable until it went extinct.

Eldredge documented three basic patterns from the study of these species.

1 After species appear, they tend to change little. Minor oscillations in their anatomy may occur, but they are reversible and do not lead to any net change in the species.
2 New species appear when small populations of the ancestral species become isolated at the margins of where the species lived. Speciation in this case occurs by cladogenesis, which is defined as the branching of a single lineage to produce two (or more) species (one being the ancestral species, the other(s) the descendant species). Cladogenesis (Figure 5-8) is distinguished from anagenesis (Figure 5-4). Speciation that happens in a population on the margins of the species range and in a region that is geographically isolated is referred to as allopatric speciation. Scientific evidence suggests that evolution and speciation are more likely to occur in small, narrow, reproductively isolated populations than in large populations that frequently have organisms interbreeding with organisms from other such populations. This is because when genetic mutations occur and new traits appear the smaller, isolated populations are much more likely to become fixed for these few new traits. By contrast, in a large species with many interbreeding populations these new traits tend to be swamped; thus these types of species will tend to change little.

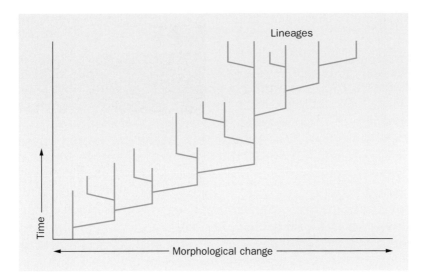

Figure 5.8 Cladogenesis. The way evolution typically occurs, by a pattern of branching speciation. Diagram from R. S. Boardman, A. H. Cheetham, and A. J. Rowell (1987) *Fossil Invertebrates*, Blackwell Publishing.

3 When speciation occurs the type and amount of change seems to be fairly modest, such as a shift in the number of lenses in the eye.

Eldredge and Gould showed that evolutionary changes typically do not comprise gradual changes across a huge population through time and from one species to another. Their findings have been supported by a great many studies. There are some exceptions, but anagenesis is not the way that life typically evolves. Rather, evolution occurs in a bunch of fits and starts.

The duration of the stability relative to the fits and starts also bears mentioning. It is believed that the speciation events discussed in Eldredge's example and in the other subsequent studies conducted since then occurred over a span of 5,000 to 50,000 years. From the perspective of an individual human life these transitions seem to be extremely slow and gradual. Compared to the duration of an individual species, however, which is typically measured in millions of years, they are quite rapid. Moreover, a 5,000 to 50,000 year transition in the fossil record will often appear rapid or instantaneous: because of the nature of the fossil record, events that occur in less than 50,000 years or so may not typically leave a trace.

When scientists working with modern animals study the types of small-scale evolutionary changes over years to decades that can lead to speciation, they observe gradual change. However, when we look at larger changes associated with speciation preserved in the fossil record, all that we typically see is the sudden appearance of a new species. Thus, evolutionary changes that may appear gradual to a person studying modern animals appear sudden to a paleontologist, because of the nature of the fossil record.

The fossil record suggests that the history of life is broken up into discrete, relatively rapid speciation events punctuated by long periods of stability or stasis. That evolutionary change frequently occurs in this way is not surprising given our discussion of uniformitarianism in Chapter 1, where we suggested that rates of change in the geological record usually are not uniform. In a way, this may be analogous to the way change happens in general. For instance, consider the communist regime in the Soviet Union. It appeared remarkably stable, yet it crumbled in a few short years.

Now, imagine if this book could send you on a virtual paleontological excursion and field trip that would allow you to walk forward in time past a set of rock layers. The classic view from Darwin predicts that as you walk through time one species would slowly and gradually evolve into a different species. In reality, what you would see, because of the prevalence of punctuated equilibria, is that the species you started with as you walked through time would not change, but abruptly another new species would appear; the original species may or may not persist along with the new species.

The recognition that species tend to persist unchanged throughout much of their history has important implications. It suggests that much of the history of life is not about lineages becoming continually more refined and better adapted to environmental conditions. This does not mean that species are not adapted; clearly they are. It also does not mean that species are never influenced by natural selection; again, clearly they are. It is simply that species are not continually changing their features, and natural selection is not necessarily the only force shaping the history of life. This is one reason why competition between and within populations, known to occur and to be important in modern populations, may not explain many of the large-scale patterns in the history of life.

In conclusion, it is worth mentioning one of the best examples of stasis comes from the lineage *Homo erectus* that is closely related to our own species, *Homo sapiens*. Throughout its roughly one million year history *Homo erectus* shows little if any change in its skeletal morphology, including its brain size, matching in detail the pattern predicted by punctuated equilibria. Similarly, within our own species, since it originated in Africa perhaps 100,000 years ago, there have been relatively few changes in skeletal morphology, and the changes that have occurred are principally related to increasing size associated with an improving diet. Science fiction writers and cartoonists often present fanciful images of our own species in the future, and these fanciful reconstructions usually involve humanoid like forms but with giant heads and tiny arms. Based on what we know about how evolution happens, however, these reconstructions are simply science fiction. Our species has changed little since it first evolved, and the current state of humanity on Earth, with large populations with the potential for nearly instantaneous global travel, are precisely the types of conditions that will prevent future evolution. The only possibilities for future substantive evolutionary change in our own lineage would occur if small populations of humans become isolated, perhaps on another planet thanks to the exigencies of space travel. Again, at least for now, this idea belongs to the realm of science fiction.

Natural Selection Operating Above and Below the Level of the Individual Organism

Natural selection does indeed act to shape populations, as Darwin and Wallace emphasized, but more and more evidence is accumulating to suggest that natural selection does not just operate at the level of individual organisms. Instead, things as distinct as genes and also species can be selected. For example, it has been shown by molecular biologists Doolittle, Sapienza, Crick, and Orgel that selection also operates at the level of genes inside of our body. In particular, gene selection explains why there are thousands of copies of some genes in our body. About

20 percent of the human genome (and this is true for other species as well) is made up of these types of genes, yet these genes appear to have no function.

How is it possible that a substantial part of our genomes are effectively junk? Current scientific understanding suggests that what has happened is that through time the number of copies of these genes has increased relative to other genes. This is possible because these genes do not have any specific function and their proliferation is not detrimental to the organisms they are inside of. Our bodies simply do not notice them, and the number of copies of these types of genes increases through time, just as certain individual organisms produce more offspring through time. In effect, because of selection at the level of the gene (gene selection) the representation of these genes is increasing in future generations, relative to other genes. As long as these genes remain invisible to the organisms carrying them, by not having any negative effects, the number of copies of these genes will continue to increase.

Just as there are things inside organisms that can be selected, there are things at larger scales, that contain organisms, that can be selected. For example, above the level of organisms, species are also potential targets of selection. Species selection was first identified as a potentially significant process by paleontologists Steve Stanley, Stephen Jay Gould, and Niles Eldredge, who recognized that, because of punctuated equilibria, species are stable entities for millions of years. Furthermore, the paleontologist Elisabeth Vrba recognized that species selection can be defined in a way that is analogous to organismal natural selection. The premise is that the number of species within a group can increase or decrease through time. However, sometimes this number will increase not necessarily because the organisms within these species are more fit or better adapted, but rather because the species themselves are more likely to speciate. If species selection is an important process, and thus far the number of definitive examples supporting species selection is very limited, it would really change views about why particular groups become dominant in the fossil record through time. Perhaps such groups proliferate not because the organisms in those groups are more complex, more fit and better adapted, but rather due to the speciation process itself. Perhaps they were simply more likely to speciate.

As an example, consider modern insects and in particular beetles. There are nearly 400,000 species of beetles known to science. Are there so many species because individual beetle organisms are incredibly fit? Certainly they must be doing something right and are well enough adapted to survive; however, this may not be the only reason why there are so many species of beetles. The number of species of beetles could have increased simply because they might have an inherently greater tendency to speciate. If so, and unfortunately this has not been tested yet, it would mean that species selection could explain why there are so many species of beetles. In addition to this hypothetical example, more concrete examples of species selection have been offered from the study of the fossil record.

Before these examples are introduced, it is worth briefly discussing speciation by expanding the discussion from the section earlier in the chapter on punctuated equilibrium. There are really two ways that speciation can occur. The first was already described and is termed anagenesis. This is when one species, over time, is converted into another species (Figure 5-4). The other way speciation occurs is by cladogenesis, which results when one species fragments to produce two or more species (Figure 5-8).

Certainly each of these types of speciation occurs, and the question is one of frequency. Current evidence suggests cladogenesis is the primary way in which speciation occurs. In reality, this should not be too much of a surprise because today more than 1 million species are known; many more undescribed species (perhaps tens of millions more) exist. This does not even include the many species that have gone extinct. All of these species are descended from one or a few ancestral species that lived long ago. Anagenesis is not a process that creates additional species; it only transforms pre-existing species. Thus, to generate the great increase in diversity known to have occurred, there must have been extensive cladogenesis.

The Fossil Record and Species Selection

Some of the best potential examples of species selection involve fossil snails from the Cretaceous and Cenozoic of the mid-Atlantic coast and the Gulf Coastal Plain. In these rocks, there is a very diverse record of fossil marine snails (Figure 5-9) and other organisms. One feature that has made snails particularly attractive for studies of species selection in the fossil record is that they possess a distinct larval phase. Snail's pass through a larval phase before they metamorphose into adult gastropods, and in modern marine snails there are basically two types of larvae, though there is some intergradation between these two types. In the first, planktonic larvae, larvae float around in the oceans, feeding for several weeks or even months before they settle down and metamorphose into adult snails. During this time they can drift with ocean currents hundreds of miles. Modern snail species with this larval type tend to have very broad geographical ranges.

By contrast, there are also nonplanktonic species. These species develop inside of eggs hatched by their mother. They feed on egg material and then metamorphose into adults around the time they hatch. Modern snail species with this larval type tend to have relatively narrow geographical ranges. The difference in the average geographical ranges between planktonic and nonplanktonic species principally results from differences in the larval phase, because adult snails travel only small distances throughout their lifetime, at least when compared to the thousands of kilometers that larvae can float in ocean currents.

A convenient feature of gastropods is that it is often possible to infer the larval type of a species without studying it in the wild or raising it in a tank in the lab. This is because in living species, and in well-preserved fossil species, a small larval shell is preserved at the apex of the snail's shell (Figure 5-10). By examining the shape and size of this larval shell, scientists can often infer the larval type of a long-extinct species.

Studies by paleontologist Thor Hansen on fossil snail species from the Cenozoic of the Atlantic and Gulf Coastal Plain revealed interesting patterns. First, just as with the modern biota, snails with a planktonic larval type have a very broad geographical range, whereas species with a nonplanktonic larval type tend to have a narrow geographical range. Furthermore, it was found that differences in larval type were associated with differences in resistance to extinction. Species with a nonplanktonic larval type (and the associated narrow geographical range) were much more likely to go extinct than species with a planktonic larval type. This actually reflects a more general pattern in the fossil record identified by the paleontologist Steve Stanley. Species with narrow geographical ranges typically have higher extinction rates than species with broad geographical ranges. This is thought to be because the narrower the geographical range of a species the more susceptible it is to environmental perturbations and other chance events that lead to extinction;

(a)

(b)

Figure 5.9 Cenozoic snails. Fossil snails from deposits along the coast of the USA. Images from J. Hendricks.

Figure 5.10 Fossil snail showing the larval shell. The larval shell is preserved at the top of the shell, with the transition between larva and adult occurring at the faint line marked by the arrow. Image from W. Allmon.

such narrowly distributed species also typically have a smaller population size, which again generally makes them more likely to go extinct.

Hansen and subsequently paleontologist Dave Jablonski also recognized other salient patterns. For example, they found that in spite of their greater likelihood of going extinct, species with a nonplanktonic larval type outnumber species with a planktonic larval type by ratios of about two or three to one. This result did not necessarily make sense (given the higher extinction rate of nonplanktonic species) until it was combined with the results from genetic studies of modern nonplanktonic snail species. In particular, it was shown that nonplanktonic species typically have more genetic variation across their range than what is found in planktonic species. This is because different populations of a nonplanktonic species do not disperse broadly as larvae and are therefore more likely to become isolated from one another and hence become genetically distinct. Through time, such species are likely to have divergent populations that will ultimately speciate, because geographical isolation of populations is one of the important ways in which speciation occurs.

In contrast to nonplanktonic species, the different parts or populations of planktonic species are constantly interbreeding with one another, because their component organisms can disperse over great distances, effectively preventing populations of the species from becoming isolated. This suggests that a nonplanktonic larval type and the increased genetic variation it provides may be a motor leading to a greater tendency for such species to fragment and speciate, explaining why nonplanktonic species outnumber planktonic ones.

Thus, what Hansen and Jablonski proposed is that the relatively greater diversity of nonplanktonic snail species (relative to planktonic species) exists because different populations of their species are more likely to become isolated and give rise to new species. If so, this is an example of species selection. The individual organisms in these nonplanktonic species are not necessarily any more fit than organisms in planktonic snail species. In fact, they may be less fit in some respect given that nonplanktonic species seem to go extinct at a faster rate; they just are more likely to speciate. In the end, this suggests that entire species can be selected, not just individual organisms.

Progress and the History of Life

For a long time it was thought that evolutionary change would produce more progressive lineages, and especially views on evolution that pre-dated Darwin, including those of Lamarck, placed a significant emphasis on the notion that evolutionary change is progressive. Some of these Lamarckian views persisted into the early part of the 20th century. Even pre-evolutionary views have some relevance here. In these views it was often asserted that God had created all the Earth's species; this was based on the supposition that all of the Earth's creatures were perfectly designed and interacted with one another harmoniously. This viewpoint was referred to as the argument from design. Darwin felt this argument was fallacious because numerous cases of poor design exist; furthermore, whatever good design existed he explained as a byproduct of life being descended from a common ancestor, along with the added mechanism of natural selection.

Reasons for Darwin's hesitancy to use the term evolution in his *On the Origin of Species* have already been presented in the previous chapter, but an additional reason is relevant here: there was the equation, at least in vernacular usage, that evolution was understood to mean progress. Darwin did not believe that descent with modification necessarily implied progress. To him, natural selection was about adaptation to local environments, not organisms becoming more advanced and progressive. Some features or adaptations that allow an organism to do well in one environment would not necessarily be beneficial in another type of environment. For instance, having the ability to climb trees might be beneficial for an organism that lives in a forest, but not if it lives in a desert. Furthermore, such concepts as progress can be subjective and therefore difficult to define.

In spite of this reasoning, Darwin did not completely reject the notion that progress had occurred during the history of life. He felt that natural selection was not only about adaptations to local environments, but it also related to the ability of organisms to outcompete one another. If those organisms that were most competitively fit survived, it was conceivable that through time fitter and more progressive lineages might prevail. This is why Darwin felt that modern organisms, if put into an environment with organisms from the Eocene, would quickly outcompete them. Thus, although Darwin made a partial break with the notion of progress, he still believed that progress could occur, and he still felt that in some respects the history of life was about the evolution of more progressive and fitter lineages. Based on discussions presented in Chapter 4, it is clear that our modern society has not given up the idea that evolution and progress are synonymous. Why is that? Probably it partly stems from the fact that some people want to believe ours is an ordered, purposeful world.

Thus, there is the notion in the scientific and popular literature that progress has occurred throughout the history of life. Lamarck accepted it early in the

19th century. The famous 20th century evolutionary biologist Julian Huxley, grandson of Darwin's bulldog, Thomas Henry Huxley, also accepted it. Darwin was not as sure, and he waffled on this point. Our modern society still partly believes that the evolution of life is best described as being progressive. A key element of science is that propositions and hypotheses can and indeed must be tested. How does the proposition that evolution is about progress fare?

The Evidence for Progress at the Grand Scale of the History of Life

When thinking about progress and whether evolution is progressive it can be recognized that there is substantial difficulty pursuing this debate or even defining what progress means. Most scientists would be hard pressed to argue that a panda is more or less complex or more or less progressive than a polar bear, although clearly most scientists would rather meet a panda bear in a dark alley. At this scale, arguments about evolution being progressive in the evolution of species within the bear lineage, seem to hold little water. What about at a larger evolutionary scale? Can other aspects of the history of life be best characterized as progressive?

If we look at the history of life we can say that through time there is an actual increase in the complexity of organisms. The earliest organisms that appear in the fossil record, around 3.5 billion years ago, were bacteria (see Chapter 12) and today there are oak trees, blue whales, and humans. Does this simple pattern mean that a trend towards progress pervades life? Stephen Jay Gould, incorporating work from paleontologist Dan McShea, has presented a detailed analysis of this question. Gould argued that if one were to single out a few of life's many lineages, for example humans and blue whales, there might be the tendency to say that their evolutionary origins record a trend towards increasing progress. To consider this issue in a rigorous way, however, it is necessary to consider the variation present within the entire system of life and recognize that some of the same types of bacteria that evolved long ago are around today and persist side by side with humans and blue whales. Gould argued that past views about increasing complexity in the history of life were biased because they defined complexity using a single measure, such as average complexity, or they considered usually just one best example of complexity, like humans.

Gould's fundamental argument about why the history of life is not about the evolution of more complex or progressive lineages, or at least why there is not an active mechanism that favors more complex and progressive lineages, considered the significance of the existence of organisms of minimal complexity, such as bacteria. In his view, effectively the first bacteria that evolved could not be significantly less complex and still qualify as organisms. That is, these bacteria lie against a wall of minimal complexity; imagine this wall being a left wall marking the point where complexity has a zero value. Then, the only way for life to expand outward is to move away from this wall; thus Gould argued that evolution was naturally disposed to produce more progressive and complex lineages (Figure 5-11). In essence, the only way to evolve was upwards and outwards.

Today, the most abundant types of organisms are still bacteria: organisms that may lie near that left wall of minimal complexity. For example, Gould argued that we now live in the age of bacteria, and our planet, at least for the last 3.5 billion years, has always been in the age of bacteria. This is a sobering thought given that

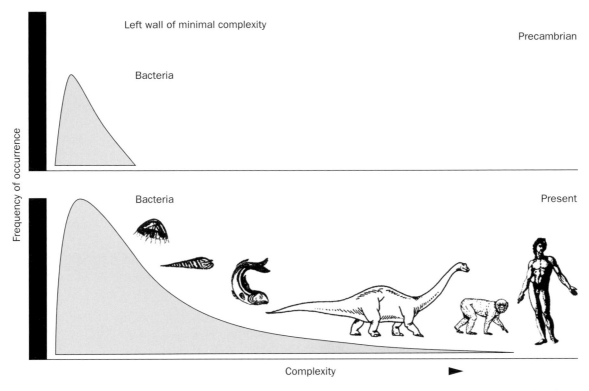

Figure 5.11 **The evolution of complexity.** Life began at a left wall of minimal complexity, as in the top panel. The only way for evolution to proceed was for life to become more complex, as in the bottom panel. Diagram from *Full House* by Stephen Jay Gould, copyright © 1996 by Stephen Jay Gould. Used by permission of Harmony Books, a division of Random House, Inc.

some scientists refer to the last 65 million years as the age of mammals and the last ten thousand years as the age of humans. It may make sense, however, given there are swarms of bacteria floating through the air and covering our bodies (not to mention the pages of this book) at densities greater than 100,000 bacteria per cm^2. The number of *E. coli* bacteria in any human's gut far exceeds the total number of people to ever have inhabited the Earth. Typically 10 percent of a human's dry body weight is made up of bacteria; many of these bacteria people need in order to live. Bacteria, unlike all other organisms, are ubiquitous, occurring at oceanic vents that are 480° F; they also may live within the Earth's crust down to depths of several kilometers, while surviving at the top of Mount Everest and inside clouds in the atmosphere. If we take into account bacteria in the Earth's crust and in the ocean their weight, or biomass, is greater than that of all the living plants and animals combined.

Gould argued cogently that the fact that this minimal level of complexity has persisted for so long and still thrives is good evidence that there is no active trend driving the evolution of complex and progressive units. The increase in complexity seen is due to a passive drifting away from a wall: life was moving in the only direction available to it. Further, he argued that just because a small component of life has increased in complexity does not mean it is accurate to say that as a whole there was an increasing tendency for life, or an active mechanism driving life, to become more complex. This could have just happened by passive diffusion, in a manner analogous to what happens in a tank of water, divided by a barrier, where

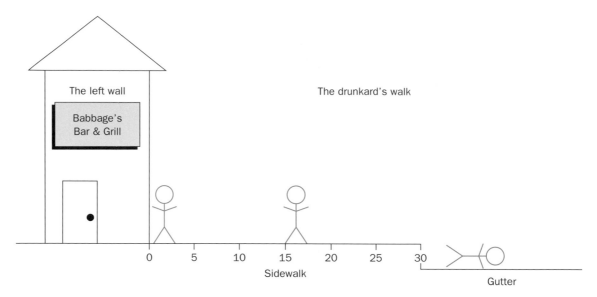

Figure 5.12 **The drunkard's walk.** The diagram shows how a person leaving a bar can end up in the gutter, solely by moving at random; life became more complex in an analogous fashion. Diagram from *Full House* by Stephen Jay Gould, copyright © 1996 by Stephen Jay Gould. Used by permission of Harmony Books, a division of Random House, Inc.

one side has a colored dye, and the other side lacks it. Remove the barrier and slowly the dye will expand to the other side, even though there is no particular force actively driving it in that direction.

Gould offered a humorous example to show how, at random, complexity may have increased during the history of life. He imagined a drunk person staggering out of a bar. On one side is a wall, and on the other side lies the gutter. Imagine that the person staggers at random, a few feet in either direction, towards the wall or away from it. For simplicity assume the person is not moving up or down the street. Just by staggering at random, with a 50 percent chance of moving to the left and a 50 percent chance of moving to the right, eventually that person is going to end up in the gutter because he or she will simply bounce off the wall on the other side (Figure 5-12). There is only one direction open for continuous movement.

Similarly, consider the case for life. There is a wall of minimal complexity on one side. Even if there is no preferred direction of movement, towards or away from the wall of minimal complexity, eventually life could move in the direction of increased complexity even if there is no evolutionary advantage. Remember, life began with minimally complex single celled organisms and today there are quite complex organisms with billions of cells, such as humans.

Gould's ideas on this subject are provocative. Although the subject should not be considered settled, it seems that he has provided compelling evidence as to why we should not view evolution as a process that necessarily favors progressive, complex lineages.

The Evidence for Progress in Individual Groups

Gould also considered the evidence for progressive change in more narrowly constrained groups. One example, derived from the work of paleontologist Richard Norris, involves the evolutionary history of a set of single celled fossils called

Figure 5.13 Foram. A fossil foram. Diagram from R. S. Boardman, A. H. Cheetham, and A. J. Rowell (1987) *Fossil Invertebrates*, Blackwell Publishing.

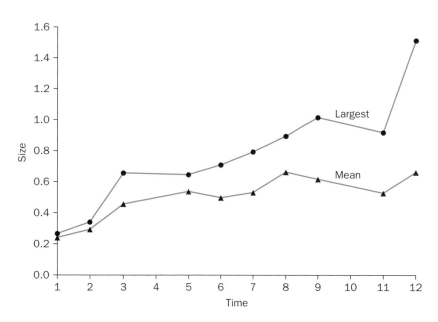

Figure 5.14 Foram body size through time. Average foram body size, and the size of the largest foram, increases through time. Diagram from *Full House* by Stephen Jay Gould, copyright © 1996 by Stephen Jay Gould. Used by permission of Harmony Books, a division of Random House, Inc.

Foraminifera or, for short, forams (Figure 5-13). When Norris looked at the mean or average size of all forams and also the size of the largest forams, through time, he identified an interesting pattern. The average size of forams and that of the largest forams tended to increase through time (Figure 5-14). This change in size may be significant to the debate about whether evolution is progressive because size is often used as a proxy for complexity: large organisms are typically more complex than small organisms. For example, elephants are held to be significantly more complex than bacteria.

On the face of it, the results from the forams might be seen as providing clear evidence that the increase in size and perhaps complexity is being favored evolutionarily. However, Gould demonstrated that when the pattern is examined in greater detail, that argument falls apart. This is because it is inaccurate to examine simply what happens to the average organisms or the largest organisms, and then conclude that there is some actively driven trend. For instance, look at this breakdown of size in forams from the different periods in the geological time scale (Figure 5-15). Although the size of the largest species increases, the size of the smallest species stays constant. This is evidence that there is a wall of minimum size and thus minimum complexity in forams, below which a foram could not function effectively. Forams are still around and presumably flourishing at these smallest sizes. By analogy to the situation observed across all life, some foram species may have passively expanded from this size minimum without actively being driven to do so.

Furthermore, Gould argued that when one considers the overall size change at any given evolutionary transition in the forams, relating to the predisposition of ancestral species to give rise to smaller or larger descendant species, a picture similar to that of the drunkard's walk emerges. When new species evolved from ancestors, they were just as likely to be smaller than their ancestors than larger. This is further evidence that there was no selection for larger size and more complex organisms.

Again, although this issue is likely not yet settled and results are needed from other types of organisms, certainly Gould has provided an interesting and compelling argument.

Is the Left Wall Continually Rescaled?

Paleontologists Andy Knoll and Richard Bambach are two others who have considered whether life's evolution has shown a tendency towards increasingly complex and progressive lineages. They recognized that Gould's arguments about life diffusing away from a wall of minimal complexity have merits but may need refining. In particular, they argued that sometimes evolution causes the position of the minimal wall to be rescaled. For example, the more complex eukaryotic cell evolved from a series of less complex prokaryotic ancestors. (The nature of this transition is discussed in detail in Chapter 12.) There is no record, however, of the eukaryotic cell re-evolving back into a more simple, prokaryotic cell type. In effect, the minimal complexity of the eukaryotic cell had been reset relative to its prokaryotic ancestors. Now there was a new left wall beneath which new types of organisms cannot evolve. Such rescaling of the evolutionary wall may happen not only at the major evolutionary transitions in the history of life but could conceivably happen at various smaller-scale transitions too.

This is a compelling argument, though some have challenged it because there may be cases of evolutionary degeneration when more complex organisms give rise to simpler organisms. For example, there are several parasitic organisms that have evolved from free-living ancestors. Evolutionary biologists Dan Brooks and Deborah McLennan, however, have demonstrated that there is no innate tendency for parasitic organisms to become more simple.

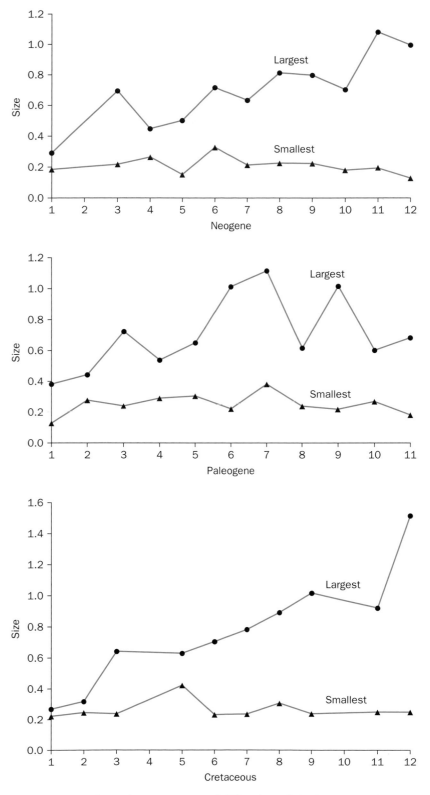

Figure 5.15 Foram body size through time *continued*. The size of the largest and smallest forams through time; significantly, the size of the smallest forams remains unchanged. Diagram from *Full House* by Stephen Jay Gould, copyright © 1996 by Stephen Jay Gould. Used by permission of Harmony Books, a division of Random House, Inc.

Conclusions: Patterns and Processes of Increasing Complexity

In summation, a fundamental distinction emerges between the pattern of increasing complexity (and progress) and the processes that produce that pattern. On the whole, more complex and possibly more progressive organisms have evolved through time. There appears to be no evidence, however, that more complex and more progressive organisms are favored evolutionarily. In short, the trend towards increasingly complex life over the past 3.5 billion years may not have been actively driven, and instead the effect may be a fallout of passive diffusion away from the earliest, minimally complex organisms.

Why the history of life should not be thought of as synonymous with a history of increasing progress and complexity is continued in the discussions of mass extinctions in Chapter 6 and the Cambrian radiation and the Burgess Shale in Chapter 13.

Additional Reading

Eldredge, N. 1995. *Reinventing Darwin: The Great Debate at the High Table of Evolutionary Theory*. John Wiley & Sons, New York; 244 pp.

Eldredge, N., and S. J. Gould. 1972. Punctuated equilibria: an alternative to phyletic gradualism. In *Models in Paleobiology*. T. J. M. Schopf (ed.). W. H. Freeman, San Francisco; pp. 82–115.

Gould, S. J. 1989. *Wonderful Life*. W.W. Norton, New York; 347 pp.

Gould, S. J. 1996. *Full House*. Harmony, New York; 244 pp.

Stanley, S. 1998. *Macroevolution: Pattern and Process*. Johns Hopkins University Press, Baltimore, MD; 370 pp.

Vrba, E. S., and Eldredge, N. (eds). 2005. *Macroevolution: Diversity, Disparity, Contingency: Essays in Honor of Stephen Jay Gould*. The Paleontological Society, Washington, DC; 210 pp.

Chapter 6

Extinctions: The Legacy of the Fossil Record

Outline

- Introduction
- Contingency
- Boundaries in the Geological Time Scale and the Nature of Extinction
- The Cretaceous–Tertiary Mass Extinction
- How has the Existence of Mass Extinctions Influenced the History of Life?
- Were Most Extinctions Caused by Asteroid Impact?
- The Permo-Triassic Mass Extinction— Causes and Consequences
- The Ordovician–Silurian Mass Extinction
- Other Mass Extinction Events: The Late Devonian and the End of the Triassic
- Habitat Degradation and Mass Extinctions
- The Sixth Great Mass Extinction: The Current Biodiversity Crisis
- Conclusions: Lessons from the Past and Future Prospects for Humanity
- Additional Reading

Introduction

In the two preceding chapters we introduced and discussed evolution. Here extinction, which is effectively the opposite side of the coin, is considered. The type of species alive on Earth today is of course related to their evolutionary history: how they have evolved and changed through time. It also depends, however, on the

Prehistoric Life: Evolution and the Fossil Record. 1st edition. By Bruce S. Lieberman and Roger Kaesler. Published 2010 by Blackwell Publishing.

pattern of extinction that life has experienced. In many ways, the stamp of extinction on life is as profound as the stamp of evolutionary history. This is because the presence of any modern species on Earth, including our own species, presupposes a long unbroken evolutionary history of the lineage to which it belongs: a history without extinction. For example, if a pivotal hominid species living in Africa and ancestral to our own species had been extinguished millions of years ago, then of course we would not be here today. This is true for the same reason that conservation biologists say, "Extinct is forever" when they speak of animals like the tiger or the panda that are in dire danger of extinction. Once a species of animal or plant has gone extinct it cannot re-evolve; this is why we must be particularly concerned when humans drive a species of animal or plant to extinction. Something roughly similar in appearance might evolve in the distant future, but the exact same species cannot evolve again.

One of the primary lessons of the fossil record is the primacy of extinction. As mentioned in Chapter 1, the great paleontologist Georges Cuvier presented the first evidence of extinction: he found that there were forms of mammals, including some species of elephants, that once roamed Europe but that are no longer living today. Estimates from the fossil record now suggest that perhaps 99.99 percent of all species that have ever lived are extinct. A cynical view of this staggeringly large percentage might be that the history of life overwhelmingly preserves a record of abysmal failure; however, an optimist might try to learn from it. Because of the tremendous abundance of fossilized extinct lineages, the fossil record is the best place for scientists to study the phenomenon of extinction.

Some key information that has been learned about extinction from the study of the fossil record includes the recognition that extinction events, when they happen, sometimes occur in bunches. For example, several times in the last 500 million years a large percentage of all the species on Earth may have gone extinct in a relatively short time period, during what paleontologists call mass extinctions. Although some of these extinction events may not have been truly instantaneous, instead requiring a few hundred to a few hundred thousand years, effectively over geological time they were instantaneous. The most well known, though not the most catastrophic, of these mass extinctions was the event that eliminated the dinosaurs 65 million years ago. Information about the various mass extinctions, what they share in common, and how they differ from one another will be introduced here. A discussion of the modern biodiversity crisis, a current mass extinction caused by our own species, will also be presented. In addition to these mass extinction events there are many other smaller scale, yet still significantly large, extinctions preserved in the fossil record.

Contingency

It is not easy to predict which lineages will survive mass extinctions (or the smaller extinctions) and which lineages will die off. In short, extinctions, especially mass extinctions, impart a certain element of chance or randomness to the history of life by toppling once dominant lineages and allowing others to take their place. This element of chance has been called contingency by paleontologist and evolutionary biologist Stephen Jay Gould. For example, as we mentioned above, our existence today presupposes the existence of an ancestral species back in Africa some millions of years ago, and indeed an unbroken series of ancestral species extending back to the dawn of animal life. If any one of these pivotal ancestors had gone extinct our own species would not be here today. Contingency is the idea that some event way back in the history of life determines the type of species that are alive today. In short, the major extinction events set the course of subsequent evolution. A famous poem by Robert Frost encapsulates nicely this principle of how contingency sets a course: "Two roads diverged in a wood, and I—I took the one less traveled by, And that has made all the difference." The famous Yankee baseball player Yogi Berra said it another way: "When you come to a fork in the road, take it."

Gould has argued convincingly that the idea of contingency means it is very likely that if we could run life's experiment again the composition of the modern biota would be very different. He argued that the survival of lineages through time is not just a matter of which were the fittest, but also which were the luckiest, because oftentimes what determines survival during times of mass extinction does not seem to be related to whether a species was particularly successful or well adapted. Gould focused specifically on conscious thought in our own species and the likelihood that it could evolve ever again on this or any other planet. He suggested that the large number of steps that needed to occur for intelligence and conscious thought to evolve and the fact that it has not evolved in any other group, vertebrate or otherwise, means that it is extremely unlikely that in some alternate-Earth experiment intelligence would ever evolve again. He suggested that it is a mere fact of contingency that we are around today and not a testament to the superiority of the chordate phylum. We shall explore this issue in greater detail in this chapter.

An opposing point of view to Gould's has been presented by the well-known paleontologist Simon Conway Morris. He argued that the history of life is replete with examples of similar types of organisms evolving independently. He suggested that porpoises may comprise an example of conscious thought that has evolved independently from humans, but within the chordate phylum. This is one of those debates that is not yet resolved in paleontology. Its resolution will depend partly on the extent to which adaptation before a major extinction determines survival during that extinction, but it also assuredly depends on defining difficult concepts like conscious thought and to what extent similar features that evolve in different organisms really are the same.

The perspective that we shall emphasize here is that in an important way extinctions do introduce into the history of life an element of chance or randomness. This element of chance means it is very difficult to make predictions about the types of animals and plants that will survive over long periods of time.

Boundaries in the Geological Time Scale and the Nature of Extinction

18th century geologists and paleontologists, who practiced science before evolution was accepted, sometimes called catastrophists, thought that all living beings were completely wiped out at each of the boundaries in the geological time scale (see Chapter 1), only to be recreated again. We know now that this is not the case. Fossils from rocks of Ordovician age look different from Cambrian fossils, yet they are evolutionarily related to one another: some are close kin; some are more distant cousins; not all species went extinct.

Background Extinction and Mass Extinction

In addition to the more severe extinction events at geological boundaries, extinction has been occurring also all the time at a more muted level: a species disappears here or a species disappears there, perhaps because the environment changed slightly or a new predator appeared. This phenomenon of the slow, regular disappearance of species is referred to as background extinction.

There are many times in the fossil record, however, when a large number of species went extinct at once, and some of the boundaries in the geological time scale mark more severe extinction horizons than others. For instance, there are five times that stand out as times of truly elevated or mass extinction, where from 70 to 90 percent of all the species on Earth went extinct. Those five intervals are called the Big Five mass extinctions. In sequence they occurred at:

1 the end of the Ordovician period;
2 during the Late Devonian;
3 at the end Permian or Permo-Triassic boundary, which is also the Paleozoic–Mesozoic Era boundary;
4 at the end of the Triassic;
5 at the Cretaceous–Tertiary boundary, which is also the Mesozoic–Cenozoic Era boundary.

Some of these extinction events may have taken as little as 1 to 100 years, or as much as a million years, to run their course. Paleontologists have good evidence for the causes of some of these mass extinctions.

The paleontologist Norman Newell made some of the earliest and most important contributions to the understanding of mass extinctions. The significance of his contributions was recognized, and research on mass extinctions was greatly expanded upon by paleontologists David Raup and J. J. (Jack) Sepkoski. In order to recognize mass extinction events and how extinction varies through time and at different geological boundaries, Newell, Raup, and Sepkoski studied the diversity of various fossil groups and how this changed through time. They used the *Treatise on Invertebrate Paleontology* as a starting point, and tabulated where and when such marine invertebrate fossils as trilobites, corals, and clams occur. Some information available for species of marine vertebrates like types of fish was also used.

Marine organisms have the best and most complete fossil record. We know less about the record of diversity of organisms that lived on land, such as insects, birds, and plants. Thus, most of our understanding of diversity and how it has changed through time is based on the fossil record of marine animals. Using these tabulations Newell, Raup, and Sepkoski identified times of dramatic mass extinction of marine life and also times when the total diversity of life was relatively stable. How paleontologists measure diversity and how it actually varies through time is a topic much debated by paleontologists; this topic and how it is resolved will partly influence our ability to recognize mass extinctions, and we shall take it up in greater detail in Chapter 7.

The Cretaceous–Tertiary Mass Extinction

The most well-understood mass extinction, the Cretaceous–Tertiary mass extinction, happens to be the most recent, and thus it is worth considering first. Paleontologists are close to reaching a consensus as to what caused this extinction, but it took a long time to reach this advanced state of knowledge, and there are a few paleontologists who still question the nature and cause of this event.

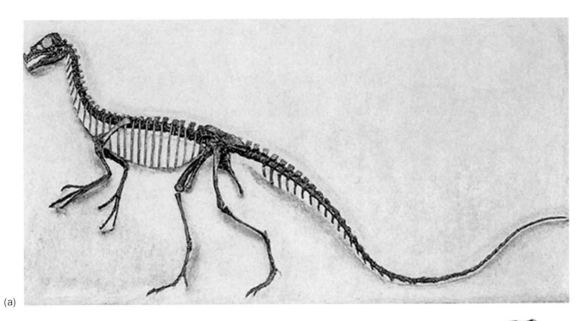

(a)

(b)

Figure 6.1 Dinosaur. *Deinonychus*, a vicious, carnivorous dinosaur (the inspiration for the raptors in the film *Jurassic Park*): (a) its skeleton, and (b) a reconstruction by Bob Bakker. Images from R. Cowen (2005) *History of Life*, 4th Edn, Blackwell Publishing; courtesy of J. Ostrom.

The Cretaceous–Tertiary extinction, which occurred 65 million years ago, is often treated as synonymous with the extinction of the dinosaurs (Figure 6-1), and it is true that these animals did go extinct at this time, but they are certainly not the only things that did. In terms of total numbers of species, they represent only a tiny part of the extinction. Other vertebrate groups that went extinct include various reptiles such as the flying pterosaurs (Figure 6-2), which were closely related to, but distinct from, dinosaurs and also a great diversity of marine invertebrate species including the ammonoids: shelled relatives of the modern squid and *Nautilus*. The ammonoids (Figure 6-3) were once abundant swimming predatory forms that lived throughout the oceans, and some reached huge sizes of up to three meters in diameter. The group had a 220 million year history before it went extinct.

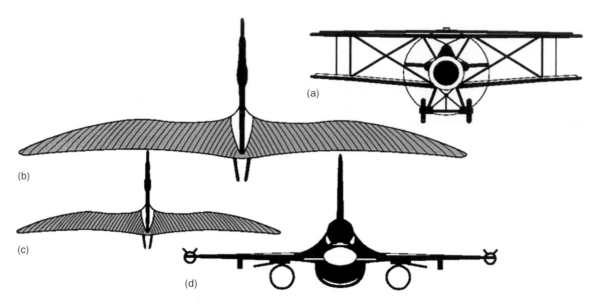

Figure 6.2 Pterosaurs. A group of fascinating, and in some cases very large, flying reptiles that went extinct at the end of the Cretaceous Period, with some aircraft for scale. Diagrams from R. Cowen (2005) *History of Life*, 4th Edn, Blackwell Publishing.

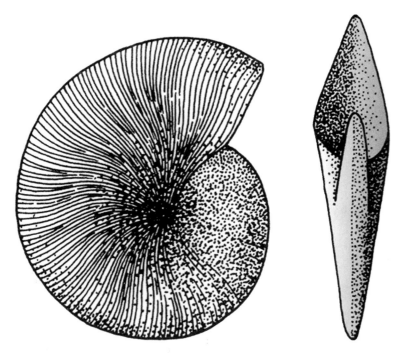

Figure 6.3 Ammonite. An example of this highly diverse group of marine molluscs (shown in two different views). Their closest living relatives are the squids; ammonites went extinct at the end of the Cretaceous Period. Diagrams from E. N. K. Clarkson (1998) *Invertebrate Palaeontology and Evolution*, 4th Edn, Blackwell Publishing; original by Bull.

Initial Explanations Focused Only on Dinosaurs

One of the problems with some of the early attempts to explain this extinction event was that scientists focused only on the problem of the extinction of the dinosaurs. An early theory posited that the extinction of dinosaurs was due to slowly increasing temperatures. This was based on studies from the modern world showing that temperature variations can cause sterility in alligators, large reptiles perhaps analogous to the dinosaurs. It was suggested that climate changed at the end of the Cretaceous caused the dinosaurs to become sterile. Birds evolved from the large nonflying dinosaurs, and they did not go extinct, but such temperature changes do not cause sterility in birds.

The second theory could be called the plant theory. It was based on the fact that many of the flowering plants, called angiosperms, the group that includes many of the tree species and effectively all of the crop species we are familiar with, contain compounds that can be poisonous. The angiosperms first appeared in abundance in the fossil record shortly before the extinction of the dinosaurs, so, the theory goes, perhaps the herbivorous dinosaurs ate them, were poisoned, and died. Then, the carnivorous dinosaurs had nothing to eat, and they died soon after, too.

There are problems with each of these ideas. In particular, neither explains the fact that many other groups of organisms, such as the ammonoids, went extinct. In fact, a great percentage of marine species went extinct, and since none of these species would have had to deal with the temperature problem or the plant problem in the same way as the dinosaurs did, their extinction effectively rules out both explanations.

Asteroid Impact at the End of the Cretaceous

A new and fascinating explanation for the extinction has emerged that can explain why these other organisms went extinct. It postulates that the Earth was struck by an asteroid and that this caused the extinction (Figure 6-4). Fortunately we have not experienced such large-scale events in modern times, although the Hollywood movies "*Deep Impact*" and "*Armageddon*" are fictional accounts of the damage such an impact would spawn were it to happen today. There are well-known, small asteroid strikes that have been documented, and there is even evidence of larger asteroid strikes that happened within the last 100,000 years (Figure 6-5), although they were small compared to the asteroid that caused the Cretaceous–Tertiary mass extinction. Even the relatively small asteroid that produced Meteor Crater in Arizona still would have inflicted significant damage locally.

The initial work that suggested an asteroid impact caused the Cretaceous–Tertiary mass extinction came from the geologist Walter Alvarez and his father, the Nobel Prize winning physicist Luis Alvarez. At first they were not actually studying what might have caused the mass extinction. Instead, Walter wanted a way of determining how quickly rocks are deposited in the fossil record. This process is called a sedimentation rate. Luis Alvarez reasoned that there is a way to calculate sedimentation rates using the fact that there is a steady stream of microscopic cosmic debris raining down on Earth from outer space. This debris, and also such larger particles as meteorites, are enriched in the element iridium (Ir), an element that is extremely rare at the Earth's surface, though it occurs deep within the Earth. The Alvarezes reasoned that if they measured the concentration of iridium in sediments they could get an idea of how rates of sedimentation varied through time.

Figure 6.4 Killer asteroid. A large asteroid and associated debris hurtling towards Earth. Image from iStockPhoto.com based on NASA.

When they did this analysis on some rocks in Italy that happened to be from around the time of the Cretaceous–Tertiary boundary, what they found was totally at odds with what they were expecting. Right at the Cretaceous–Tertiary boundary, at the interval of the mass extinction, they found a very high concentration of iridium, hundreds of times higher than normal. This increase in iridium was referred to as an iridium anomaly. There is no way that a simple decrease in sedimentation rates, allowing more time for cosmic dust to accumulate, could explain it. The Alvarezes calculated that the amount of iridium found in the anomaly could be delivered to Earth in a short time only by an asteroid that was 10 km in diameter. Although the anomaly was first found in rocks in Italy, it has since been found around the world at the Cretaceous–Tertiary boundary.

Figure 6.5 Meteor crater. From the famous impact crater site in northern Arizona. Image from iStockPhoto.com © S. Hoerold.

How might this collision have caused the mass extinction? First, the asteroid would have been traveling at roughly 50,000 km/hr. Combining this speed with the hefty mass, its collision with the Earth would have generated enormous heat and energy, more energy in fact than would be produced by tens of thousands of atom bombs detonating simultaneously. For thousands of kilometers around the site where an asteroid of this size landed everything would have been incinerated. If it struck in the ocean it also would have generated huge tidal waves that would have traveled many miles onto land, crushing many large animals and plants. The explosion and the resulting fires would have thrown up a large cloud of soot and dust into the atmosphere blocking sunlight. The decrease of sunlight reaching the Earth's surface would have inhibited greatly the plants' ability to produce energy by photosynthesis, cutting off their growth and thus also food for animals that feed on plants. The decline in sunlight would also have caused global temperatures to fall dramatically, producing a situation analogous to what is known now as a nuclear winter. This winter would have persisted for many months or even years. In short, we can see readily how all of these cascading effects working together might have produced disastrous consequences for life on land and in the oceans.

Although the iridium anomaly was the first significant piece of evidence supporting the impact hypothesis, other kinds of evidence have also emerged. For example, glass microspherules are found at the Cretaceous–Tertiary boundary (Figure 6-6). They are produced when rocks are suddenly heated to great temperatures by some sort of explosion. Such microspherules are produced when rocks first turn molten and then cool rapidly. Glass microspherules have also been found at more recent meteor impact sites, including the crater shown from Arizona

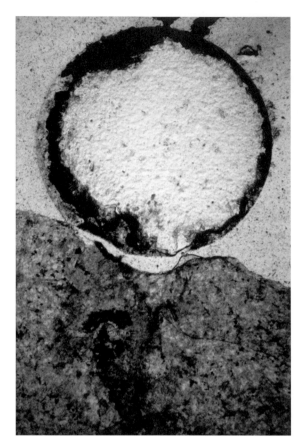

Figure 6.6 **Glass microspherules**. Further evidence of asteroid impact. Images from C. Wylie Poag, U.S. Geological Survey.

and also at nuclear weapons testing sites. High-pressure forms of the mineral quartz, called shocked quartz, which describes how they are produced, have also been found at the boundary. These forms of quartz are created under immense pressure, the type of pressure that might be caused by an asteroid impact. All this evidence was important in support of the asteroid-impact hypothesis, but the most important piece of the puzzle came into place when the actual crater caused by the meteor was found in the Gulf of Mexico, just off the Yucatan peninsula of Mexico (Figure 6-7). The crater has been buried by younger rocks, but it has a circular form, and is more than 100 km in diameter; it was created, based on radiometric dating, right at the Cretaceous–Tertiary boundary. Paleontologists call this the smoking gun left over from the asteroid impact. Based on the size of the crater, the asteroid that caused it was probably about 10 km in diameter.

The case of the Cretaceous–Tertiary mass extinction makes it clear that there are large-scale events that have profoundly influenced the history of life. If scientists simply studied modern time scales they could not see or predict these events. In short, this is further evidence, in addition to that presented in Chapter 1, that a substantive uniformitarian approach to the study of the history of life, which assumes that the study of the present is the key to the past, does not always work.

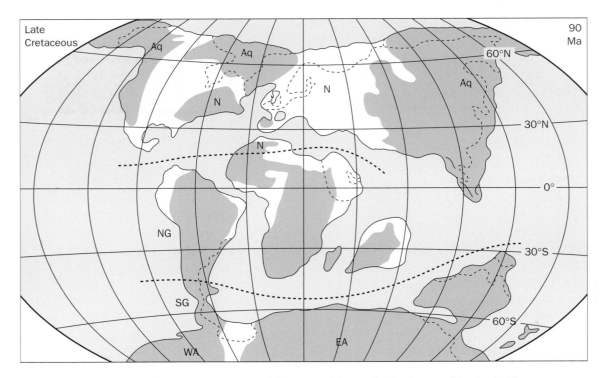

Figure 6.7 The Earth in the Late Cretaceous. The asteroid struck Earth near Mexico's Yucatan peninsula, visible near the southern tip of North America. Diagram from C. B. Cox and P. D. Moore (2005) *Biogeography: An Ecological and Evolutionary Approach*, 7th Edn.

How has the Existence of Mass Extinctions Influenced the History of Life?

It turns out that the existence of mass extinctions changes greatly the way scientists think about evolution and the history of life.

The Limits of Adaptation

Recall how Darwinian natural selection works (see the discussion in Chapter 4). Natural selection is the process whereby through time organisms become better adapted to local environments or become more competitively fit. No matter how competitively fit or locally adapted an organism is, however, it cannot survive having a 10-km-wide asteroid land on its head. Furthermore, adaptations that may have made an animal more competitive during normal times of existence do not necessarily work well during times of mass extinction. Life cannot see a mass extinction coming, because they happen quite infrequently, on time scales of millions of years. Thus, organisms cannot possibly adapt to survive them. In the Cretaceous–Tertiary extinction, more than half of all the species of animals went extinct effectively overnight. Many of these species were very fit and adapted to their local environments prior to the extinction, but this did not seem to help them survive the extreme environmental changes caused by such a cataclysmic event.

What survives a mass extinction may be determined by chance. Species may have survived because they lived on the opposite side of the world from where the asteroid struck or because they happened to possess a trait that evolved under a different kind of selection pressure that allowed them to survive during the time of the catastrophic event. An excellent example of this was documented by paleontologist Jennifer Kitchell and colleagues. There is a group of single-celled marine organisms called diatoms that are abundant fossils in Cretaceous and Tertiary sediments (Figure 6-8).

Diatoms as a group happen to have relatively low extinction rates during the Cretaceous–Tertiary mass extinction. This can be attributed to a specific aspect of their life cycle: it turns out that these diatoms can form a resting stage that allows them to go dormant, that is, to shut off all physiological activity for several months at a time. The ability to do this probably is an adaptation for when conditions in the ocean that these organisms need to sustain them change, and the diatoms typically go dormant during some seasons of the year. This type of resting stage, however, may have also allowed them to survive the major environmental changes that followed the asteroid collision 65 million years ago. The resting stage was an adaptation for surviving local perturbations in their environment that by chance happened to make diatoms more likely to survive the Cretaceous–Tertiary interval.

Mass Extinctions Reset the Evolutionary Clock

Paleontologists Stephen Jay Gould, Dave Raup, and others have argued that the evolutionary effects of mass extinctions are important because they may eliminate once dominant groups and thereby allow what were marginal groups to expand. A good example of this is the case of the dinosaurs and mammals. Prior to the end of the Cretaceous, mammals were present, but they were quite rare. Most of them were also small. These predominantly small, shrew-like animals lived in a world that was ruled by dinosaurs. It was not until the dinosaurs were wiped out that the number and variety of mammals increased. Shortly after the dinosaurs went extinct, a whole series of new mammal groups evolved and appear in the fossil record. By the Eocene Epoch, some 50 million years ago, and 15 million years after the mass extinction, mammals had diversified from the mostly small, shrew-like forms of the Cretaceous into varieties as different as whales and bats.

What happened at the end of the Cretaceous was not the story of the competitively superior, more fit, more complex mammals taking over from the inferior dinosaurs, as some used to portray it. Dinosaurs were the dominant large terrestrial organisms for 180 million years, with mammals or mammal-like forms living in their shadow for almost all of that time. If the Cretaceous–Tertiary mass extinction had not eliminated the dinosaurs, which were otherwise competitively fit, then the diverse groups of mammals we see today, including humans, would not have evolved. The asteroid impact at the end of the Cretaceous caused a cascade of changes that altered fundamentally the types of vertebrates that walk the Earth and dominate it today. This phenomenon, contingency, is how mass extinctions have influenced the course of evolutionary history.

Let us return to the subject of progress in the history of life that we introduced in the last chapter. Stephen Jay Gould has argued convincingly that the existence of mass extinctions is another reason why progress is unlikely to be a prominent theme in the history of life. Mass extinctions will eliminate any progress or evolutionary change that has occurred by resetting life's clock, as it were. Whatever progressive evolutionary change that occurs (and remember, this may not be much anyway), including the accumulation of new adaptations that make organisms more fit

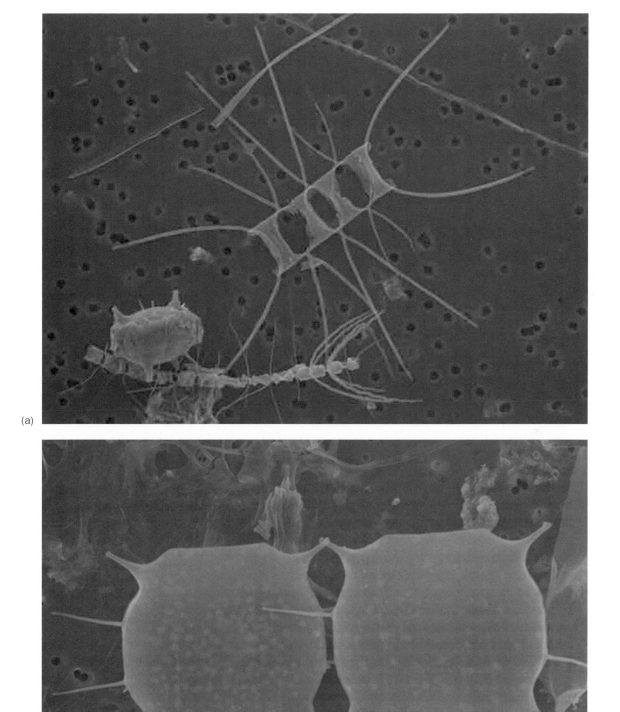

Figure 6.8 **Diatoms**. Modern, marine examples of these micro-organisms. Images from S. Morton, NOAA.

competitively, will often be wiped out by these global-scale events of mass extinction because traits that were adaptive may not help during a mass extinction.

Were Most Extinctions Caused by Asteroid Impact?

Asteroid impact provides a good explanation for mass extinction at the Cretaceous–Tertiary boundary, but were the other mass extinction events caused by asteroids or comets (which are likely to have similar effects) colliding with the Earth? Posing such a question is part of the general scientific process. In the history of life there are recurrent patterns, like mass extinction events, and we try to link them by invoking similar causes. The more phenomena scientific ideas can explain, the more powerful the ideas are considered to be, but although the idea that asteroid impacts caused mass extinctions is exciting and provocative, it may not work in general. For example, there are times, as in the late Eocene, when there is an iridium anomaly, there are high pressure polymorphisms of quartz, and there are glass microspherules. There is even evidence that a large crater formed at this time, probably comprising what is now Chesapeake Bay along the coast of the eastern United States (Figure 6-9). Paleontologists, however, do not see any sign of

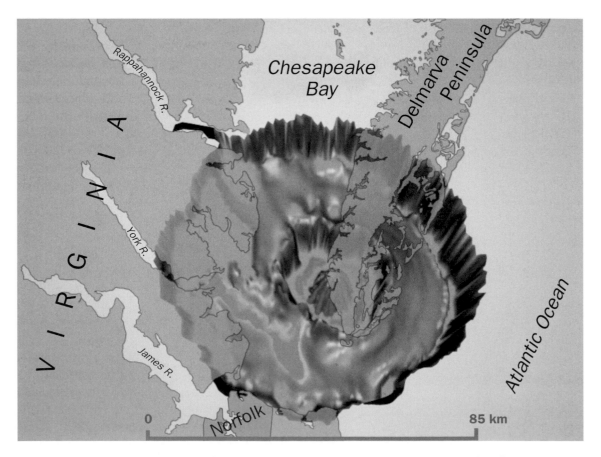

Figure 6.9 Eocene impact crater. This crater partly straddles the site of present-day Chesapeake Bay in the eastern USA. Diagram from C. W. Poag, U.S. Geological Survey.

a profound mass extinction at that time (only smaller, local scale extinctions). It appears that not all asteroid impacts cause mass extinctions. Moreover, not all mass extinctions are caused by asteroid impacts, as we shall discuss below.

The Permo-Triassic Mass Extinction—Causes and Consequences

Of the five mass extinctions the Permo-Triassic extinction is the largest: as many as 90 percent of all species may have been eliminated. Did an asteroid or comet impact play a role in this extinction? At this time the consensus appears to be no, because there is no good evidence for an iridium anomaly nor for some of the other signatures characteristic of an asteroid impact. In 2003, however, paleontologist Luann Becker and colleagues found some types of chemical compounds from the time of the mass extinction that are known only from asteroids and comets. Then, in 2004 Becker and colleagues reported finding a deep structure in the ocean buried off the coast of Australia that might be a crater, and it dates from the end of the Permian Period. Thus, what really needs to be said here is "Stay tuned!" It is intriguing that a 250 million year old event is the focus of new and ever changing ideas and thus the work of science is not finished.

There is some controversy about whether the Permian mass extinction represents one massive extinction or two very big extinctions separated by a few million years. The latter view has been suggested by paleontologists Steven Stanley and X. Yang. Another thing notable about the Permian, and also the Cretaceous, is that both the end of the Permian and the end of the Cretaceous correspond respectively to times of the largest and the second largest outpourings of volcanic lava during the last billion years. Whether these contributed to the mass extinctions awaits to be determined by future scientific research. For now, we shall present what seems to be the best-supported cause of the end-Permian extinction.

Ocean Overturn, Carbon Dioxide, and the Permo-Triassic Mass Extinction

Part of the reason there may have been such a large extinction at the end of the Permian is that the Earth's continents were in a very different configuration from that of today. Due to plate tectonics the continents have been in nearly continual motion for at least the last 2 billion years (see Chapters 10 and 11), and during the Permian the continents were joined into a single supercontinent called Pangaea (Figure 6-10). Because there was a single, large supercontinent during the Permian, there was also one large ocean. Warm water at the surface of the ocean is less dense than cooler water that is deeper, heavier and more dense. In the modern oceans, there is a profound circulation that mixes the upper and lower layers of the ocean, keeping the bottom layers of the ocean well oxygenated. This is the case because warm water from the equator moves to the poles, where it becomes cool and oxygenated, due to the cooling action of the polar ice caps, and also more dense, thus sinking; this water then migrates to the Earth's equator, because of the rotation of the Earth (Figure 6-11). This circulation is due partly to the presence of ice at the poles and also to the geometry of the continents and the relatively small intervening oceans.

In the late Permian major polar ice caps were not present. Thus, it is unlikely that there was extensive circulation between the surface and the deep oceanic waters. The result was that the ocean's deep waters were poorly oxygenated, what is termed anoxic. Waters very low in oxygen are toxic for most animal life. (An unfortunate

Figure 6.10 Pangaea. The Earth's continents were joined into a single supercontinent at the end of the Permian. Diagram from C. B. Cox and P. D. Moore (2005) *Biogeography: An Ecological and Evolutionary Approach*, 7th Edn.

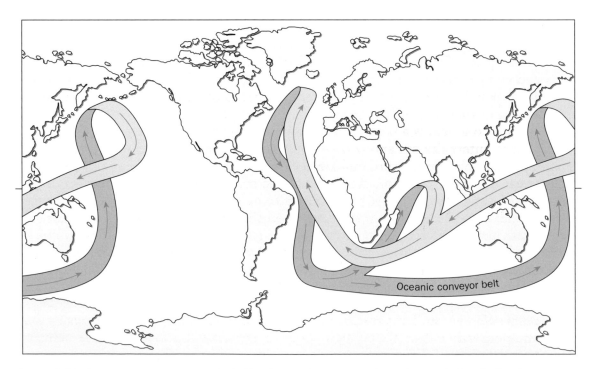

Figure 6.11 Ocean current conveyor belt. In the modern ocean there is good circulation between shallow and deep water and polar and equatorial water. Diagram from C. B. Cox and P. D. Moore (2005) *Biogeography: An Ecological and Evolutionary Approach*, 7th Edn.

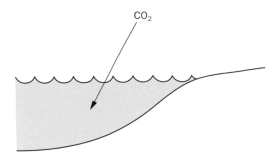

Figure 6.12 Oceans in the Permian. As organic matter is buried in the oceans, CO_2 moves from the atmosphere to the oceans. Diagram by M. Smith.

example is the conditions that occurred in Lake Erie in the 1970s, albeit for different reasons. There, due to intense pollution, the lake became anoxic and inhospitable for most life forms.) Paleontologists Tony Hallam and Peter Wignall have argued that these anoxic waters may have spread into shallow water at the end of the Permian, wreaking havoc on much of the marine life, and it is likely that this oceanic anoxia played some role in the Permian mass extinction. Anoxia also causes another phenomenon: a large amount of organic matter begins to be buried at the bottom of the ocean. Paleontologist Andy Knoll and colleagues have suggested the effects of this may have played a fundamental role in driving the extinction. These effects are related to the concentration of carbon dioxide (CO_2) in the oceans and the atmosphere. The buried organic matter is rich in carbon and also carbon dioxide. As the amount of CO_2 in the oceans increases, the amount in the atmosphere concomitantly declines (Figure 6-12).

CO_2, the Greenhouse Effect, and Global Warming

Introducing how changing CO_2 concentrations in the atmosphere affects climate necessitates a brief digression about topics you may be somewhat familiar with: the greenhouse effect and global warming. The greenhouse effect is caused by the fact that carbon dioxide and some other compounds like water and methane allow the Sun's energy to enter the atmosphere, but when it turns to heat and is scattered back off the Earth's surface, it cannot escape the atmosphere as easily. The result is that the atmosphere is heated. The process is in some ways analogous to what happens when one leaves a car parked outside on a sunny day with the windows closed. Light enters the car and is then converted to heat, which cannot easily escape. Even if the temperature outside is cool, the sun will make the inside of the car heat up. If it's very hot, the inside of the car will become scorchingly hot.

As atmospheric carbon dioxide concentrations rise, the greenhouse effect becomes more marked, and temperatures rise. Abundant evidence suggests that over the last few hundred years CO_2 levels in our atmosphere have increased, due in large part to the activities of humans, especially the burning of coal, oil, and gas, and temperatures have already begun to increase due to these changes. This is a topic that we will take up in greater detail in Chapters 10 and 11.

An extreme example of what happens to planets with high CO_2 levels is the case of the planet Venus: the temperature at the surface is 700° C. Venus is closer to the Sun than the Earth is, but not enough closer to explain the great difference between the temperature of the Earth and the temperature of Venus. The reason Venus is so hot is that its atmosphere is about 99 percent CO_2 whereas Earth's is less than 1 percent CO_2. The result: Venus has a much more prominent greenhouse effect than Earth. Reducing CO_2 levels in the atmosphere of our planet (or that of Venus)

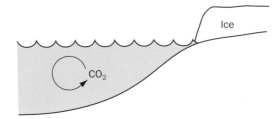

Figure 6.13 Oceans in the Permian *continued*. As temperatures cool, ice forms. When ocean circulation resumes CO_2 moves from deep to shallow water. Diagram by M. Smith.

will, all things being equal, reduce the greenhouse effect and cause a decline in global temperatures.

CO_2 and the Permo-Triassic Mass Extinction

Because of the effects Knoll and colleagues described, at the end of the Permian CO_2 levels were dropping in the atmosphere, while they increased in the deep ocean. With a decrease in the greenhouse gases, there was a concomitant drop in temperature. Significant increases or decreases in temperature can change circulation patterns in the oceans. At the end of the Permian, as temperatures fell, oceanic circulation, which was previously absent or very limited, increased suddenly. At the end of the Permian the point may have been reached when deep waters started to circulate to the surface. As a result lots of carbon dioxide was brought from deep water to shallow water at the surface (Figure 6-13). CO_2, like carbon monoxide, is a substance that is toxic to most animals, and it turns out that some types of animals are much more likely to be affected by carbon dioxide poisoning than others. Specifically, organisms that are relatively inactive throughout their lives and do not or cannot move around much, like corals or brachiopods, are particularly negatively affected by increased CO_2 levels. By contrast, active, mobile groups like the vertebrates (fish) or the molluscs (clams and snails, which are active at least when compared to corals!) are not as prominently affected by CO_2 poisoning. When Knoll and colleagues studied the Permo-Triassic mass extinction, they found that although many species went extinct, it is the inactive groups that were the most negatively affected at this time: precisely the groups that were predicted to be most susceptible to CO_2 poisoning.

The Ordovician–Silurian Mass Extinction

The extinction at the Ordovician–Silurian boundary coincides with a time of major glaciation, leading some paleontologists to implicate the glaciation in this mass extinction. Still, it is somewhat puzzling why a major glaciation occurred at this time because it followed a period of very warm conditions. It is also not clear why a glaciation would necessarily trigger a mass extinction because the relatively recent Ice Ages that occur during times of extensive glaciation do not coincide with a mass extinction. For this reason, Adrian Melott, a physicist at the University of Kansas and Bruce Lieberman, a paleontologist at the University of Kansas (and coauthor of this book), along with their colleagues, proposed that there may have been a gamma-ray burst that occurred at the end of the Ordovician that helped trigger the mass extinction.

Gamma-ray bursts are tremendously powerful explosions produced by stars as they collapse and die. Astronomers have recognized that some have occurred within

Figure 6.14 Gamma-ray burst. What would happen to Earth if there was a nearby gamma-ray burst. Diagram from NASA.

our own galaxy and in nearby galaxies; if a gamma-ray burst occurred close enough to Earth, the negative effects on our biosphere would have been profound (Figure 6-14). In particular, a gamma-ray burst would have produced an initial blast of radiation that damaged the ozone layer, greatly reducing its strength and effectiveness. The ozone layer shields life at the planet's surface from the dangerous effects of ultraviolet (UV) radiation produced by our own Sun (see Chapter 12 for a discussion of how the ozone layer formed and also how significant it is for life on the planet). Reducing the effectiveness of the ozone layer would have allowed a much greater amount of UV radiation to penetrate our atmosphere. The effect of this would have been particularly catastrophic for life on land; however, in the Ordovician there were few terrestrial organisms, and most organisms lived in the oceans (see discussion of how the land was colonized in Chapter 17). Marine organisms that lived in shallow water would be most affected by UV radiation. As a secondary affect, a gamma-ray burst could have produced a dramatic decrease in global temperature and thus could have caused the short-lived global scale glaciation that occurred at the end of the Ordovician. Melott, Lieberman, and colleagues argued that the weakening of the ozone layer and the sudden fall in temperatures was a one-two punch for life on the planet, precipitating the mass extinction.

At this time, the precise causes of the Ordovician–Silurian mass extinction are being debated, although it is likely that glaciation played some role. What is clear is that it was the second largest mass extinction in the history of life. Paleontologist Peter Brenchley and colleagues have shown that many species, especially those living in the tropics, died out at the end of the Ordovician. The tropics are in a belt around the Earth's equator where the temperature is always warm. Furthermore, most of the Earth's species live in the tropics, so an extinction event that disproportionately strikes the tropics will be particularly severe. Global cooling associated with glaciation will affect negatively the species that live in the tropics. In fact, paleontologists Steven Stanley and Niles Eldredge have argued that glaciation may be one of the fundamental factors causing mass extinctions. Glaciations have a deleterious effect on tropical species because all organisms depend on a specific habitat and environment. In the face of a changing environment, organisms will migrate to follow their preferred environment. There are some times, however, when this becomes difficult, especially when a species' preferred environment starts to contract and shrink in size; then, the animals and plants that need that environment have no place to go. If a species' preferred habitat lies in the tropics, as temperatures start to fall, the size of this preferred habitat will constrict as regions where temperatures are relatively colder start to expand (Figure 6-15). As we discussed earlier, the smaller the geographical range of a species, the more susceptible it becomes to extinction. This is true for many reasons. One of the most important reasons is that in general as the geographical range of a species declines, its population size also declines. The smaller the population size, the easier it is for some chance event to eliminate the species. Thus, the way global cooling might have acted to cause a mass extinction at the end of the Ordovician is that it first would have greatly restricted the geographical ranges of tropical species, thereby making them much more likely to vanish.

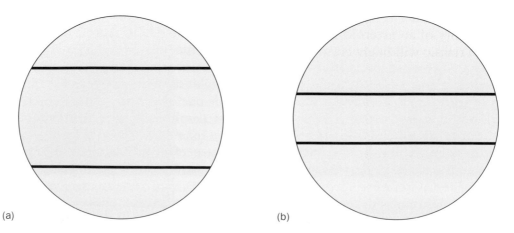

(a)   (b)

Figure 6.15 **Shrinking tropics.** As global temperature drops the size of the tropics, represented by the space between the bands in (a) and (b), becomes more restricted. There is concomitantly less habitat available for tropical species.

Other Mass Extinction Events: The Late Devonian and the End of the Triassic

Our knowledge of mass extinctions in the fossil record is far from complete, but knowledge is expanding. One mass extinction event that has emerged as particularly interesting occurred in the Late Devonian. Part of the reason this mass extinction may be so interesting is because the term mass extinction may be a bit of a misnomer. This mass extinction has been considered comprehensively by paleontologists George McGhee and Alycia Stigall. They found that one cause of the Late Devonian event might actually have been a decline in speciation rates, and therefore McGhee suggested that the term biodiversity crisis might be more appropriate for the Late Devonian.

Originally, the five mass extinctions recognized in the fossil record were identified on the basis of an apparent decline in total diversity. The net diversity of life at any one time, D_n, is made up of the total number of species on the planet at a previous time, D, plus any new species that are produced, S, minus any species that go extinct, E, or, in equation form:

$$D_n = D + S - E$$

A decline in D_n could be caused either by an increase in the amount of extinction, E, or by a decline in speciation, S.

New research has also emerged on what was perhaps the least well known of all the mass extinctions: the end-Triassic extinction. Paleontologist Paul Olsen and colleagues have shown that there may be an iridium anomaly at the end of the Triassic, suggesting that an asteroid or comet may have struck the Earth at that time, triggering a mass extinction like the one at the end of the Cretaceous. Geologist Paul Renne and colleagues have by contrast argued that the extinction event was caused by a protracted period of volcanism that produced extensive lavas. These lavas are still visible today along the eastern seaboard of the United States. The famous Palisades, a landmark visible across the Hudson River from New York City, are one example, although extensive lava deposits are also known from other places as far north as Canada and as far south as Brazil. Notably, the end Triassic, like the end Cretaceous, seems to show both evidence of extensive volcanism and evidence of an asteroid impact during a time of mass extinction. (Remember that we discussed earlier the prodigious lava flows at the end of the Permian, along with the possibility of an asteroid impact at this time.) The debate about the nature of the end Triassic will likely continue as our knowledge of this event expands, but it seems that now there are two well-supported but competing hypotheses for why there was a mass extinction at that time.

Habitat Degradation and Mass Extinctions

As can be seen from this discussion, it is clear that the study of mass extinctions and their causes is ongoing. In many cases, the causes for the extinction events, especially the Ordovician–Silurian, Permo-Triassic and Cretaceous–Tertiary may have been different, but they do share commonalities. In particular, the mechanism of the mass extinctions are different, but the results engendered by the mechanisms were the same: habitat degradation. If habitats are degraded quickly enough, organisms will go extinct. During these three different extinction events habitat degradation was caused by different things. In the case of the Cretaceous–Tertiary it was the global conflagration from a massive asteroid and the subsequent global cooling. In the Permo-Triassic, there may have been the spread of oxygen-poor and carbon-dioxide-rich water into shallow water. Finally, in the Ordovician–Silurian there may have been a gamma-ray burst that eliminated the ozone layer and further, the range of species in the tropics became restricted as temperatures cooled. Each of these extinction events occurred on different time scales. The environmental changes at the Cretaceous–Tertiary happened extremely suddenly. In the case of the others, the environmental changes may have taken longer (perhaps tens or hundreds of thousands of years for their ultimate results to be felt). In the end, however, the results were the same: mass extinction.

The Sixth Great Mass Extinction: The Current Biodiversity Crisis

In addition to the big five mass extinctions there is a sixth mass extinction event that is worth addressing because we are right in the middle of it. It is called the current biodiversity crisis. The rates of extinction today are high enough to rival the extinction rates during any of the other big five events. When just an acre of tropical rain forest is paved over or burned to the ground, a few hundred species are likely to go extinct. Further, land is being lost in tropical rain forests at a pace of many thousands of acres each year.

In addition to the criterion of rate, there are other similarities between the sixth mass extinction and the other five, because they are all caused by habitat degradation. Humans (*Homo sapiens*), through their profound influence on the planet, have become an agent of habitat degradation and thus resemble previous agents of mass extinction. In order to elucidate this, it is worth considering how the influence of humankind has changed since our species first evolved. Humans were not always a scourge.

Humans as Hunters of Large Mammals

The earliest fossil humans are known from African rocks that are roughly 100,000 years old. Considering that bacteria have been around for more than 3.5 billion years (see discussion in Chapter 12) that provides some perspective on our own species' place in the history of life: we are a very recent phenomenon. Notably, there's no real evidence that humans had a negative effect on other species or caused any extinctions until around 40,000 years ago. These extinctions that began 40,000 years ago occurred shortly after the development of new hunting weapons and the expansion of the human species in great numbers out of Africa. The expansion out of Africa was probably triggered after the population size of *Homo sapiens* passed a critical threshold.

A spate of extinction events that affected principally large mammals began around 40,000 years ago and lasted until around 10,000 years ago. These extinctions actually continue today (note the plight of the modern panda and tiger) but have been overwhelmed by another set of extinctions. Numerically relatively few species went extinct during the interval 40,000 to 10,000 years ago, so that paleontologists do not typically classify them as a true mass extinction. Still, the effects of these extinctions can be seen today in many places, including modern zoos, which are some of the few places we can still encounter many species of large mammals.

If there were such a thing as a Pleistocene zoo it would have been much more exciting than anything present today. This is because 40,000 years ago North

Figure 6.16 Giant ground sloth. An extinct ground sloth, from the famous La Brea tar pits in Los Angeles, in the University of Kansas Natural History Museum. Image by B. Scherting.

America was home to several species of giant ground sloths, some as large as 15 feet tall (Figure 6-16); saber-toothed cats with vicious front teeth (Figure 6-17); cave bears bigger than the modern grizzly; woolly mammoths (furry relatives of the modern elephant) (Figure 6-18); giant beavers that stood as much as six feet tall; the dire wolf (Figure 3-3), which was quite a bit larger than the modern wolf; and the list goes on and on. Today, however, they are all extinct.

The reasons for these extinctions still are debated by some, and in 2008 Firestone and colleagues have argued that a comet impact and its associated climatic effects 10,000 years ago may have been the coup de grâce for some large mammal species, but most scientists accept the prevailing theory: as humans spread out of Africa they hunted or killed these mammals or the prey that they depended on, exhausting the supply of animals in one region and then moving on to another. One of the driving forces behind the recognition of this was the research of paleontologist Paul Martin. It appears that the extinctions were especially severe in North America and South America because the mammals in these places were not familiar with humans and quickly succumbed. In Africa, where animals had been around ancestral lineages of our own species for millions of years, they were used to hominid hunters, and because of this familiarity the extinctions at this time in Africa were not as dramatic as elsewhere.

Figure 6.17 Saber-toothed tiger. An extinct giant cat, from the famous La Brea tar pits in Los Angeles, in the University of Kansas Natural History Museum. Image by B. Scherting.

The effects of how animal naiveté (concerning humans) affects their survival can be illustrated using a poignant example. Charles Darwin in his book *Voyage of the Beagle* described how people collected the Falkland Islands fox. The Falklands were effectively devoid of humans until the 19th century, and Darwin was one of the first westerners to encounter what even then was a rare animal; today it is extinct. Darwin recognized their scientific value and sought specimens to bring back with him to natural history museums. Darwin apparently collected them using what seems an item ill adapted to capturing foxes: a geological hammer. Because they were naïve animals concerning humans and the danger they signified, to collect them he simply walked up to a fox and bonked it over the head with his hammer. The species is extinct today because of its initial low population size and also its naiveté. Consider how difficult it is today to catch even a glimpse of a North American red fox, unless it is road kill, never mind to approach within striking distance. Albeit a different species, the red fox shares similarities with the Falkland Island fox.

A variety of evidence supports the notion that humans caused the extinctions of the large mammals between 40,000 and 10,000 years ago. First, the timing of the extinctions of the large mammals of each continent matches the time that humans arrived at those continents. For example, humans reached North America before they arrived in South America, and the extinctions in North American mammals happened first. By 10,000 years ago, humans had spread to every continent but Antarctica. There is not only a match in the timing of the extinctions with the

Figure 6.18 Woolly mammoth. A skull of this extinct relative of the modern elephant in the University of Kansas Natural History Museum. Image by B. Scherting.

timing of human habitation, there is also a fossil record of this early hunting preserved in the Americas, Asia, and Europe. There, at various sites, huge caches of bones have been found that preserve evidence of butchering. The animals found butchered are large herbivores that would have been sources of food. Other mammals not known from such sites may have been killed because they were dangerous to humans (like the saber-toothed cat, but not the giant beaver!), or some of these species may have died out simply because we had eliminated their prey.

Humans as an Agent of Habitat Degradation

Our species has had a negative effect on the life of our planet for quite some time. The extinctions before 10,000 years ago, however, were really just the tip of the iceberg. Around 10,000 years ago a profound innovation in human society occurred that altered the relationship between the human species and the entire biosphere. This innovation was the development of agriculture: humans settled on the land and began to cultivate crops, primarily focusing on one or a few plant species. Later, such animals as cows, pigs, and horses were domesticated. To do this, it was necessary to clear land and alter habitats.

The development of agriculture is perhaps the greatest success story in the history of the human species. The number of people that can be supported by a hunting society is fairly small, while farming allowed population size to expand greatly because many more people could be fed. Thus, from our perspective, agriculture is extremely important and beneficial. Modern society benefits greatly from the fact that we can go to the supermarket and buy a steak instead of needing to go outside and hunt a woolly mammoth with a spear. As the population size of humans expanded and the demand for food increased, however, larger and larger areas either came under cultivation or had to be used for human habitation. The necessary consequence is that humans are taking away the space where other plants and animals formerly lived. In a sense we began to act exactly like the glaciers at the end of the Ordovician, although now not only the species in the tropics were having their ranges geographically restricted. We began to eliminate the preferred habitats of countless species and continue to do so while expanding the range we occupy as well as the ranges of domesticated animals, farm crops, and also the pests that feed off our civilization—rats, pigeons, bedbugs, and cockroaches. As we restrict the geographical ranges of many plants and animals and degrade their habitats through pollutants, the natural result is extinction. As we degrade more and more habitats, the result has become mass extinction.

Conclusions: Lessons from the Past and Future Prospects for Humanity

The profound difference that agriculture meant to humans is that we shifted from hunter gatherers that were moderately successful and had a limited effect on other living beings to being wildly successful while having a very negative effect on other organisms. Probably none of the great technological advances of society would have been possible without the development of agriculture because we would all be so busy procuring food day to day that there would be little time for making new inventions. Being successful and expanding the geographical range that we occupy and modify, however, comes at the expense of other living beings. We are now causing habitat degradation, and thus functioning in the exact same manner as other agents of mass extinction. The real question is if we will be willing to compromise and slow the effects of habitat degradation or continue on as before. In the short term, our species may continue to be successful. It will be our children and grandchildren who will have to deal with this most urgently if we do not take action now, and by then it may be too late, because extinct is, after all, forever—and this includes our own species.

What paleontologists have recognized based on life's distant past is that there were several episodes of global mass extinction where entire ecosystems collapsed, but even though these events occurred in the distant past they still have relevance today. It is clear that this ongoing mass extinction, if it continues, will have a profoundly negative effect on us: if not by driving us to extinction then by leaving us in a world of rats, roaches, pigeons, and maybe a few crop species. It is true that after every one of the mass extinction events of the past the world's animals and plants eventually recovered, but the time scale for recovery is on the order of at least five to ten million years. This length of time means that any recovery will be small comfort for us or our children and grandchildren, if our species even survives the throes of this mass extinction. (The long recovery time is partly attributable to the fact that it takes a lot longer for new species to evolve than for them to go extinct, just as it takes much longer to build a building than to tear one down.)

Thus, even if our species survives this mass extinction, it would live in a biologically impoverished world. The odds favor our extinction, however, because the fossil record of other mass extinctions suggests that if enough species or links in the global ecosystem vanish then the entire global biosphere collapses. Think of the many contingencies that were necessary for our species to be here today, including the fact that some ancestral lineage made it through each of the five succeeding mass extinctions, including the Permo-Triassic when as many as 90 percent of all Earth's species went extinct, and the Cretaceous–Tertiary mass extinction, when the mighty dinosaurs were felled. Yet what if we were to fall victims to the very mass extinction of which we are the agents? That would be the type of irony that hopefully can be avoided.

Additional Reading

Benton, M. J. 2005. *When Life Nearly Died: The Greatest Mass Extinction of All Time.* Thames & Hudson, London; 336 pp.

Eldredge, N. 1994. *Miner's Canary.* Princeton University Press, Princeton, NJ; 272 pp.

Eldredge, N. 1997. *Dominion.* University of California Press, Berkeley, CA; 190 pp.

Gould, S. J. 1996. *Full House.* Harmony, New York, 244 pp.

Hallam, A., and Wignall, P. B. 1997. *Mass Extinctions and Their Aftermath.* Oxford University Press, New York; 328 pp.

Knoll, A. H., Bambach, R. K. Canfield, D. E., and Grotzinger, J. P. 1996. Comparative earth history and late Permian mass extinction. *Science* **273**: 452–457.

McGhee, G. R., Jr. 1996. *The Late Devonian Mass Extinction.* Columbia University Press, New York; 378 pp.

Raup, D. 1992. *Extinction: Bad Genes or Bad Luck?* W. W. Norton, New York; 224 pp.

Stanley, S. M. 1987. *Extinction.* Scientific American, New York, 242 pp.

Ward, P. D. 2004. *Gorgon: Paleontology, Obsession, and the Greatest Catastrophe in Earth History.* Viking, New York; 257 pp.

Chapter 7

Systematics and the Fossil Record

Outline

- Introduction
- Methods and Approaches in Systematics
- The Growth of Molecular Biology and Improvements in DNA Sequencing Technology
- The Spread of Computers and Computer Programs Used to Study Evolutionary Relationships
- Systematics and How to go About Identifying Species in the Fossil Record
- Systematics and its Relevance for Identifying Patterns of Mass Extinction
- Systematics and the Meaning of Adaptations
- Concluding Remarks
- Additional Reading

Introduction

As already described in Chapters 3 and 4 some of the fundamental evidence that evolution has occurred is the fact that life is arrayed in a hierarchical pattern: some groups are more closely related evolutionarily, while others are more distantly related. This was in fact recognized even before Darwin published his *Origin of Species* and was codified in the Linnaean hierarchy. This hierarchy, which we discussed in greater detail in Chapter 3, allows different organisms to be classified into groups showing more or less evolutionary relatedness or affinity. For example, two species in the same genus, like *Crocodilus niloticus* and *Crocodilus porosus*, the Nile River (of Africa) and Australian saltwater crocodiles, respectively, are closely

Prehistoric Life: Evolution and the Fossil Record. 1st edition. By Bruce S. Lieberman and Roger Kaesler. Published 2010 by Blackwell Publishing.

Figure 7.1 Nile crocodile. This large carnivore basks on the bank of the Nile River. Image from iStockPhoto.com © N. Avivi.

related species in the class Reptilia (Figure 7-1). Species in different classes, like our own species *Homo sapiens*, which is in the class Mammalia, are more distantly related to species of *Crocodilus*, albeit they all belong to the same phylum, the Chordata. There is in fact an entire scientific discipline called systematics that is devoted to reconstructing the history of life. It includes the study of life's hierarchical pattern and the classification of organisms into different groups.

One of the important ways that paleontologists study the fossil record is using information from systematics. For example, to study and recognize mass extinctions (considered in the last chapter), we need information about the diversity of species and how this changed through time. A dramatic drop in the number of species at a particular time in the fossil record might imply a mass extinction and can be identified only through systematic studies. To consider what was happening during the Cambrian radiation (to be discussed in detail in Chapter 13) we need to know how rapidly the diversity of animals may have been increasing shortly after they evolved; again, systematic knowledge of the types of species and the number present is required. This systematic information also has to be carefully collected, because as described more fully below, differences in how we classify various groups might cause differences in the interpretation of how big a mass extinction truly was. Information from the field of systematics is useful even if scientists want to know whether what is called a species by biologists studying the modern world is equivalent to what paleontologists studying the fossil record call a species. Finally, an understanding of information from systematics is necessary if we wish to hypothesize that an organism has some kind of trait or morphological feature that can be called an adaptation. This is because it is crucial to understand when the particular traits called adaptations evolved. Each of these topics will be introduced in this chapter, along with examples that illustrate various points.

Significantly, some of the most exciting advances in paleontology and evolutionary biology over the last few years have occurred in the area of systematics. These advances have occurred in three parallel areas:

1 improved methods to study the evolutionary relationships of organisms;
2 advances in our ability to sequence the DNA of living and rarely fossil organisms;
3 the development of computers and computer programs that make it easy to analyze the large amounts of data that come from DNA studies and also from detailed studies of the anatomy of organisms.

Relevant information about these advances will also be introduced later in this chapter.

Methods and Approaches in Systematics

The way scientists determine the evolutionary relationships of organisms is to compare the various features of organisms, be it their anatomy, such as bones, shells, and cellular structures, or their DNA sequences (discussed more fully below). Building on the example from Chapter 4, jaguars and humans, being mammals, resemble each other more closely than they resemble snails (some talk-radio personalities aside). The jaguar and humans share teeth, hair, the fact that they can provide their young with milk, not to mention their common vertebrate skeleton with backbone and other bones. All of these characters and also a large number of commonalities in their DNA sequences separate the jaguar and the human from the snail and indeed from many other nonvertebrate organisms. (Of the over 1,000,000 species of animals that have been described by scientists only around 5,000 are mammals.) In general the features unique to some groups and that separate them from other groups, such as the ones that separate jaguars and humans from snails in this example, are called shared derived characters. They indicate that the jaguar and the human share a closer evolutionary history than either shares with the snail. As discussed in Chapter 4, the last common ancestor of humans and jaguars may have lived around 60 million years ago while the last common ancestor of jaguars, humans, and snails lived perhaps 550 million years ago (Figure 7-2).

Let us suppose now that we are interested in refining further our understanding of evolutionary relationships as we add in another species to the mix: the modern chimpanzee. It too, like the jaguar and human has characters that identify it as a mammal. For example, it has hair, teeth, and chimp mothers can provide milk to their young. Based on the information currently discussed it would fit onto the previous evolutionary tree as shown in Figure 7-3. Considering the characters on this tree thus far, there is really no way without having additional characters to determine the relationships among these three mammals. They are thus

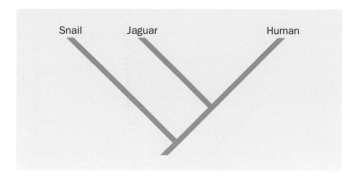

Figure 7.2 Mammals and snails. A cladogram relating jaguars, humans, and snails. Diagram by F. Abe, University of Kansas.

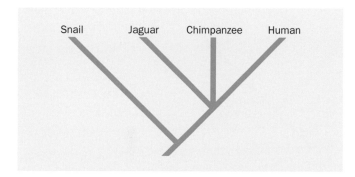

Figure 7.3 Mammals and snails *continued*. A cladogram relating jaguars, chimps, humans, and snails, created without considering any shared derived characters that might allow for resolution among the different mammal groups. Diagram by F. Abe, University of Kansas.

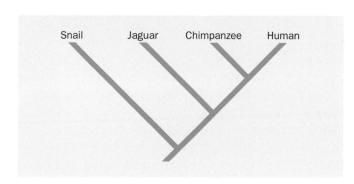

Figure 7.4 Mammals and snails *continued*. A cladogram relating jaguars, chimps, humans, and snails. By considering shared derived characters the relationships among the different mammal groups can be resolved. Diagram by F. Abe, University of Kansas.

placed in an unresolved position on the tree relative to one another. This is because the available characters discussed thus far are what are called shared primitive characters: the characters like teeth, hair, and the ability to provide milk characterize the ancestor of all of these species. There are, however, many other characters that chimps and humans have in common but that jaguars (and also snails and certain radio talk show hosts aside) lack, which allow us to further resolve this tree. To name just a few, both chimps and humans have a relatively large brain and a five-digit grasping hand with a prominent thumb, and they also share the placement and shape of the teeth, especially the molars. These characters are the aforementioned shared derived characters, which means they evolved recently in a closely related common ancestor. Shared primitive characters evolved a long time ago in a distant common ancestor. When attempting to determine the evolutionary relationships of species, the shared derived characters are the only characters that matter. In principle and for the purposes of this text, the distinction between shared derived and shared primitive characters is straightforward, but in practice it can be more complicated (Figure 7-4).

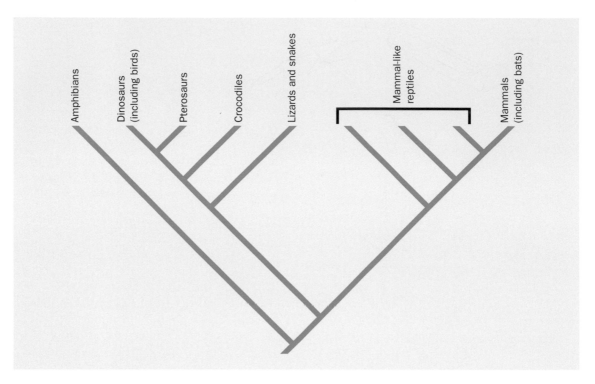

Figure 7.5 Flight evolved independently. A cladogram showing the relationships of several vertebrate groups including bats and birds. It is clear that flight evolved independently in these two groups. Diagram by F. Abe, University of Kansas.

The approach to systematics that considers evidence only from shared derived characters is referred to as cladistics. The development of cladistics was motivated by systematist Willi Hennig in the middle of the 20th century, and cladistics came to be used increasingly by other scientists at the end of the 20th and beginning of the 21st century. The development of cladistics, with the recognition that only shared derived characters are important for determining evolutionary relationship, was an important conceptual advance in the field of systematics.

One important lesson learned from systematics is that not all characters shared in common between species are shared derived or shared primitive characters. There are also some characters that are analogous or convergent. For example, both bats and birds have wings that they use to fly, but these wings evolved independently or convergently in each group: bats are mammals, while birds evolved from reptiles (Figure 7-5). The fact that bat and bird wings are convergent is pretty easy to recognize (Figure 7-6). As can be seen, the structure of the two wings and the bones that comprise them, and thus the mechanics of how they operate in flight, are very different.

There are times, however, when convergences are much more tricky to identify. One or several characters in two different species might be so similar that scientists could mistakenly treat these species as being closely related and having evolved from a recent common ancestor. An example of this type of convergence occurs in two Devonian trilobites *Bouleia* and *Phacops*. The former is from South America, the latter from North America. These two species are very similar, especially in the overall inflation and shape of the head and the position of the eyes (Figure 7-7). Because of these similarities, it was originally thought that they were closely related,

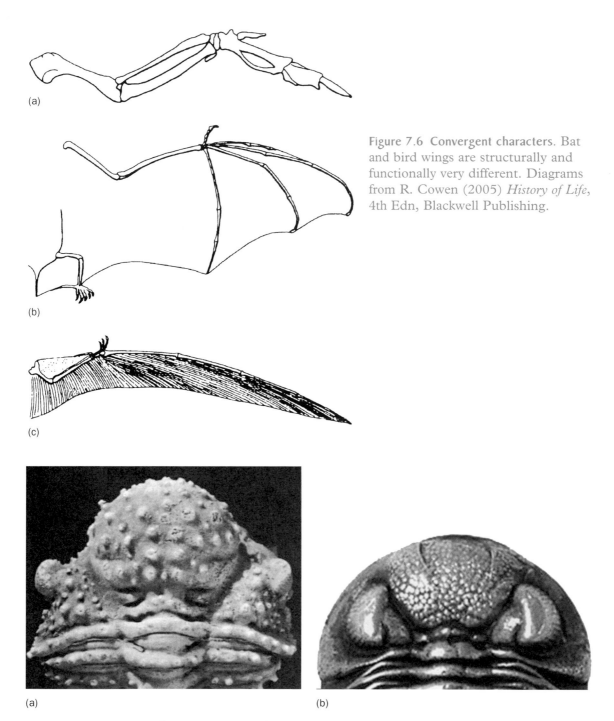

Figure 7.6 Convergent characters. Bat and bird wings are structurally and functionally very different. Diagrams from R. Cowen (2005) *History of Life*, 4th Edn, Blackwell Publishing.

(a)

(b)

Figure 7.7 Convergent characters *continued*. Heads of two species of trilobites. They were once thought to be closely related but are now known to be convergent. (a) *Bouleia*, image from Eldredge, N. (1971) Patterns of cephalic musculature in the *Phacopina* (Trilobita) and their phylogenetic significance. *Journal of Paleontology* 45, 52–67. (b) *Phacops*, image from N. Eldredge.

but paleontologist Niles Eldredge, using a comprehensive study of all of the characters of these species showed that they were in fact much more distantly related.

Another example might involve the same mutation along the same stretch of DNA in two different species. This in fact happens quite commonly, and given that this part of these species' DNA will be chemically identical, the case is different from the bat and bird wing, which really represent only superficial convergences. It might be asked how can systematists hope to determine if characters are convergences when this happens? It turns out that the best way systematists can determine if one character, or several characters, that species share are convergences is by comparing all of their characters. If two species differ from one another in most of their characters, and are shown not to be derived from a relatively recent common ancestor, then any of those characters held in common can be identified either as primitive characters that they share or as convergences.

The Growth of Molecular Biology and Improvements in DNA Sequencing Technology

Technical advances in molecular biology now mean that it is relatively easy, even routine, to sequence an organism's DNA. The DNA molecule governs heredity, contains the genetic material, and codes in large part the appearance of organisms. The genetic instructions are encoded in the relative order of nucleotides found along the DNA strands. Nucleotides possess one of four bases called adenine (A), guanine (G), thymine (T), and cytosine (C). The relative order of the bases is called the DNA sequence and is what determines an individual's genetic code. Closely related individuals share a more similar order of A, G, C, and T bases (DNA sequence) than more distantly related organisms because they inherited this DNA from a more recent common ancestor; more distantly related organisms differ more in their DNA sequence because they share a more ancient common ancestor and their DNA sequences have had more time to evolve independently.

These advances in molecular biology mean that systematists no longer have to be content with just comparing the anatomies of modern organisms to see how they are related to one another. Instead, they can decipher and compare the DNA sequences. Still, it is quite laborious to sequence the entire genetic code of any organism because their DNA is composed of literally billions of smaller submolecules called nucleotides. There are only a few hundred species, human beings are one of these, for which the complete DNA sequence is known. One of the advances that made DNA sequencing relatively easy was the development of a technique that is termed polymerase chain reaction (PCR) technology. PCR is used to amplify tiny samples of DNA to produce billions of copies of a stretch or fragment of DNA (a few hundred to a few thousand bases long), providing abundant material for sequencing. Nowadays there are even automated sequencers that greatly reduce the amount of "hands on" time required for sequencing. DNA sequencing was possible before the invention of PCR and automated sequencers, but each of these new technologies has made it much easier to collect DNA sequences from large numbers of different species. (This is the same type of technology that police labs use to solve crimes by matching either the victim's or the perpetrator's DNA.)

Systematists can use these differences in the genetic code to determine how organisms are related. Organisms that share a large component of their genetic code are closely related. More distantly related organisms have fewer parts of their sequences in common, although the same basic principles from cladistics apply: systematists line up equivalent stretches of DNA from different organisms and then look for shared derived similarities (the same bases) in their genetic code (Table 7-1). Just as with anatomical characters, as already mentioned, there can be convergences. DNA sequences provide further evidence, in addition to the anatomical traits already described that they share, that chimps and humans are

Table 7.1 Hypothetical short stretch of DNA sequences for four different species with similarities, relative to humans, highlighted in bold

Snail	**CC**ATG**TT**A**CC**
Jaguar	**CCTT**AA**TAGC**
Chimp	**CCTTACTAGC**
Human	**CCTTACTAGC**

very closely related evolutionarily: we share almost 99% of our genetic code in common with chimps.

DNA can be used to consider patterns of evolutionary divergence within individual species. For example, it can be used to reconstruct the order in which different regions of the Earth were colonized after our own species migrated out of Africa. DNA can also be used to answer larger-scale systematic questions such as how the various phyla of animals are related to one another.

Ancient DNA

There are indications that some long dead, fossilized organisms can be sequenced, and there is an entire field devoted to the study of what is called "ancient DNA." For instance, sometimes animals can be preserved by chance in a manner that resembles the process of mummification (see Chapter 1). A famous example involves a specimen of an extinct giant ground sloth that died about 10,000 years ago that was preserved in the very dry environment of a cave.

The skin and body are preserved, resembling a mummy, without the wrappings of course. The well-known woolly mammoths preserved frozen in the permafrost of Siberia (Figure 1-16) has also had a fragment of its DNA sequenced, and some paleontologists have even seriously proposed bringing mammoths back from the dead by using this DNA along with that of modern elephants.

One thing that scientists who study ancient DNA have discovered is that DNA is a very sensitive molecule that tends to fall apart or degrade with time. The longer an organism has been dead, the more degraded its DNA becomes, and if the DNA is too degraded it becomes very tricky to extract it, amplify it using PCR, and sequence it. Currently the oldest known, definitive, examples where ancient DNA has been successfully isolated and sequenced all involve organisms that lived less than about 11,000 years ago; these examples are all fairly straightforward and uncontroversial.

For some time there has been the even more intriguing goal of discovering and sequencing ancient DNA from species that lived on Earth much further back in deep time. This idea first gained broad currency with the publication of Michael Crichton's book *Jurassic Park*. The premise in that science fiction book was that scientists could recreate dinosaurs by isolating the DNA from their blood which might be contained inside the guts of mosquitoes preserved inside amber. Amber is a yellow, hard substance that is the fossilized remains of tree resin. Insects that get trapped in the sap of trees are sometimes preserved in beautiful condition, looking as if they just died (Figure 7-8).

Although aspects of *Jurassic Park* are completely far-fetched, there are mosquitoes and other biting insects preserved in amber that formed when

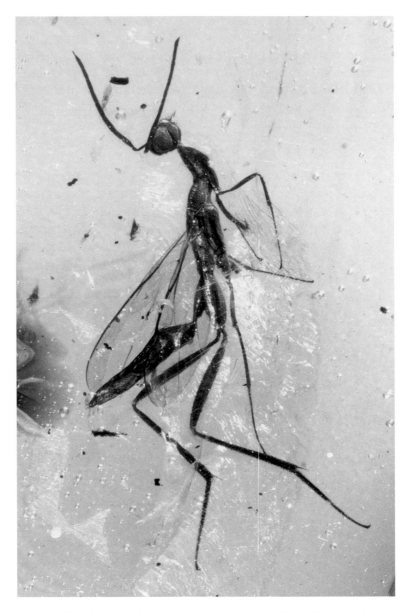

Figure 7.8 Amber. A fossilized insect (a wasp) preserved in amber. The specimen is roughly 30 million years old. Image from M. Engel, University of Kansas.

dinosaurs walked the Earth. There can also be tiny air bubbles preserved within these amber nuggets like time capsules preserving the constituents of ancient atmospheres. What is still controversial at this point is whether DNA can be isolated successfully, amplified, and sequenced from insects preserved in amber that is tens of millions of years old.

In what was an exciting and potentially groundbreaking result Rob DeSalle, John Gatesy, and others reported in the journal *Science* that they had isolated and sequenced DNA from a roughly 30 million year old termite preserved in amber. They even used a cladistic analysis to compare their sequence with sequences from several other modern termite species: they found it assumed a relatively basal position within the tree of modern termites and was thus the likely ancestor of many modern groups of termites.

Unfortunately, other scientists have had a hard time reproducing these scientists' results, and it remains controversial as to whether they actually isolated the DNA of the fossilized termite or instead may have inadvertently sequenced some modern-day contaminant. Still, irrespective of the ultimate verdict by scientists on the validity of these results, it points out the potential of the new technologies that will be available to paleontologists in the 21st century. Contamination can be a problem in any analysis of DNA sequences because if only the smallest bit of some other organisms' DNA inadvertently enters the original sample it will be amplified many times over during PCR. Contamination is, however, a particular problem in analyses of ancient DNA because the samples that these scientists must work with often contain extremely tiny quantities of DNA. The smaller the initial quantity of DNA the harder it is to amplify and the more likely it will be overwhelmed by some contaminant.

Returning to the theme from *Jurassic Park*, it should be noted that even if scientists were able to isolate dinosaur DNA from the gut of a mosquito they would still be a long way from making a dinosaur. Indeed, isolating dinosaur DNA might be the simplest (albeit still very difficult) step in the complex chain of steps necessary to bring the large terrestrial dinosaurs back to life. Thus, bringing back an extinct organism is still farfetched, but it is theoretically possible to analyze fragments of ancient DNA for systematic purposes.

It is still too soon to say whether elements of the story of Jurassic Park have moved from science fiction to science fact. It is likely, however, that ancient DNA will continue to be an exciting research area for many years to come. This means that some paleontologists, instead of picking up a rock hammer and heading off to some exotic locale to collect fossils, may don a lab coat and enter a modern molecular biology sequencing facility.

The Spread of Computers and Computer Programs Used to Study Evolutionary Relationships

The final development that has helped make systematics an exciting research area is the easy access scientists now have to high-speed computers. Consider that scientists have gathered DNA sequence data from various species and also identified many anatomical characters that various species share. Because some characters are primitive and others are the result of convergence, it will not be easy to just gaze at the animals, or their DNA sequences, and somehow come up with an answer as to which species are each others' closest relatives. Instead, scientists need some way to process large amounts of data, and this is where computers come in. Using various computer programs systematists can analyze large amounts of data to come up with well-supported evolutionary trees. Easy access to computers has made it possible for systematists all over the world not only to analyze their data but also to share it via the worldwide web. For example, the time may come in the not too distant future when paleontologists will not have to board a plane and travel thousands of miles to see an important fossil specimen housed in a distant museum. Instead they will be able to call up an image on the internet and analyze the fossil without leaving their own office. (In fact, one of us [RLK] helped make 3D images of fossils in our museum at the University of Kansas and these are available on the web.)

Systematics and How to go About Identifying Species in the Fossil Record

For the past 250 years a tempestuous debate has raged among systematists that has little relevance to the world at large. It is a debate about what are species and how do we define them, a topic we first introduced in Chapter 3. In reality the debate about definitions may not matter all that much in practice. For instance, the famous systematist and evolutionary biologist Ernst Mayr showed that the various bird species he and other scientists identified in the tropical rain forests of New Guinea were almost precisely the same as the various types of birds the aboriginal hunting peoples that occupied these islands identified. Many scientists argue that species represent groups of organisms that can interbreed with one another in the wild. Other scientists argue that species are groups of organisms that share in common unique features of their anatomy or DNA. It turns out that these definitions are related because organisms that are interbreeding will pass down their common appearance and similarities in their DNA.

Although details of this debate are not essential, it is true that subscribing to aspects of both of these definitions could be potentially problematic for paleontologists studying the fossil record. For instance, paleontologists cannot determine whether extinct organisms were interbreeding with one another, although in practice biologists have nearly as much trouble with breeding tests in nature. Nor of course can paleontologists sample the DNA of extinct fossil organisms, except perhaps very rarely as already mentioned above. The one thing that paleontologists do have is information about the anatomies of their fossil organisms. Remember, as already described in Chapter 5, these tend to persist relatively unchanged for many millions of years, as part of the pattern described as punctuated equilibria. It turns out that there is considerable evidence that the anatomy of organisms serves as a good proxy for whether organisms can interbreed with one another; it also is an underlying reflection of their genotype.

Among the evidence supporting this is an analysis by paleontologists and biologists Alan Cheetham, Jeremy Jackson, and Nancy Knowlton that was based on work with modern marine animals. They found that a group of animals in the phylum Bryozoa that resemble one another very closely, what we might call a species if we only had their fossilized remains, is also the same group that can interbreed with one another and the same group that shares such a close complement of their DNA. Thus, they would be called the same species by all these different criteria. This result is crucial for paleontologists because it means that even though they do not have a precise way of measuring the genetic content of organisms, or a way of verifying which organisms could and did interbreed with one another, we have a good way of inferring it. By looking at the fossil shells preserved in the face of a cliff we are in effect staring across a set of beaches that persisted for countless millennia. Each layer of fossils in the face of the cliff is analogous to the dead shells we would see if we were to stroll on the beach today. But today those dead shells are a relict of the animals that are living and interbreeding deeper out beneath the waves. The shells that are identical or closely similar represent the same species. Perhaps several species are visible as dead shells, revealing several species that lived farther out in the water. The fossil record is directly akin to that, except when we see the fossil shell we know its other counterpart existed under the waves millions of years ago.

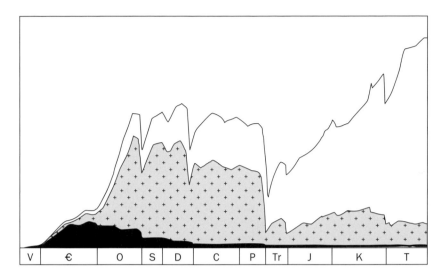

Figure 7.9 Diversity curve. The diversity of marine animals from the late Proterozoic (far left) to the Pleistocene (far right). This curve is derived from the work of J. J. (Jack) Sepkoski. Beneath the curve, the different colors represent the diversity of different types of organisms that contribute to the total diversity. For instance, the black curve approximates the diversity of trilobites. Diagram from E. N. K. Clarkson (1998) *Invertebrate Palaeontology and Evolution*, 4th Edn, Blackwell Publishing.

Systematics and its Relevance for Identifying Patterns of Mass Extinction

Our understanding of the nature of several of the concepts that were already discussed, for example, mass extinctions (see Chapter 6) depends on the existence of detailed systematic data. This is because mass extinctions were originally recognized and to a large degree are still recognized today by taking compilations of diversity through time. The result is a diagram like this, derived from the work of paleontologist J. J. Sepkoski (Figure 7-9). The particularly large jags down represent mass extinctions, while large jags up might represent times of extensive evolution. For example, the Cambrian radiation, which we shall discuss in more detail in Chapter 13, is the relatively large jag up on the left-hand side of the diagram. It represents the time when animals first evolved and diversified in the fossil record, and it may have been a time of relatively rapid evolution.

The way diversity curves like the one shown above are compiled is by determining the number of species, genera, or families that lived during broadly defined geological time intervals. Ultimately this information is derived from a large number of systematic studies by paleontologists, some conducted recently, others a long time ago. Our knowledge of diversity is not static: it is continually improving and changing through time. One reason is that paleontologists continue to find new fossils in new places and from new time periods. For example, imagine that a genus whose last occurrence was thought to be at the end of the Jurassic was found in Cretaceous rocks. This would subtly change our knowledge of overall diversity: instead of regarding the genus as being one that disappeared as part of the extinction event at the end of the Jurassic, it would be tabulated now along with other genera that went extinct sometime during the Cretaceous. The diversity curve shown above is built up from data on literally thousands of genera, so changing the stratigraphic range of one genus has little effect on the overall shape of the curve. If, however, there was enough new systematic

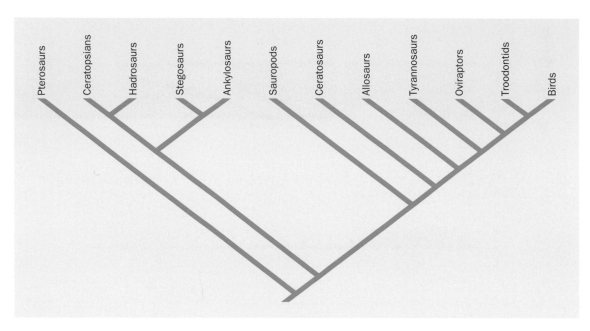

Figure 7.10 Bird–dinosaur relationships. A cladogram relating pterosaurs, the different major groups of dinosaurs, and birds. Diagram by F. Abe, University of Kansas.

information from the fossil record that altered the stratigraphic ranges of many taxa substantially then the overall shape of the curve could change, and along with it, perhaps, our understanding of mass extinction events of the past.

Another reason the shape of this curve could change is that there may be new systematic studies that alter pre-existing ways that paleontologists classify animals. A classic example involves birds and dinosaurs. Originally and through much of the 20th century paleontologists and biologists treated birds and dinosaurs as distinct taxonomic groups, and it was thought that all the dinosaurs went extinct at the end of the Cretaceous. More recently, however, paleontologists have conducted more detailed systematic studies using computer algorithms to consider a comprehensive set of anatomical characters from extinct dinosaur groups such as *Tyrannosaurus rex* and extinct and modern species of birds (Figure 4-9). These analyses have shown that extinct and modern birds evolved from the stock of organisms that we have referred to broadly as dinosaurs (Figure 7-10). We still call them dinosaurs, but the term has become a little more complex.

Based on this evolutionary history and also recent beautifully preserved fossils discovered from China it turns out that some fossils we call dinosaurs, including even *Tyrannosaurus rex*, may have actually been covered with feathers (Figure 7-11), and depending on how one defines the term bird, some of these specialized dinosaurs (but not *T. rex*) may actually have been primitive birds. (Nevertheless, as Figure 7-11 shows, there is something quite different between these carnivorous "dinosaurs" and modern birds.)

Dinosaurs will be discussed in greater detail in Chapter 17. The key point for now is that dinosaurs as an entire group did not go extinct at the end of the Cretaceous. Yes, it is true that many species of dinosaurs, especially the large, nonflying ones including *Tyrannosaurus rex*, went extinct right at the end of the Cretaceous. Some lineages of dinosaurs, however, especially the winged dinosaurs, that is, the birds, survived that mass extinction and are still with us today.

Figure 7.11 Feathered dinosaur. A dinosaur, covered with feathers, and closely related to *T. rex*, from the Cretaceous of China. Image from Zhonghe Zhou, Institute of Vertebrate Palaeontology and Palaeoanthropology, Chinese Academy of Sciences, with assistance from Desui Miao, University of Kansas.

This improvement in our understanding of the relationship between birds and dinosaurs was derived from systematic studies that could ultimately subtly change our understanding of the total diversity of animal life and how it changed throughout the Phanerozoic. This is because it means that the extinction at the end of the Cretaceous was slightly more muted than originally thought, although its effects and the total diversity loss were still prodigious. Again, one such change is unlikely to greatly modify Sepkoski's diversity curve. If, however, enough systematic studies were to be conducted that showed that many Cretaceous (or other) groups thought to have gone extinct actually survived into the Cenozoic, it would lessen the magnitude of the end Cretaceous extinction. Indeed, paleontologists Colin Patterson and Andrew Smith argued that many of the mass extinctions appear more dramatic than they actually were because of this phenomenon just described. They argued that although we may have some basic knowledge of the types and numbers of genera and families in the fossil record, without detailed studies on evolutionary relationships using modern computer based methods this knowledge should be considered only sketchy. The jury is still out as to whether Patterson and Smith are correct about this, but minimally their study represents an important argument that additional systematic studies aimed at teasing apart the evolutionary relationships of fossil taxa should be conducted. We can clearly see that systematic studies, by focusing on evolutionary relationships, provide important information that paleontologists need in order to study mass extinctions.

Systematics and the Meaning of Adaptations

Systematic studies also provide crucial information about the meaning and nature of adaptations, and given that one of the critical ideas from Darwin's *Origin of Species* was natural selection and the related notion of adaptations, discussed at length in Chapter 5, this is given added significance. Darwin in the *Origin of Species* spent much time discussing adaptations: traits that perform a specific function that were honed by natural selection. Like many other concepts and questions in paleontology and evolutionary biology, we need systematic studies to test whether traits that serve important functions are in fact adaptations. Systematic studies are important for determining if a structure is an adaptation because these studies allow us to determine the evolutionary history of a trait. Paleontologists and evolutionary biologists Elisabeth Vrba and Stephen Jay Gould were among the first to emphasize how important systematic studies might be for the study of adaptations. Consider the example of eyebrows in humans. It has been argued occasionally that eyebrows are an adaptation to keep sweat out of our eyes. This is certainly possible, but what might be wrong with this claim? One potential problem with it is that many other organisms aside from humans have eyebrows, yet they might not be used to keep sweat out of their eyes. For example, if you have a cat you can see that they have fine, long whisker-like hairs coming out above their eyes exactly where eye brows should be, although BSL's cat was never caught sweating. Seals also have similar types of eyebrows and again these clearly do not keep the sweat out of their eyes. It turns out that many other kinds of mammals have eyebrows, too. Our understanding of the evolutionary history of mammals, derived from systematics, means that the state of having eyebrows evolved long before humans evolved, some time back during the early history of mammals and perhaps more than 50 million years ago. Therefore, they are primitive characters. Yes, perhaps eyebrows may occasionally prove useful to us when we are working up a sweat, although still this may not help us leave more descendants in the next generation, but eyebrows certainly did not evolve to keep sweat out of our eyes. They evolved in one of our distant ancestors, perhaps as a result of natural selection for something else.

The Spandrels of St Mark's Cathedral and their Relevance to the Question of Adaptation

The example of eyebrows as adaptations in humans is one of many such examples, and it turns out that since the publication of the *Origin of Species* some scientists have been far to willing to say that any trait found in any organism is adaptive without adequately justifying the proposition. Often discussions of adaptations deteriorate into a set of what are called just-so stories. This style of argument was rightly ridiculed by the famous French author Voltaire who remarked sarcastically, in French, that "Things cannot be other than they are. . . . Everything is made for the best purpose. Our noses were made to carry glasses, so we have glasses. Legs were clearly intended for pants, and we wear them."

Stephen Jay Gould and evolutionary biologist Richard Lewontin have brilliantly shown how these types of adaptive just-so stories can be easily applied to architecture, thus exposing how ludicrous they are. They concentrated on St Mark's Cathedral in Venice, Italy. If you're ever fortunate enough to go to Venice Italy, you will likely go to the Piazza San Marco, where St Mark's cathedral is located (Figure 7-12). This cathedral has a great central dome held up by arches with beautiful illustrations on the inside of the dome. The arches are also covered with illustrations. Where the arches meet there is a small triangular space that has to exist because any time there are two rounded arches supporting a dome,

Figure 7.12 St. Mark's cathedral. The beautiful façade of the cathedral, at the time being renovated, in Venice, Italy. Image by B. S. Lieberman, University of Kansas.

where they meet there has to be a slight space (Figure 7-13). Each of these triangular spaces is typically called a spandrel, and they are beautifully and harmoniously illustrated with paintings of the four apostles, Matthew, Mark, Luke, and John. One could argue that the design of the spandrels is so elaborate and fits in so harmoniously with the rest of the cathedral that they were specifically built for that purpose, but this does not actually describe why they exist. There was a structural architectural constraint, a spandrel, created any time a builder placed a dome on rounded arches. The 15th century Venetians did a good job of decorating these spandrels with religious allegories to get them to blend in with the rest of the paintings in the cathedral. They were not made, however, specifically to convey that allegory, even though there are four of them and there are four authors of the New Testament gospels. They had to exist, and the decorators put them to good use.

When biologists since Darwin have been asked to characterize the evolution of life, they have often emphasized that organisms possess many traits that are adaptations. But maybe most traits are not adaptations. Maybe many traits are like the spandrels in St Mark's cathedral. They are present for some other reason or because some other structure exists, and nature takes advantage of them.

Figure 7.13 **Spandrel**. The structure labeled "A", which is also sometimes called a pendentive. Diagram by E. Viollet-le-Duc (1856) *Dictionnaire raisonné de l'architecture française du XIe au XVIe siècle*. Librairie-imprimeries reunites, Paris.

One prominent case of where this revised mode of thinking might be warranted is in the example of the dinosaur *Tyrannosaurus rex* (Figure 7-14). People often question the purpose of the diminutive arms of *T. rex* and what they could have been an adaptation for. There have been some pretty farfetched explanations including mating courtship. Even if they were used for that purpose, it is unlikely that the small arms in *T. rex* evolved for this reason. They may have eventually come

Figure 7.14 *Tyrannosaurus rex*. The skeleton of this awesome dinosaur in side profile, highlighting the large skull and the diminutive arms. Image from iStockPhoto.com.

to play this role, but there is no evidence that active natural selection led to the establishment of this trait. It is important to remember that just because something has utility does not mean it originated for this reason. There is excellent evidence that all life has evolved and follows a pattern of common descent. We cannot assume though that every trait that evolved during that history was actively shaped by natural selection to serve that purpose. (We should mention in this context that the tiny arms may be like our appendix—present as a vestigial trait. After all, with six-foot jaws, who needs forearms? *Tyrannosaurus rex* did not need its forearms for feeding any more than modern birds do.)

Another excellent example of structures that have been called adaptations but are really more likely spandrels are the antlers of the extinct Irish elk. Stephen Jay Gould provided important information about the evolutionary history of antlers, using his understanding of systematic patterns in the group that includes the Irish Elk; this shed light on the question of whether the Irish Elk's huge antlers were indeed adaptations. The Irish Elk is known from fossils throughout Europe, including Ireland of course, and has a dramatic and remarkably large set of antlers (Figure 7-15). The Irish elk went extinct around 10,000 years ago probably due to human hunting pressure (see discussion in Chapter 6). It turns out that the Irish Elk is not in fact an elk but rather a deer (American Elk are deer, and the British use the term elk where we would use moose), and this fact, plus the recognition that it was not restricted to Ireland, means that the name Irish Elk is a bit of a misnomer. Still, the Irish Elk has the largest set of antlers known among modern or extinct deer (or elk); not only are its antlers large, but the overall body size of this deer is quite large as well. It is also the largest living or fossil deer (or elk). While its body size is only somewhat larger than the body size of its closest relatives, the antlers are much larger in comparison. The truly prodigious size of the antlers had

Figure 7.15 Irish elk. A skeleton of this giant extinct deer, with a paleontologist for scale. Image from M. J. Benton and D. A. T. Harper (2009) *Introduction to Paleobiology and the Fossil Record*, Blackwell Publishing.

led some scientists to suggest that the large antlers were an adaptation, perhaps used in courtship battles. Modern male deer use their antlers to fight other male deer in order to gain more mates. Perhaps the truly large antlers in Irish Elk evolved to facilitate the battle for mates.

Although this explanation is certainly possible, it does not seem likely for several reasons. First, Gould recognized that there is a consistent relationship between the sizes of various species of deer and the size of their antlers. As species of deer increase their relative sizes, the sizes of the antlers increase much more rapidly. Relatively larger species of deer have relatively much larger racks of antlers. If you scale the size of the Irish Elk relative to the size of other deer species, its antlers are exactly the right size. Given that the Irish Elk is a large deer, it should have very, very large antlers: it turns out that the antlers of the Irish Elk are not unusually large, all things considered. Moreover, the Irish Elk lived during the time of the Ice Ages, when climatic conditions were unusually cold, at least relative to today or relative to the last 60 million years of earth history. Many other types of animals were unusually large during the Ice Ages; a well-established relationship exists between temperature of the environment and the body sizes of mammals. As the average temperature falls, the average body size of mammal species increases. Maybe the relatively low temperatures during the time when the Irish Elk first evolved and subsequently lived favored its large body size. Because it was a large deer, it would be expected to have unusually large antlers. Thus, the large antlers did not evolve to facilitate courtship battles for mates, though undoubtedly they were subsequently used for this purpose. To this extent they are like spandrels. Our knowledge of systematic patterns in deer and the relationship between body size and antler size puts this potentially murky case of adaptation in a clearer and less adaptive light.

Concluding Remarks

Although the field of systematics has been around for a long time, research in this area continues to make crucial contributions to paleontology and evolutionary biology. Research in this area is still significant because the truth is that although we have a good idea of the range of species that have existed on the planet over the past several hundred million years our knowledge continually grows and expands with each new fossil discovered. Ultimately it is the scientist on the ground, the systematist who makes these new scientific discoveries and integrates them with our pre-existing knowledge, that helps our scientific understanding of the history of life to increase through time. The field also continues to have relevance, however, because of new technological advances. We predict that studies of the DNA sequences of living and fossil species will also continue to increase our knowledge of how life has evolved through time. It is encouraging and noteworthy that one of the disciplines that early on made the fundamental contributions to ideas on evolution, and helped to convince people that life had indeed evolved, still continues to be a relevant scientific discipline in the 21st century.

Additional Reading

Brooks, D. R., and McLennan, D. 1991. *Phylogeny, Ecology, and Behavior*. University of Chicago Press, Chicago, IL; 441 pp.

DeSalle, R., Yudell, M., and the American Museum of Natural History. 2004. *Welcome to the Genome: A User's Guide to the Genetic Past, Present, and Future*. John Wiley & Sons, Chichester; 204 pp.

DeSalle, R., and Lindley, D. 1997. *The Science of Jurassic Park and the Lost World: Or, How to Build a Dinosaur*. Basic Books, New York.

Eldredge, N., and Cracraft, J. 1980. *Phylogenetic Patterns and the Evolutionary Process*. Columbia University Press, New York; 349 pp.

Gould, S. J., and Vrba, E. S. 1982. Exaptation—a missing term in the science of form. *Paleobiology* **8**: 4–15.

Gould, S. J., and R. C. Lewontin. 1979. The spandrels of San Marco and the Panglossian paradigm: A critique of the adaptationist programme. Pp. 147–164 in J. Maynard Smith and R. Holliday (eds.), *The Evolution of Adaptation by Natural Selection*. The Royal Society, London.

Grimaldi, D. A. 2003. *Amber: Window to the Past*. Harry Abrams, New York; 216 pp.

Martin, A. J. 2006. *Introduction to the Study of Dinosaurs*. John Wiley & Sons, London, 560 pp.

Poinar, G., and R. Poinar. 2001. *The Amber Forest*. Princeton University Press, Princeton, NJ, 292 pp.

Wiley, E. O. 1981. *Phylogenetics: The Theory and Practice of Systematics*. John Wiley & Sons, London, 456 pp.

Wiley, E. O., D. Siegel-Causey, D. R. Brooks, and V. A. Funk. 1991. *The Compleat Cladist*. University of Kansas Museum of Natural History Special Publication No. 19.

Chapter 8

Principles of Growth and Form: Life, the Universe, and Gothic Cathedrals

Outline

- Introduction
- Galileo's Principle
- Galileo's Principle and its Relevance to the Biology of Living Organisms
- Galileo's Principle and Constraints on the Evolution of Large Body Size
- Galileo's Principle and its Relevance to Medieval Architecture
- Galileo's Principle and its Relevance to Cratering Density in our Solar System
- Concluding Remarks
- Additional Reading

Introduction

When paleontologists and evolutionary biologists study the history of life they hope to uncover general unifying processes and principles. Without these their results deteriorate into a simple narrative: this happened, then this happened, then this happened. Not only is such a treatment rather boring, but it's hardly synthetic either. You may recognize that the existence of historical patterns and the search for unifying processes and principles partly describes the relationship between

Prehistoric Life: Evolution and the Fossil Record. 1st edition. By Bruce S. Lieberman and Roger Kaesler. Published 2010 by Blackwell Publishing.

time's arrow and time's cycle considered in Chapter 1. One of the most exciting discoveries a scientist can make is when he or she not only uncovers a general unifying principle that applies in one scientific area, like paleontology, but when they are able to relate a similar principle to a different scientific discipline, or even a completely distinct area of human intellectual inquiry. Here, we will consider one such general unifying principle whose relevance was elucidated by paleontologist and evolutionary biologist Stephen Jay Gould and evolutionary biologist John Tyler Bonner. This principle not only helps us understand the nature of some of the key evolutionary events on this planet, like the origin of multicellular life and the emergence of complex body plans, but it also appears to help explain aspects of planetary geology and medieval architecture. Some of the key evolutionary events we consider here occurred during the Archean and Proterozoic eons, which are informally referred to as the Precambrian, and will be documented more fully in Chapter 12.

Galileo's Principle

The unifying principle we refer to here governs the size and shape of organisms, the size and shape of medieval, gothic cathedrals, and the amount of craters on the planets and moons of our solar system. It is caused by a phenomenon recognized by the famous scientist Galileo Galilei. Galileo, among the greatest scientists of the Middle Ages, was born in Italy in 1564. Not only did he build one of the first telescopes and use it to make important observations about the nature of the stars and our solar system, he also recognized what is today referred to as "Galileo's Principle." This principle specifies that as the size of an object increases, its surface area or outer covering increases as a power of 2 or a square, and its internal volume increases as a power of 3 or cube. Galileo himself used this observation to explain various features of living organisms.

It turns out that this relatively simple principle constrains in an important way the manner in which organisms grow. It is worth demonstrating the reality of Galileo's Principle using a simple example. Imagine a cube one centimeter on a side that could somehow be uniformly inflated to a cube ten centimeters on a side (Figure 8-1). The smaller cube's volume is just its height multiplied by its width multiplied by its depth; if we multiply accordingly (1 cm × 1 cm × 1 cm) we obtain a value of 1 cm^3. The surface area of this cube can be calculated as the total area on all six of its faces. The area of each face of the cube is the height of the respective square multiplied by its width; if we multiply accordingly (1 cm × 1 cm) we obtain a value of 1 cm^2; then the total surface area of the smaller cube is 6 cm^2. In this smaller cube the ratio of its surface area to its volume is 6 to 1.

Now consider the larger cube; its volume is much greater than that of the smaller cube: following the calculations above (10 cm × 10 cm × 10 cm), we obtain a volume of 1,000 cm^3. The surface area of the larger cube is also considerably bigger than that of the smaller cube: (10 cm × 10 cm) × 6 which is 600 cm^2. But note that the surface area of the bigger cube has not grown proportionately with the growth of its volume. In particular, in the larger cube the ratio of its surface area to its volume is now 6 to 10 or 0.6: much smaller than the ratio in the smaller cube. Thus, as the cube has expanded or "grown" its volume has increased much more rapidly than its surface area.

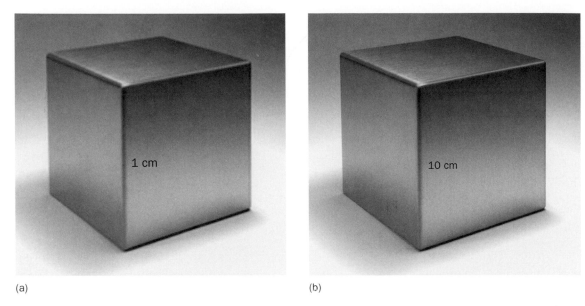

(a)   (b)

Figure 8.1 **Two cubes of different size.** These are used to illustrate Galileo's principle. Images from iStockPhoto.com © A. Prill.

Galileo's Principle and its Relevance to the Biology of Living Organisms

One reason "Galileo's Principle" matters to biology is that in many living things volume can be thought of as a proxy for weight. As animals and plants grow larger their volume increases more rapidly than their surface area. Considered a different way, simply by growing larger, the surface area of an organism decreases relative to its volume or weight. Why might this matter? There are many important biological functions that depend upon surfaces and thus available surface area; further, these functions that occur at a surface must serve the entire volume of the organism.

Feeding in Organisms

For instance, there are many organisms that obtain food through their surfaces: one well known example involves single celled bacteria. When a single celled bacterium feeds the food enters across the external cell membrane (Figure 8-2). The amount of food that a bacterium can consume is limited by the available external surface area of the cell. Further, all the nutrients must be distributed to all parts of the cell, i.e., the entire volume of the cell. In a simple single celled organism this is accomplished by a process like diffusion. Critically, any part of the cell that did not receive adequate nutrients would start to die, ultimately killing the entire organism. As we shall explore shortly, because of Galileo's Principle, there are limits determining how big a single celled organism can get; in particular, how big a cell can be while still maintaining an adequate surface area to take in enough food while transporting it throughout the cell. This is why more than 99.99% of all single celled organisms are microscopic and extremely small.

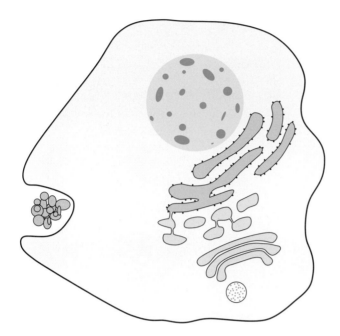

Figure 8.2 **Feeding in a bacterium**. The larger cell is engulfing the smaller cells on the left. Diagram from J. and J. Giannini, St Olaf College.

One thing we see is that as organisms became larger, they evolved a way of dealing with thc constraints imposed by Galileo's Principle. In particular, they employed various means of maximizing their surface area relative to their volume. This point can be well illustrated by comparing humans to bacteria. We might tend to think of human's as much more complex than bacteria, and rightly so, yet even humans are constrained by Galileo's Principle. For example, we must rely on surfaces and thus surface area to digest food. Although this surface is much more complex than the simple external cell membrane of a bacterium, ultimately our mouth, stomach and gut region are surfaces where digestion takes place (Figure 8-3). In addition, the nourishment we take in at these surfaces must be circulated to all of the cells in our body, that is, to our internal volume; how this is done will be explained shortly, but again recognize if any part of our body did not receive this nourishment it would die. Last time we checked, gangrene was still a very bad thing. Thus, for humans to survive as large organisms, sometime during the evolutionary history of our lineage a strategy was necessary to maximize our surface area, and thus take in as much food as possible, relative to our volume.

We can actually see such a strategy at work when we examine the small and large intestines of humans, where most digestion takes place. First, notice the size of our intestines, upwards of 30 feet in length. If you ever wondered why our intestines are so big recognize that ultimately the answer is Galileo's Principle. The size of our intestines is a way of dealing with the fact that as an organism grows its volume increases more rapidly than its surface area; in effect the substantial size of our intestines is a way of maximizing the surface available to take in food. If we zoom in on the surface of the intestine we find even further support for the notion that Galileo's Principle is at work. The surface of the intestine is highly convoluted. All of these convolutions greatly increase the effective surface area where digestion can occur. Were it not for our massive intestines, with the concomitant complex folding pattern, there would not be sufficient surface area to take in enough food to support the substantial volume contained within our bodies.

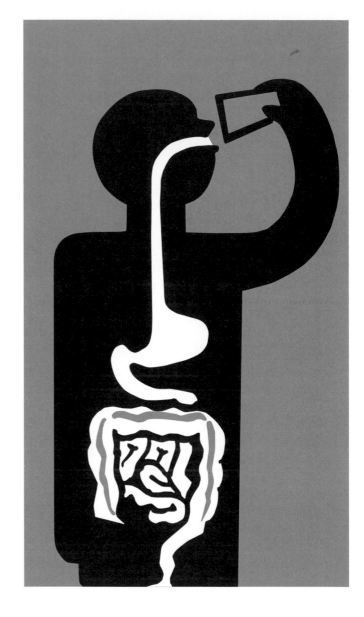

Figure 8.3 Feeding in a human.
A diagram showing how a liquid
passes from the mouth, to the
stomach, and on to the intestines.
Diagram from iStockPhoto.com
© D. Yemelyanov.

Respiration in Organisms

Just as most organisms have to take in food, a process that ultimately occurs at their
surfaces, they also have to take in oxygen to respire; to use the colloquial term
they need to breath. Galileo's Principle also comes into play when we consider
how living organisms breathe. Bacteria that require oxygen take it in through an
external membrane, the same site where they take in food. The oxygen needs to
reach all parts of the cell, or those parts of the cell will die; in a bacterium oxygen
is transported by diffusion. Humans of course also need oxygen to live, and this
oxygen needs to reach all cells in the body or these cells will die; how this transport
happens in humans is of course different from what happens in bacteria. Before we
focus on the interesting phenomenon of how oxygen is transported throughout the
body, consider again how surface areas first come into play, this time in human
respiration. We take in air, which contains oxygen, through the nose and mouth;
oxygen ultimately reaches the lungs. In effect, the lungs act like a large surface, in
a manner directly analogous to the external membrane of a bacterium. However,

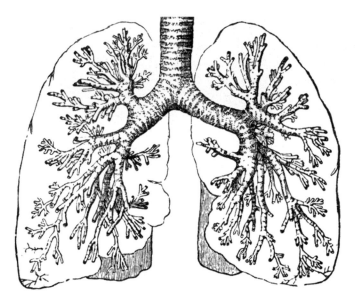

Figure 8.4 Human lungs. The interiors of our lungs are complexly branched. Diagram from iStockPhoto.com © M. Strozier.

given that a human is so much larger than a bacterium, it has a much larger internal volume (both absolutely, but also relatively) that it needs to distribute this oxygen to. With these constraints we might predict that just as with the intestines the lungs should show certain features that will increase their effective surface area: in the human lung we do see just such a phenomenon. A close up of a human lung shows that it is complexly branched with a myriad of tiny chambers (Figure 8-4). Again, just as with the intestine, this is a remarkable but necessary adaptation required in large animals like humans that need to maximize their surface area relative to their volume.

Strength of Bones

Strength of bones in vertebrates is another case where the importance of Galileo's Principle is manifested. It turns out that the strength of a bone depends on its cross-sectional area (Figure 8-5). We shall see that ultimately this sets limits on the size that terrestrial vertebrates can attain, and it also determines what the legs of large vertebrates will look like. Again, recognize that as a vertebrate grows in size the volume (or weight) of their bones increases more rapidly than the surface area

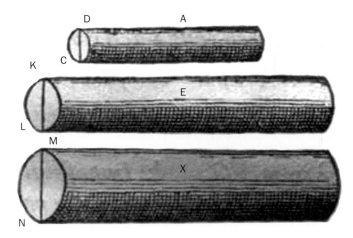

Figure 8.5 Bones and cross-sectional area. The strength of bones, and cylinders, is related to their area, but as the size of a bone, or cylinder, increases its volume or weight increases much more rapidly than its cross-sectional area. Diagram from Galileo.

Figure 8.6 **Elephant and humans**. Walking side by side, with some other humans along for the ride. Image from iStockPhoto.com © S. Podgorsek.

of those bones. Further, the surface area of those bones is necessary to support the weight of the animal.

Consider the example at the start of the chapter where we simply inflated a smaller cube to create a bigger cube. Imagine we were deranged scientists who decided to tinker with human form; in particular, suppose we wanted to create an incredibly tall human being in order to rule the world. Further, suppose we were somehow capable of inflating the dimensions of the human leg bone, increasing its length just as we did with the sides of the cube, by a factor of 10. Imagine our experiment was successful and we finally decided to unleash our giant, who had been lying on a table, on the world. As we the push the large lever used to tilt the table holding our giant to the floor we wait with eager anticipation; as soon as the

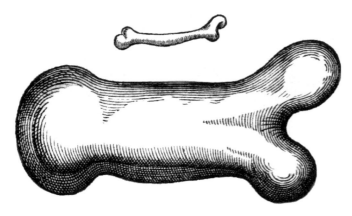

Figure 8.7 Human and elephant femurs. Notice that the thigh bone of the elephant is somewhat longer, but proportionately much wider, than the thigh bone of a human. The great increase in the cross-sectional area of the thigh bone is needed to support the much greater weight of the elephant. Diagram from Galileo.

giant's feet hit the floor her legs crumble beneath her. Oh well, plans for world domination squelched and back to the drawing board.

Even though our plans were foiled, which was a good thing, you might wonder why this happened? The answer is simply Galileo's Principle. The weight of our giant grew at a much quicker rate then the ability of her leg bones to support her. In tall animals that stand on their legs, bones have to become comparatively thicker to support their great weight. To create our giant there would have to be a real expansion in the cross-sectional area of her femurs to support her weight. This principle in action can be illustrated by a real example. Compare a human to an elephant (Figure 8-6). Of course the elephant is taller than the human. However, compare their leg bones (Figure 8-7). The femur of the elephant is proportionately much, much wider than the femur of the human, whereas overall its length is not that much longer. The answer again lies in the fact that as an animal grows in size its weight or volume increases much more rapidly than its surface area.

Galileo's Principle and Constraints on the Evolution of Large Body Size

Thus far we have considered two important phenomena related to Galileo's Principle: the important role of surface areas in obtaining food and oxygen; and the fact that as an organism grows its volume increases more rapidly than its surface which it ultimately depends on for food and oxygen. Evolutionarily this principle matters because when life first evolved all organisms were very small; one of the important evolutionary patterns that occurs, which was discussed in Chapter 5, is that as organisms have evolved over time there has been a general increase in size and complexity (although small and simple organisms have persisted). Here we will consider how the constraints of Galileo's Principle influenced the evolution of large size.

Given Galileo's Principle there are only two fundamental ways that organisms can become large; we see that both strategies are employed in living and fossil organisms. One strategy is that adopted by many animals alive today, including humans. That is, they evolved complex internal organs. These complex internal organs include the lung: it has a highly convoluted wall that increases its surface area and thereby its ability to take in oxygen; and the intestines, which are long, highly coiled and have a highly convoluted wall to increase the available area to absorb food. Organisms that follow this strategy, particularly animals, also evolved

Figure 8.8 Tapeworm. A large specimen preserved in alcohol. Image from K. Jensen, University of Kansas.

a circulatory system, for example blood with vessels, a heart, etc. This blood system distributes and transports the oxygen and food absorbed at the organs to all the parts and cells of the body.

There is a second strategy that is also seen in some modern animals. This strategy is exemplified by the tapeworm which can become very large. For example, tapeworms can grow to 30 feet in length, yet they do not possess the complex internal organs used to process food or breathe oxygen that other organisms have (Figure 8-8). Instead, a tapeworm is very thin, a fraction of an inch thick, so thin that they can absorb food and oxygen (they need very little of the latter) directly through the skin; these products in turn are passed to all parts of the body by diffusion. By becoming large yet staying thin the tapeworm has managed to maximize its surface area relative to its volume. The tapeworm is a highly specialized parasitic organism that depends on its host; it could not persist for long outside of the host's body. Thus, in tapeworms this strategy evolved secondarily from an organism that had greater internal complexity.

Intuitively, however, we might suspect that this second strategy seems easier to evolve. In fact, if we consider the sequence of evolutionary events in the fossil record, some support is leant to this notion. For example, sponges are perhaps the earliest diverging animal group. They also were the first definitive large animals to appear in the fossil record, with evidence from fossils suggesting they evolved before

Figure 8.9 Ediacaran. A large specimen of *Dickinsonia* from the Proterozoic of South Australia. Image by B. Lieberman, specimen in the Yale University Peabody Museum of Natural History.

600 million years ago. Evidence has recently been found that rocks 700 million years old contain sponge-specific compounds or biomarkers. In sponges Galileo's Principle is in full effect and they follow the second strategy mentioned. In particular, they lack complex organs and a transport system, instead relying on the surrounding water to obtain food and oxygen directly through their internal body cavity. As we might expect, the inside of large sponges becomes extremely convoluted to maximize the surface area exposed to water.

The Ediacarans—an Early Multicellular Experiment?

A similar phenomenon is also seen in an enigmatic group of organisms called the Ediacarans (Figure 8-9) or vendobionts (we shall use the former term here). Ediacarans may have been the first large multicellular organisms, aside from sponges, to have evolved. Ediacarans are now wholly extinct. The earliest of these date from around 610 million years ago; few are younger than 540 million years old, making them a largely Proterozoic evolutionary phenomenon. Their name is derived from the place in South Australia where they were first found.

The precise nature of these fossils is enigmatic and paleontologists love to argue about what kinds of organisms they were. Some paleontologists have argued that they were akin to modern jellyfish; others, including paleontologist Dolf Seilacher, have suggested they represent a separate evolutionary origin of large size by a lineage distinct from modern animals. In any event, it is true that the Ediacarans were among the first organisms to evolve large size, and further that in many ways Ediacarans were unlike most animals alive today because they were large, flattened sheets, resembling a big and thin pancake. They also lacked visible mouths, limbs, and other organs. Taken together, these suggest that the Ediacarans may have evolved large size while employing the "tapeworm solution" to the constraints imposed by Galileo's Principle: they were very thin in order to absorb nutrients and oxygen through their skin and possessed no complex organs.

Immediately after most of the Ediacarans disappeared the Phanerozoic Eon (and Cambrian Period) begins. Then, roughly 540 million years ago, we encounter the first abundant remains of undoubted animals (exclusive of the sponges). Most Cambrian, and Phanerozoic, fossils are very different from the Ediacarans because they clearly represent animal remains (Figure 8-10). It is true that sometimes paleontologists argue about the precise affinities of some Cambrian organisms, as we shall see in our discussion of the Burgess Shale in Chapter 13; however, irrespective of that debate, they clearly were animals that had complex organs, limbs, and a circulatory system. Thus, perhaps this solution to the constraints imposed by Galileo's Principle, which in principle seems more difficult, evolved later.

Body Size in Insects or How Big can a Bug Get? No Need to Worry, Thanks to Galileo's Principle

As already mentioned Galileo's Principle acts to limit the size that certain organisms can become, at least without a significant change in overall body form. It also sets concrete limits on the size that certain types of animals can be. An excellent example involves terrestrial arthropods, particularly insects. Insects have holes in their exoskeleton. Oxygen enters these holes and passes through a series of tubes that attach to living cells; the oxygen is transferred via diffusion. As an insect becomes larger, oxygen has to diffuse over a much larger area and volume. This sets a critical limit on the size insects can become: given current oxygen concentrations, an insect cannot become much larger than about a foot in length. In the distant past, some terrestrial arthropods were larger than the largest terrestrial arthropods alive today. This was particularly true during the Carboniferous Period, when dragonflies had a body length of one and a half feet and wingspans of two and a half feet (Figure 8-11). The reason these larger sizes were possible then was because during the Carboniferous oxygen levels were perhaps twice as high as they are today (we will discuss this topic in greater detail in Chapter 11). Millipedes during this interval may have grown to a bit more than three feet. Still, no terrestrial arthropods exceeded this size limit. Like it or not, the stuff of 1950s Hollywood screenwriters' dreams and our nightmares, the massive, lion-sized (or bigger) arthropods stalking human prey today just isn't possible, thanks to Galileo's Principle, but we suppose that's why they call it science fiction.

Galileo's Principle and its Relevance to Medieval Architecture

The paleontologist and evolutionary biologist Stephen Jay Gould also documented how Galileo's Principle applies to the architecture of cathedrals in medieval Europe. Just like organisms, cathedrals vary greatly in their size; also, like organisms, the first cathedrals were very small, and over time architects built larger and larger

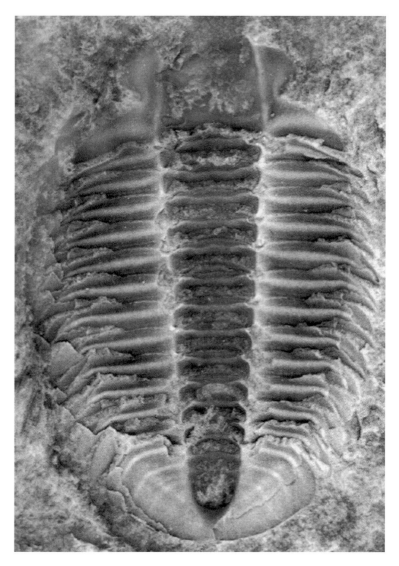

Figure 8.10 Cambrian trilobite. A fossil from Antarctica. Image by B. S. Lieberman, specimen in the University of Kansas Museum of Invertebrate Paleontology.

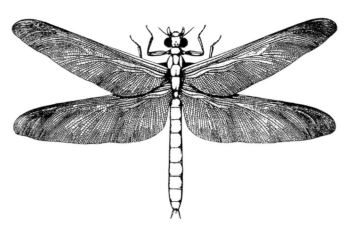

Figure 8.11 Carboniferous dragonfly. The largest flying insect that ever lived. Diagram from R. Cowen (2005) *History of Life*, 4th Edn, Blackwell Publishing; after Handlirsch.

cathedrals. Although of course medieval architects did not need to worry about their cathedrals feeding and respiring, there were various other architectural principles they needed to take into account, especially as they built ever larger cathedrals. This was because during the Middle Ages several constraints were imposed on architects because of the absence of steel girders, electric lighting, and air conditioning. These constraints implied that the only way to support the cathedral, to light it, and to maintain sufficient air flow was through the external walls of the cathedral: in effect, their surfaces. In particular, since steel girders did not exist there was no way of gaining extra strength for the building other than the strength of the stone walls themselves. Further, light and air could only come through windows in the external walls of the building. These principles mattered in cathedral building because as architects sought to make ever larger cathedrals (Figure 8-12) they had to deal with

Figure 8.12 Large gothic cathedral. The front of Notre Dame, in Paris, France. Image by B. S. Lieberman.

the following issues that arise directly from Galileo's Principle: the surface area of the cathedral, i.e., the means of lighting, ventilating, and supporting the cathedral, increases more slowly than the interior, i.e., the volume or weight of the cathedral.

The architectural issues Galileo's Principle necessitated can be visualized nicely by comparing the plan of a small church with that of a larger cathedral. The 12th century church from Essex, England in Figure 8-13 has a very simple plan compared with the much larger, and taller cathedral from Norwich, England shown in Figure 8-14. In comparison to the small church, the large cathedral is

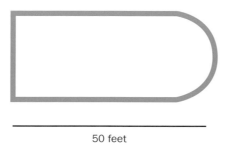

50 feet

Figure 8.13 Simple 12th century church. A schematic floor plan of a small church in Essex, England.

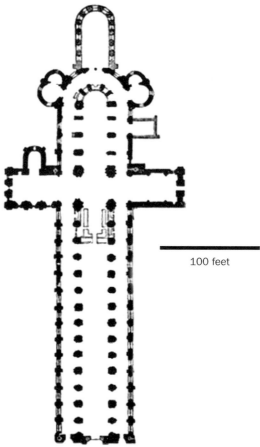

100 feet

(a)   (b)

Figure 8.14 The Norwich Cathedral. (a) An image from iStockPhoto.com © R. Laurence. (b) A schematic floor plan, diagram by R. King with assistance of B. Thayer, of this large 12th century English cathedral.

(a)

(b)

Figure 8.15 Flying buttresses. These struts help support the walls of large gothic cathedrals. Shown are examples from: (a) the Cathedral of Jerez de la Frontera in Andalusia, Spain © S. Morris; (b) the backside of Notre Dame Cathedral © J. L. Gutierrez. Both images from iStockPhoto.com.

somewhat taller but really relatively much narrower than the small church. (This partly is a consequence of the absence of steel girders in Medieval Europe: these today can span large spaces whereas in the Middle Ages this had to be accomplished by stone; widely spaced stone walls would demand a widely spanning stone ceiling which would be difficult to support.) It is also an important consequence though of the fact that a wider cathedral, especially a tall one, would be tremendously heavy. All other things being equal, a narrow cathedral will weigh much less than a broad cathedral. This matters because the only support for this tremendous weight of the cathedral in Figure 8-14 are the walls themselves. Again, it is the issue of volume or weight increasing more quickly than surface area or strength. Remember we encountered a similar situation in the world of living organisms: the need for feeding and breathing by organisms required that as they grew larger they needed to maximize their surface area relative to their volume.

Another distinctive difference between the plan of the church shown in Figure 8-13 and the plan of the cathedral in Figure 8-14 is the many outpocketings or projections that are visible in the latter. These are chapels that are used in the religious rituals held in the cathedral. These chapels also serve a utilitarian (from an architectural perspective) role as well because each outpocketing or chapel increases the surface area of the cathedral, and that in turn provides more support for the walls and ceilings of the cathedral. Without these added outpocketings or chapels the cathedral in Figure 8-14 would not be able to stand up for long. These outpocketings are directly analogous to the outpocketings in the walls of our lungs and intestines that were discussed above and increased the surface area available to take in oxygen and digest food.

Flying Buttresses: A Dramatic Architectural Solution to Galileo's Principle

Throughout the Middle Ages architects sought to build ever larger cathedrals. As with living organisms, there are distinct constraints on the size a cathedral can reach. In fact, the surface area to volume relationship described by Galileo meant that cathedrals could not become much bigger than the Norwich cathedral (Figure 8-14), at least without an architectural innovation that would greatly increase the relative surface area of the cathedral. This innovation came in the form of flying buttresses. These are dramatic stone supports that sit on the outside of the cathedral wall, and a particularly famous example comes from Notre Dame cathedral (Figure 8-15). By touching the walls, buttresses serve to greatly expand the effective surface area and thus the strength of the outside of the cathedral. In effect, they are directly akin to the convolutions in the walls of our intestines or lungs. Unlike the chapels discussed above, flying buttresses play no functional role in the interior or religious aspects of the cathedral, although some may find them decorative in a "kitschy" sort of way. They did allow, however, medieval architects to design and build truly colossal cathedrals.

Galileo's Principle and its Relevance to Cratering Density in our Solar System

If you examine images of the inner rocky planets of our solar system like Mercury, Venus, Mars, and also our Moon, one thing you notice is that their surfaces are

(a)

(b)

(c)

(d)

Figure 8.16 Craters on planetary bodies. (a) Mercury, (b) Venus (this planet is shrouded in clouds so its craters are not directly visible), (c) Mars, and (d) the Moon. Images from NASA.

covered by craters (Figure 8-16). This is very much unlike the situation on Earth where, except in rare cases, few signs of cratering are prominently visible at the surface (Figure 8-17). This is not to say that craters cannot be found on Earth. For example, recall Figure 6-5, which shows a crater that formed about 100,000 years ago in Arizona, when a relatively small meteor roughly 100 meters in diameter struck the Earth; however, such prominently visible craters are fairly rare. The reason for their rarity is not because the Earth wasn't being hit by meteors. In fact, as meteors struck the planets all around the Earth, and also our Moon, they were also striking the Earth, yet why the absence of visibly expressed meteor craters

Figure 8.17 The Earth from space. A classic image from NASA.

today? The answer lies in Galileo's Principle. Again, the paleontologist and evolutionary biologist Stephen Jay Gould was among the first to recognize the role of Galileo's Principle in this area. We find it fascinating that this principle can explain such disparate phenomena: the shape and size of organisms; the form of gothic cathedrals; and aspects of planetary geology.

Most of the large craters on the Moon, Mercury, Venus, and Mars were created prior to 3.5 billion years ago. (Note, only the rocky planets can be considered in this discussion. Other planets further from the Sun like Jupiter, Saturn, Uranus, and Neptune are largely gaseous and thus they are very different from the Earth, Mars, the Moon, etc.; in particular, their more gaseous nature means that their surfaces are being continually reworked.) Thus, these planets were struck by meteors deep in the past, yet their surfaces have changed relatively little. The Earth was also struck by meteors deep in the past but its surface has changed dramatically since then. How can this difference be explained by Galileo's Principle?

Again, the difference relates to the relative size of the planetary bodies. There is an inverse relationship between the relative sizes of the planets and the degree of cratering. Note that in the list of planetary bodies we are considering those with the smallest diameters, like Mercury and the Moon, are the most cratered; Mars is larger than these two and it has fewer craters; Venus is larger still and it has fewer craters than Mars; the Earth is the largest of these objects and it has the fewest craters (Figures 8.16 and 8.17). Relevant here is that the difference in the sizes of these planets influences the mass of the planet and its effect on gravitational attraction. This determines what happens to craters on a planet's surface in the long term.

The reason the size of a planet relates to its gravitational attraction derives from Newtonian physics. Gravity is described as a force, F, where

$$F = (Gm_1m_2)/d^2$$

with G a constant, m_1 the mass of the first object (let's say a planet), m_2 the mass of the second object, and d the distance from the first to the second object. The reason we need be concerned with this equation is that the mass is proportional to the volume of the planet and since the Earth, the Moon, Mercury, Venus, and Mars all have similar densities their respective masses, and the gravitational force or pull they exert, will be directly related to their diameter.

Planetary Size, Atmospheres, and Erosion

The gravitational pull of a planet affects its present-day cratering density in two important ways. First, the more powerful the gravity of a planet is, the longer that planet will retain its atmosphere. For example, the Earth, because of its large size, has an atmosphere whereas the much smaller Moon and Mercury, with their weaker gravity, have little or no atmosphere, as distinguished from deep space. Earth's atmosphere is of course essential for life, but it is also important for the problem at hand. This is because wind and rain, which are associated with our atmosphere, are powerful erosive forces. One reason the Earth has few craters is because its atmosphere has eroded them away. Given enough time the crater shown in Figure 6-5 will erode away. By contrast, Mercury and the Moon lack significant atmospheres, and thus there has been little if any erosion at their surfaces over the last several billion years. We don't know how long the famous astronaut's footprint on the moon will last (Figure 8-18), but unless some other astronaut comes along and kicks Moon dust over it, it will persist on the surface of the Moon for a long time to come, because of the absence of erosion on the Moon. Footprints on our own planet's surface will not be preserved unless they are quickly buried.

Figure 8.18 Neil Armstrong's footprint on the Moon. Another classic image from NASA.

The atmosphere on Mars (which is a smaller planet than Earth but bigger than Mercury and Mars) is quite thin when compared to Earth's, yet it is much more prominent than that of Mercury and the Moon; there is also evidence that at least in the distant past significant volumes of water may have coursed over Mars' surface. Polar ice caps, comprised of frozen carbon dioxide or dry ice and also liquid ice also exist on Mars. Given its intermediate size and thus atmosphere, relative to the Earth and Mercury, we would expect to see an intermediate degree of cratering on that planet, which is exactly what is observed. Venus is larger still than Mars, nearly as large as the Earth, and it too has a more prominent atmosphere than Mars: its atmosphere happens to be very rich in carbon dioxide and sulfuric acid. This larger planetary size and more prominent atmosphere explains why there are fewer craters on Venus' surface than on Mars'. There are, however, more craters on Venus' surface than on Earth's. The reasons for this difference between Venus and the Earth are fascinating, and involve the relative absence of plate tectonics (discussed immediately below) on Venus.

Planetary Size, Plate Tectonics, and Crustal Recycling

Plate tectonics is another process that explains why there are differences in the degree of cratering on the various rocky planets. We will explore plate tectonics in greater detail in Chapters 10 and 11; however, some discussion is relevant here because plate tectonics on our planet has meant that a large amount of the Earth's crust has been recycled. This crustal recycling has also served to eliminate many of the craters that had been produced by ancient meteor impacts. Because of plate tectonics, older ocean crust is subducted back into the Earth and remelted while new ocean crust is produced as lava (Figure 8-19). Most (about 70%) though not all of the Earth's crust is oceanic crust, and because it is being continually recycled there is little oceanic crust older than about 180 million years remaining. This means that any record of ancient (e.g., 3.5 billion years old, like that preserved on the Moon) cratering on the ocean floor (70% of the Earth's surface) would have been obliterated. Continental crust like that comprising the bulk of North America, Europe, South America, Asia, Africa, Australia, and Antarctica is lighter and less dense than oceanic crust and tends not to subduct and will not be destroyed by this route, although it will of course be subject to erosion. Therefore, continental crust is the place on Earth where the record of events older than 180 million years old is

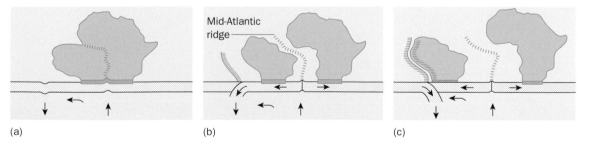

(a)   (b)   (c)

Figure 8.19 Plate tectonics and crustal recycling. New oceanic crust is generated (up arrow) causing continents to drift apart (sideways arrows). As new oceanic crust is created, older crust is destroyed (down arrows). Diagram from C. B. Cox and P. D. Moore (2005) *Biogeography: An Ecological and Evolutionary Approach*, 7th Edn, Blackwell Publishing.

preserved. Very old rocks, so common on the Moon's surface, are comparatively rare on Earth's surface.

It turns out that the size of the planet is related to the extent to which plate tectonics (and thus crustal recycling) can occur, just as it was related to the extent to which atmospheric weathering and erosion can occur. Part of the reason plate tectonics occurs on Earth is the active heat in our planet's interior. This heat is generated by radioactive decay. The amount of radioactive decay, and thus the amount of internal heat a planet generates, is related to the overall volume or size of the planet; also, the amount of heat a planet loses to outer space is related to its surface area. In fact, the degree of plate tectonics on the various planets matches the gradient in size/volume and thus internal heat they can generate: from the Moon, Mercury, and Mars, upwards to Venus and the Earth. (If, however, differences in planetary size solely explained the extent of plate tectonics then Venus and the Earth should have nearly equivalent degrees of plate tectonics: they do not, and plate tectonics is more extensive on the Earth than it is on Venus. The reasons for this difference may relate to the extreme heat on Venus; this heat baked off the planet's water and water may be a prerequisite for the presence of full, Earth-style plate tectonics.) Just to summarize, the gradient in increasing planetary size and volume is matched by a declining relative planetary surface area; further, this gradient corresponds to a gradient of increasing internal heat with a concomitant decline in the potential to release that internal heat; the more internal planetary heat the greater the prevalence of plate tectonics; finally, the more prevalent plate tectonics is, the more extensive the recycling of the planetary surface. Thus, we can see why Galileo's Principle provides an important explanation for the pattern of cratering we see on various planets in the solar system.

Concluding Remarks

Ultimately, science is about the search for generalities that explain or describe the behavior of groups of objects. Galileo's Principle is one such generality. Stephen Jay Gould demonstrated how even objects seemingly as dissimilar as the human intestine, the flying buttresses of the Notre Dame cathedral in Paris, and craters on the surface of the planet Mercury are governed by the constraints of surface area to volume. This is because at base all living and inanimate objects answer to the same physical requirements. How does this relate to the material we discussed in Chapters 4 and 5, where we considered evolution and Darwin's mechanism for evolutionary change: natural selection? Clearly many of the traits that animals and plants possess are adaptations that were shaped by natural selection. Can we call the convolutions in the lungs and intestines of humans "adaptations" to increase the surface area of these surfaces? Such a view would not be inaccurate as of course those organisms that relatively speaking had more convolutions would be better able to absorb oxygen and food and this might provide them with a competitive advantage relating to survival and reproduction. Remember, however, that in Chapter 7 we also considered how some traits in animals and plants that have been called adaptations are actually spandrels: structures that already existed that were subsequently utilized for a different purpose. To distinguish whether these convolutions are actually spandrels or adaptations we would have to determine whether these structures were pre-existing and later co-opted for a given function or instead had been continually shaped by natural selection. Clearly such structures and other related ones (that evolved to deal with the relationship between surface area and volume) evolved long ago and are today maintained because an animal lacking such structures would quickly perish. Thus, they could be thought of as maintained or conserved adaptations.

However, we suspect these convolutions on the surfaces of our lungs and intestines reflect something deeper and more profound. Consider: would we call Mercury's excessive record of cratering an adaptation to the planet's small size? Of course not. It is a simple aspect of geometry and the fact that a small planet will tend to have a weaker atmosphere and less prevalent plate tectonics, each of which conspire to leave the planet's surface covered with craters. Similarly, living organisms are subject to physical and mathematical principles, for example, those outlined by Galileo. A key point, in summation, is that life on Earth, the cratering of planet's, and even medieval cathedrals are governed by a similar scientific principle that extends across biology, geology, and architecture.

Additional Reading

Bonner, J. T. 2006. *Why Size Matters: From Bacteria to Blue Whales*. Princeton University Press, Princeton, NJ; 176 pp.

Buss, L. W., and Seilacher, A. 1994. The phylum Vendobionta: A sister group of the Eumetazoa? *Paleobiology* **20**: 1–4.

Fedonkin, M. A., and Waggoner, B. M. 1997. The late Precambrian fossil *Kimberella* is a mollusc-like bilaterian organism. *Nature* **388**: 868–871.

Gould, S. J. 1977. *Ever Since Darwin*. W.W. Norton, New York, 285 pp.

Greeley, R. 1994. *Planetary Landscapes*. Springer, Berlin, 288 pp.

Knoll, A. H. 2004. *Life on a Young Planet*. Princeton University Press, Princeton, NJ; 277 pp.

McMenamin, M. A., and McMenamin, D. L. 1990. *The Emergence of Animals*. Columbia University Press, New York: 217 pp.

Schütz, B. 2002. *Great Cathedrals*. Harry N. Abrams, New York; 472 pp.

Vita-Finzi, C. 2006. *Planetary Geology*. Terra, London; 168 pp.

Wilson, C. 2005. *The Gothic Cathedral: The Architecture of the Great Church 1130–1530*. Thames & Hudson, London; 304 pp.

Chapter 9
The Role of Fossils in the Genesis of Myths and Legends

Outline

- Introduction
- Paleontologist's Have Come from Many Different Walks of Life and Have Sported Many Different Hairdos
- Paleontology in Ancient Greece
- Native American Contributions to Paleontology
- Concluding Remarks
- Additional Reading

Introduction

Mythic beings and legendary creatures are present in virtually all cultures: stories about these go back thousands of years. Here we will explore the genesis and derivation of some of these legends. We will present evidence, in some cases incomplete yet tantalizing, in other cases more concrete, that suggests many of these mythic creatures, and the legends surrounding them, in fact derive from ancient fossil discoveries. Part of what we will focus on is how cultures tried to explain these fossil finds, and sometimes this involved devising myths and religious explanations. It is exciting and even fascinating from a historical and paleontological perspective to be able to look back on a myth, and explain it as the byproduct of an actual fossil organism meeting humankind's vivid imagination. We will present several cases where this can at least be surmised. One of the interesting examples involves the fossil oysters of the genus *Gryphaea* (Figure 1-17) already mentioned. These fossils

Prehistoric Life: Evolution and the Fossil Record. 1st edition. By Bruce S. Lieberman and Roger Kaesler. Published 2010 by Blackwell Publishing.

are abundant in England, and their relatively large, curved, and gnarly appearance earned them the moniker "Devil's toenails."

Although some of these myths may appear naïve, and their genesis may seem whimsical from our current perspective, modern society should be wary of staring down its nose at the civilizations and cultures that spawned these mythic beliefs. The author Adrienne Mayor in her book "*Fossil Legends of the First Americans*" relates an interesting story about one of the early fossils found, by Europeans, in North America. Fossil mastodon teeth were uncovered on a plantation in South Carolina in the early 1700s; the slave master who ran the plantation surmised these were the teeth of a giant human victim of the biblical flood. However, the African American slaves who worked the plantation correctly recognized that the teeth belonged to some sort of elephant; they were of course familiar with elephants, being from Africa and the mastodon is indeed a relative of modern elephants. Thus, technically, aside from the discoveries of the native Americans, that will be described more fully later in this chapter, the first fossil correctly identified in North America was identified by African American slaves; by contrast, their slave master explained the fossil through recourse to an unscientific myth.

Even more importantly, studies on this topic have also shown that not all ancient interpretations of fossils were naïve and quaint. Indeed, evidence is accumulating that at least some individuals within these ancient cultures recognized the scientific and naturalistic significance of paleontological finds: they realized they represented extinct life forms; these cultures may have even grasped a nascent concept of evolution.

Paleontologist's Have Come from Many Different Walks of Life and Have Sported Many Different Hairdos

The traditional view of the science of paleontology is that it started in Europe roughly 200 years ago, and the practitioners all looked like the subject of Figure 9-1. It is also usually assumed that the modern day paleontologist will look similar, though perhaps they will sport a hipper haircut (Figure 9-2), but in reality not all paleontologists are men (Figure 9-3). It is true that many of the important scientific contributions to paleontology have occurred during the last 100 years and were accomplished by citizens of what are referred to as "developed" countries, including the United States, Canada, the United Kingdom, Australia, and other European countries. This is not to say, however, that there was no one doing important paleontological work before this time and there were no paleontologist's who belonged to other cultures. The truth is, as we shall describe in this chapter, the Native Americans had important paleontological insights dating back at least to AD 500, and the man in Figure 9-4, who lived in the 19th century, had a deep understanding of aspects of paleontology. He may have been a better paleontologist than either of the authors of this book: he certainly had a better haircut. Adrienne Mayor, in her book on fossil legends of the Native Americans, showed that the subject of Figure 9-4, White Bear, knew where to find fossils and he knew that they represented traces of ancient life forms that are now extinct.

Mayor, and other authors cited in the additional reading listed at the end of this chapter, have shown that interest in fossils and knowledge of paleontology certainly extends back to the time of the ancient Greeks, around 400 BCE; we know this because there were ancient Greeks who wrote on this topic. It is even possible that

(a)   (b)

Figure 9.1 Early paleontologists. (a) Barnum Brown, working on the skeleton of *T. rex* at the American Museum of Natural History (AMNH). Image courtesy of Division of Paleontology, AMNH, with assistance of S. Bell and M. Ellison. (b) William Buckland, the English clergyman who published the first scientific description of a dinosaur, image from A. Martin.

Figure 9.2 Paleontologist. Roger Kaesler, holding a cast of a fossil scorpion. Image courtesy of the Paleontological Institute, University of Kansas.

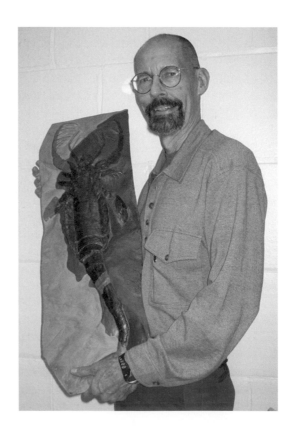

Figure 9.3 **A modern paleontology lab**. Featuring students, faculty, and staff at the University of Kansas, with from left to right Erin Saupe, Cori Myers, Curtis Congreve, Francine Abe, Bruce S. Lieberman, Wes Gapp, and Talia Karim. Image by P. Cartwright.

Figure 9.4 **Paleontologist**. Native American fossil hunter White Bear. Image from A. Mayor.

humankind has been interested in, and curious about, fossils for far longer. For example, there have been fossils found at sites where humans dwelled 30,000 years ago; moreover, some of these fossils were brought from far away. Maybe the oldest profession on Earth isn't what we think it is: maybe it's paleontology! Seriously, the precise reasons why fossils from distant places are found at 30,000 year old human dwellings may never be known, but perhaps the fossils had some religious significance, or maybe the owners simply thought they looked interesting.

Paleontology in Ancient Greece

Adrienne Mayor not only has written a fascinating book on Native Americans and paleontology, but she has another book detailing what the ancient Greeks thought about fossils; understanding of the latter topic has a longer history, although Mayor's 2002 book has shed significant new light on the subject. It turns out that the Greek countryside contains fossil remains of immense animals that once walked the Earth. Because of this, there were likely many times that ancient Greeks encountered fossils. When they did, there was a natural desire to explain what these bones represented. Typically, what the ancient Greeks did was fit the bones into their myths or used their myths to explain the occurrence of gigantic bones.

The Cyclops

For example, consider Figure 9-5. The woman on the left is the mother-in-law of one of the author's of this book (BSL); the fossil skull belonged to a woolly mammoth. The ancient Greeks had a legend about a giant one-eyed monster who dwelled in caves on islands in the Mediterranean Sea and terrorized travelers, feasting on them whenever possible. This monster is referred to in Homer's *Odyssey* and of course is not BSL's mother-in-law, who is actually a very nice woman, but rather the dreaded Cyclops (Figure 9-6). The mammoth skull in Figure 9-5 almost certainly illustrates how the legend of the Cyclops began. At one time large relatives of the elephant roamed Europe, Asia, and North and South America. These include the woolly mammoth (Figure 6-18) and also the mastodon. In fact, we have already described, in Chapter 6, how animals like the woolly mammoth and mastodon survived until as recently as 10,000 years ago, and also how humans probably drove them to extinction (turning the story of how the Cyclops terrorized ancient humans literally on its head).

If one interpreted the mammoth skull in Figure 9-5 as if it were a human skull, as the ancient Greeks probably did, the large hole in the center of the skull would be taken for the socket of a single eye. In actuality, all primates, including humans, are somewhat unusual in having their eyes at the front of the head. In most mammals (and other vertebrates) the eye sockets are on the side of the head and mammoths, mastodons, and elephants are no exception. The large hole in the center of an elephant's skull demarcates the site of the trunk. Ancient Greeks would have found skulls of mammoths and mastodons in various sites in and around the Mediterranean, including in caves, and in some of these caves they were surrounded by the bones of many other animals (this is typical of fossil cave deposits). The ancient Greeks were not familiar with living elephants, since they had been extinct in Europe for thousands of years; thus, they interpreted them as one eyed giants, and assumed that the surrounding bones were those of their victims. This is almost

Figure 9.5 The origin of the legend of the Cyclops. BSL's in-laws stand beside a fossil mammoth skull. Image by B. S. Lieberman.

Figure 9.6 A Cyclops. Note the single eye and the tusks. Diagram from iStockPhoto.com © P. Angeles.

assuredly how the legend of the Cyclops began. (Note, Alexander the Great, a very famous ancient Greek, encountered elephants when he tried to conquer what is modern day India, but the legend of the Cyclops and indeed, the *Odyssey*, dates back to a much earlier time.) Gould, Mayor, and others have argued convincingly that perhaps many of the legends of ancient Greek civilization (and probably of other civilizations) involving fantastic creatures arose from observations of fossils from unfamiliar, extinct beings.

The Griffin

Another creature from ancient Greek mythology that may derive from the discovery of fossil bones is the griffin (Figure 9-7). The griffin was supposed to have been roughly the size of lions, and was believed to sport claws and a beak like an eagle. They were said to nest on the ground in packs. Further, it was claimed that their homeland was in central Asia, where they guarded gold in deserts. Adrienne Mayor in her book "*The First Fossil Hunters: Paleontology in Greek and Roman Times*" describes the genesis of the legend of the griffin in detail. Griffins first appeared in Greek culture shortly after the Greeks encountered the Scythians, a nomadic group of people from central Asia; Scythia was an important source of gold in ancient times. The Scythians not only brought gold to Greece but also, apparently, the legend of the griffin, and the myth of the griffin was probably not derived from pure fancy. The reason we can suggest this indirectly traces to this man, Roy Chapman

Figure 9.7 A Griffin. The mythological beast sports claws, a beak, wings, and a lion's hindquarters. Diagram from iStockPhoto.com by J. Tenniel © D. Hendley.

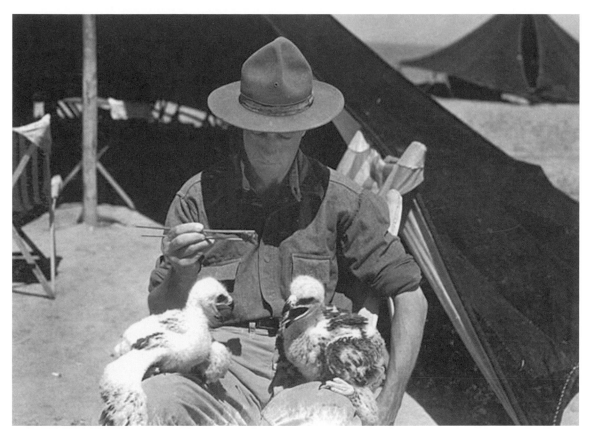

Figure 9.8 Roy Chapman Andrews. The famous paleontologist from the American Museum of Natural History (AMNH) feeding baby eagles in the Gobi Desert of Mongolia. Image courtesy of the AMNH, with assistance from S. Bell and R. Edwards.

Andrews (Figure 9-8), a scientist from the American Museum of Natural History (AMNH).

Andrews led an expedition to Mongolia's Gobi desert in the 1920s; there he uncovered not Scythian gold but paleontological gold. (As an aside, Andrews was the model used for Hollywood's *Indiana Jones*. Wherever he traveled, Andrews took his broad brimmed hat and also, because in the 1920s, and even today, the Gobi desert contains dangerous bandits, he usually carried a revolver. We are unclear as to how Hollywood came up with *Jones'* bull whip. Most puzzling, Hollywood felt that Indiana Jones would be more exciting as an archaeologist than as a paleontologist: we beg to differ.) Chapman made many important paleontological discoveries in the Gobi desert. One of these was the dinosaur *Protoceratops* (Figure 9-9) which is a relative of the more famous, and larger dinosaur, *Triceratops* (Figure 9-10); incidentally, *Triceratops* was considered the nemesis of *T. rex*, but it was doubtful if *Protoceratops* would have been anything other than a tasty snack to *T. rex*. Famously, Andrews also found some specimens of *Protoceratops* standing over or near clutches of dinosaur eggs. These were the first documented discoveries of dinosaurs eggs. Paleontologists now recognize that dinosaurs not only laid eggs but also likely guarded them; further, some dinosaurs may have even guarded their babies after they hatched from their eggs; finally, these probably were not the eggs

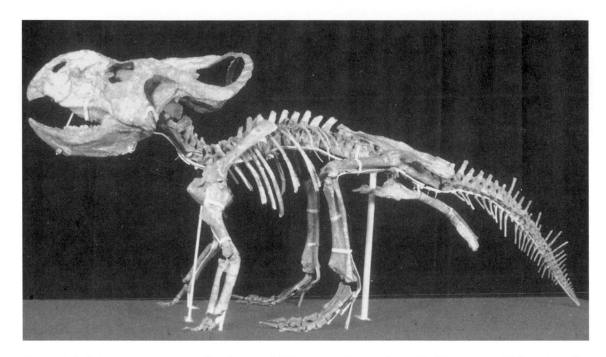

Figure 9.9 *Protoceratops*. A small relative of *Triceratops*. Image from D. Fastovsky, University of Rhode Island.

Figure 9.10 *Triceratops*. A skull of this large Cretaceous dinosaur. Image from iStockPhoto.com © C. Ermel.

of *Protoceratops* but rather belonged to another dinosaur: *Oviraptor*. This is one of the traits (these are discussed more fully in Chapter 17) that links dinosaurs with birds and also with more distant bird relatives like the crocodile and alligator: these animals also guard their nests, and stay with their young, at least for a short time, after they hatch.

Adrienne Mayor made a compelling case that the legend of the griffin derives from the animal that Chapman Andrews first brought to the light of science; if so, he clearly must not have been the first person to find one of these fossils. Note that *Protoceratops* has a beak and claws, and is roughly the size of a lion. The mythical animal's wings are more difficult to explain, but they may represent the crest on the head; it is also possible that the early Scythians, assuming that any animal with a beak and claws must be a bird, used some artistic license and added them on.

Legendary Greek Heroes: Fossil Bones Combined with a Vivid Imagination

The discovery of large, fossilized mammal bones also led the ancient Greeks to conclude that long ago humans were much larger; this may be the source for legends about the ancient Greeks, and the mythological creatures they fought. One of the important Greek myths focused on Heracles or Hercules, who was supposed to have battled the famous monster of Troy. Again, Mayor argued that this monster was an extinct fossil mammal, possibly *Samotherium*, a five million year old relative of the modern giraffe that sported a skull that was nearly a yard long. She was able to infer this because in one image of the battle on a ceremonial vessel Heracles is fighting a giant skull projecting from a cliff, and the skull looks much like the skull of *Samotherium*. Further, in modern Turkey, near Troy, fossils of *Samotherium* are relatively plentiful. The legend of Heracles and the monster of Troy not only suggests that the ancient Greeks were familiar with fossils, but also that they used fossils to create their myths. Many of the Greek temples also contained large hero's bones, which they worshipped. They were, in fact, worshipping fossil mammals.

Although most Greeks, and this is true of the Romans as well, thought that fossil bones represented the bones of giants and heroes, not all the people of these eras had such fanciful ideas. Some recognized that they were the remains of extinct beasts just as some ancient Greeks (described in Chapter 4) recognized that evolution had occurred. Not all ancient Greeks tried to explain natural phenomena (including fossilized bones) in the supernatural context of religion, myth, and legend. There were some among them who tried to use natural, scientific explanations to understand fossil bones and it allowed them to see how the world and its life forms had changed through time. Yes, the relative proportion of people accepting these scientific ideas has certainly increased through time but the science of paleontology, in some respects, has been around for millennia, not just a few hundred years. Very similar themes emerge upon analysis of Native American culture.

Native American Contributions to Paleontology

There are several lines of evidence, beautifully detailed in Adrienne Mayor's book *"Fossil Legends of the First Americans"*, that native Americans were intimately familiar with fossils and knew where to find them. In some cases fossils were incorporated into their myths and legends, or even accorded magical status. For example, some tribes thought fossil ammonites (Figure 6-3) could aid in finding buffalo or would shield the bearer from weapons. Fossils were even turned into powders that were used by medicine men.

Like the ancient Greeks, the native Americans also had a mythology explaining the bones as those of giant creatures that had been struck down by the Gods. Some tribes also believed that the monsters' bones were merely dormant and could be set

Figure 9.11 **Cave bear**. This giant bear fossil is from the Pleistocene of Europe. Image by G. Rabeder, University of Vienna.

loose if man misbehaved and mistreated the Earth. There were also cases where fossils were treated in a more sophisticated manner: as the remains of extinct animals. For example, the Seneca tribe of western New York spoke of a legendary huge, fast bear, and in that region (and elsewhere in North America and Europe) there are fossils of the now extinct giant cave bear (Figure 9-11). This animal went extinct roughly 15 thousand years ago and would have dwarfed a modern grizzly; notably it had long legs, suggesting it may have been a speedy runner.

Another case where legend may match paleontological reality involves Native American tribes from Mexico and the American southwest. Several tribes had legends of a large flying bird that would capture humans. This legend may derive from the teratorn, a bird with a wingspan of 12–17 feet, that as recently as 8,000 years ago lived in the western United States. It resembled an eagle but weighed 50 pounds and could have caught deer and perhaps even small children. Because it went extinct only 8,000 years ago it certainly co-occurred with ancestral Native Americans in the western United States, and they also could have found its bones in caves. Other potential examples involve the Plains Indian tribes who told stories of a Water Monster and a Thunder Bird that were constantly at war with one another; these regions are among the world's richest repositories of giant marine reptiles (especially mosasaurs) and flying reptiles (pterosaurs). Mosasaurs (Figure 9-12) are marine relatives of the snake that could grow upwards of 40 feet in length. Pterosaurs (Figure 6-2) are close relatives of dinosaurs and some grew to the size of an F-16 aircraft.

Figure 9.12 Mosasaur. This giant marine reptile is from the Cretaceous of Kansas and is housed in the University of Kansas Natural History Museum. Image by B. Scherting.

Some Native American tribes even had a protoevolutionary view of the history of life, perhaps based in part on their understanding of the fossil record. In particular, they posited a series of past worlds that were destroyed before this one, and they recognized extinct fossil species that might have been ancestors of living modern ones. These views of course match some of the facts described throughout the book and especially in Chapters 4–6.

Native Americans and Fossil Footprints

Exposures of dinosaur footprints are common in certain places in the United States, including the Connecticut River Valley and also the southwest and again Adrienne Mayor in her book "*Fossil Legends of the First Americans*" untangled a remarkable story relating to these. Some of the most famous dinosaur tracks are the familiar three-toed footprints (Figure 9-13) that were made by various types of carnivorous dinosaurs. Of course, the resemblance of these dinosaur trackways to modern bird footprints has not been lost on modern paleontologists, and interestingly it was not lost on the Native Americans either. In places in the American southwest, rocks near dinosaur trackways feature Indian pictograms that date back a thousand years: these depict giant birds. This is rather ironic considering that for a long time paleontologists debated the evolutionary relationships of dinosaurs and only relatively recently has it been broadly recognized that dinosaurs are in fact closely

Figure 9.13 Dinosaur footprint. Footprint of a large carnivorous dinosaur, with human foot for scale. Image by A. Walton, University of Kansas.

related to birds (as we shall describe more fully in Chapter 17). The Native Americans appear to have beaten many of these professional paleontologists to the punch and may have long recognized the kinship of dinosaurs to birds.

Another case where Native Americans correctly identified fossils before their counterparts in universities in the United States involves the famous *Daemonelix* or Devil's corkscrew fossil, known from roughly twenty million year old rocks in western Nebraska (Figure 9-14). The name Devil's corkscrew whimsically evokes the name Devil's toe nails used by the residents of medieval England: all we can say is "that would have been a mighty big bottle of wine" given the dimensions on some of the corkscrews, which can approach ten feet in length. At first, paleontologists didn't know what could have made these corkscrew-shaped fossils and some thought, incorrectly, that they might represent strange tree roots. By contrast, the Lakota Sioux called them Beavers' lodges and it turns out their identification was the correct one. (The Lakota Sioux might have found skulls inside the corkscrews that resembled modern beaver skulls or perhaps they recognized that the corkscrew diggings resembled the corkscrew pattern that beaver's used to gnaw down a tree.) Confirmation came from paleontologists who found fossilized beavers inside some of the corkscrews.

Figure 9.14 Devil's corkscrew?. The trace fossil *Daemonelix*, a fossil beaver burrow from Nebraska. Image from M. Hertig, Agate Fossil Beds National Monument, National Park Service.

Native Americans as Fossil Guides

It is important to note that most of the famous dinosaur paleontologists who explored the West for fossils in the late 19th and early 20th century had Native American guides. Although early accounts downplayed their role, it is clear that they not only helped professional paleontologists get from place to place, but they also knew where the fossils were, and took the paleontologists to the best fossil-bearing localities. These fossils were sent to various famous museums like the AMNH in New York City and the Yale University Peabody Museum of Natural History in New Haven, Connecticut. It is true that sometimes the Native Americans received little or no credit, but in some cases they did. Most noteworthy is the famous Professor Marsh of Yale University who tried to intervene to help the Native Americans keep their land from government swindlers (although Marsh may have not been so scrupulous when it came to his other paleontological endeavors: these included claiming to have spent ample time collecting fossils while braving the

Figure 9.15 *Brontotherium*. A large fossil mammal from the American West. Diagram from R. Cowen (2005) *History of Life*, 4th Edn, Blackwell Publishing; based on Osborn.

elements when it was usually his subordinates who performed these tasks). Marsh was especially close to the Sioux chief Red Cloud and apparently he named the famous dinosaur *Brontosaurus* (see Chapter 15), which means Thunder Lizard and the famous large mammal *Brontotherium* (Figure 9-15), which means Thunder Beast, in his honor, in homage to the aforementioned mythological Thunder Bird, and, of course, because of their prodigious size.

Concluding Remarks

In summation, there is some kinship between the musings of ancient Greeks and Native Americans and the scientific approach to the study of fossil record that we describe in this book; however, there clearly are differences too. We find it fascinating that for a very long time humans have been interested in explaining distinctive bones found in rocks, or shells found in places that are today thousands of miles from the sea. It is a part of the intellectual curiosity that, along with our artistic yearnings, represents one of the crowning achievements of our own species. Throughout this book we try to put into context what fossil discoveries tell us about the evolutionary history of life, the history of this planet, and the future of our own species if we continue on our current path. Interestingly, some Native American tribes hold that fossils represent the bones of beasts that had been destroyed because they harmed the Earth. Let us hope that the extinction of our own species, and the fossils we would ultimately leave behind, will not be a case where fact replicates legend: we know that the greatest threat to the survival of our species is the long-term harm we are doing to the Earth and its life forms.

Additional Reading

Colbert, E. H. 1984. *The Great Dinosaur Hunters and their Discoveries*. Dover, Mineola, NY; 384 pp.

Gould, C. 2002. *Dragons, Unicorns, and Sea Serpents: A Classic Study of the Evidence for their Existence*. Dover, Mineola, NY; 416 pp.

Gould, S. J. 2002. *The Structure of Evolutionary Theory*. Harvard University Press, Cambridge, MA; 1464 pp.

Jaffe, M. 2000. *The Gilded Dinosaur*. Crown, New York; 424 pp.

Jones, D. S., and Gould, S. J. 1999. Direct measurement of age in fossil *Gryphaea*: the solution to a classic problem in heterochrony. *Paleobiology* **25**: 158–187.

Mayor, A. 2001. *Fossil Legends of the First Americans*. Princeton University Press, Princeton, NY; 488 pp.

Mayor, A. 2001. *The First Fossil Hunters: Paleontology in Greek and Roman Times*. Princeton University Press, Princeton, NJ; 384 pp.

Wallace, D. R. 2000. *The Bonehunters' Revenge: Dinosaurs and Fate in the Gilded Age*. Mariner Books, New York; 224 pp.

Chapter 10
Plate Tectonics and its Effects on Evolution

Outline

- Introduction
- Early Ideas on Continents in Motion: Continental Drift
- Plate Tectonics: Continental Drift in a Different Guise and with a Valid Mechanism
- The Evolutionary Implications of Plate Tectonics
- Biogeography
- Concluding Remarks
- Additional Reading

Introduction

One of the most important and all encompassing advances in the field of geology was the development of plate tectonics: the idea that the Earth's surface is broken up into a series of plates in motion; the continents ride on top of these plates. Plates move episodically, at an average rate of a few to several centimeters a year; yet slowly but inexorably the plates and continents can move tremendous distances, and dramatically change their position and geometry throughout the course of Earth history. Plate tectonics led to a revolution in the way scientists studied the history of the Earth and the history of Life. This is because, as we shall outline here and in the next chapter, changing the position and arrangement of the Earth's plates and continents can influence the life of this planet in a variety of ways. First directly, either by isolating animals and plants from one another, which

Prehistoric Life: Evolution and the Fossil Record. 1st edition. By Bruce S. Lieberman and Roger Kaesler. Published 2010 by Blackwell Publishing.

spurs evolutionary change, or by bringing formerly separated animal and plant groups into contact with one another, which may spur evolution and also extinction. Plate tectonic movements can also initiate regional and global climate changes which in turn will have a secondary, indirect effect on life and evolution. Think of Antarctica, which was not always a frozen continent, but once had lush forests with dinosaurs sporting among the trees. Here we will introduce the notion that geology, through plate tectonics and other mechanisms, fundamentally influences biology and climate. The notion that climate influences biology was already introduced in our discussion of natural selection. An interesting point that we will develop in later chapters is that life has also played a role in influencing geology and climate. (We will say that climate even plays a role in influencing geology, although this subject is beyond the scope and aims of this book.) In short, there are feedbacks and the relationships between geology and biology and climate and biology are not one-way streets; the Earth, Life, and climate are all fundamentally and inextricably linked.

Early Ideas on Continents in Motion: Continental Drift

The idea that the Earth's continents have moved significant distances throughout Earth history dates back at least to the 19th century. One of the first persons to make a rigorous, well constrained argument for what was then called continental drift was the German meteorologist Alfred Wegener. He presented this idea in 1912, and offered several lines of evidence suggesting that what he called continental drift had occurred. Although not all aspects of Wegener's ideas on continental drift proved correct, they still deserve mention, not only because he presented an important kernel of what is today known as plate tectonics, but also because he summarized mountains of geological and paleontological evidence suggesting that the continents had changed their positions through time. However, in his lifetime his ideas were accepted by few scientists: the related, though not identical version of plate tectonics did not become a broadly accepted scientific idea until the 1960s. The reason provides a lesson about the scientific process, and the way humans assimilate new information and come to accept new ideas in any field, not just geology. Further, much of the evidence Wegener accumulated to support continental drift is still valid evidence for plate tectonics.

One of the important lines of evidence Wegener adduced was that there was a geometric fit between continents on either sides of the Atlantic Ocean such that the coastlines of Africa and South America fit almost perfectly together (Figure 10-1): like the once joined pieces of a jigsaw puzzle that now lie thousands of miles apart. Wegener argued that this fit suggested that Africa and South America were once joined in this position. Wegener also described how rocks that are the same age but lie on either side of the Atlantic Ocean can be very similar. For instance, rocks in Newfoundland, which lies in eastern Canada, are nearly the spitting image of rocks in England while rocks in Brazil are very similar to those in western Africa. Wegener further argued that the southern continents, including South America and Africa, were once joined into a single large supercontinent termed Gondwanaland. There was also geological evidence that this southern continent experienced a widespread glaciation in the Carboniferous period, while other continents that are today in the northern hemisphere lack evidence of a Carboniferous glaciation.

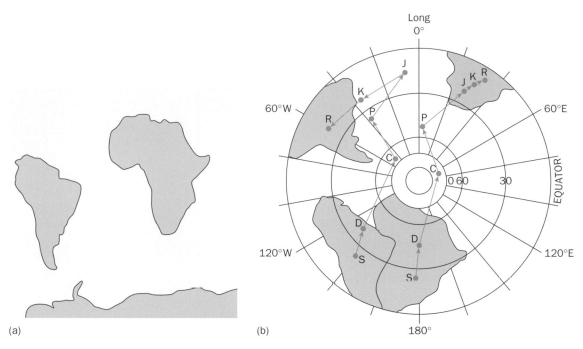

(a)   (b)

Figure 10.1 Current and former positions of South America and Africa. Showing the geometric fit between these two continents. The green lines with arrows in figure (b) refer to the motions of these continents relative to the pole, and the letters represent the relative positions at various geological periods, for example, "S" represents Silurian, "D" represents Devonian, etc. Image from C. B. Cox and P. D. Moore (2005) *Biogeography: An Ecological and Evolutionary Approach*, 7th Edn, Blackwell Publishing.

Paleontological Evidence for Continental Drift

Some of the most important evidence Wegener presented for continental drift derives from fossil organisms. Just as rocks on both sides of the Atlantic seemed to match up, the fossils they entomb also matched up. For example, Carboniferous rocks of today's southern continents (including not only South America and Africa but also Antarctica) contain unique and distinctive plant fossils belonging to the genus *Glossopteris* (Figure 10-2). Further, these fossil plants are only known from the southern continents. A possible way of explaining the distribution of this plant only in the southern continents would be to invoke that all of these continents were once connected, and isolated from the northern continents. Perhaps one of the best examples of fossil evidence supporting continental drift also comes for the Carboniferous: the small freshwater reptile *Mesosaurus* is only known from Carboniferous rocks in eastern Brazil and western Africa (Figure 10-3). How could an exclusively freshwater organism cross the wide Atlantic ocean? The answer, of course, is that it didn't have to. Carboniferous lakes in eastern Brazil and western Africa were in close association, and taken in sum the various bits of geological and paleontological evidence provide strong support that the Earth's southern continents were joined back in the Carboniferous and thus subsequently split apart.

Not all examples of paleontological evidence supporting continental drift are restricted to the Carboniferous. For instance, there are certain Cambrian trilobites that today are found in North Africa, central Europe, Scandinavia, and the eastern most tip of North America: around Boston, Massachusetts, and in eastern Newfoundland, Canada, but nowhere else on the North American continent. For

Figure 10.2 Glossopteris. A reconstruction of this Carboniferous plant. Diagram from M. J. Benton and D. A. T. Harper (2009) *Introduction to Paleobiology and the Fossil Record*, Blackwell Publishing.

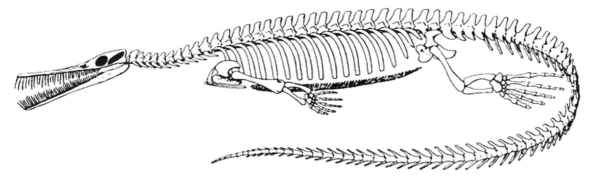

Figure 10.3 Mesosaurus. A reconstruction of this Carboniferous freshwater reptile. Diagram from R. Cowen (2005) *History of Life*, 4th Edn, Blackwell Publishing.

instance, if you search in Cambrian rocks west of Boston, in New York State, or a little bit west of eastern Newfoundland, in western Newfoundland, a totally different type of Cambrian trilobite is found. It might be asked how one type of Cambrian trilobite could come to be distributed on both sides of the Atlantic Ocean, and over much of Europe and northern Africa, yet occur no further west on the North American continent than Boston. These trilobites were discovered well before most paleontologists accepted continental drift (or plate tectonics) and it was instantly recognized that their geographical range was odd or unusual; because they presented such a paradox, they were named *Paradoxides* (Figure 10-4). This trilobite is actually famous enough to have appeared on a Canadian stamp in 1990. Their distribution would no longer seem paradoxical, however, if the continents have moved significant distances and regions today separated by great distances were once in contact back in the Cambrian (and regions today in contact were once far apart in the Cambrian). In this particular case, it turns out that parts of eastern North America, especially the Boston area and eastern Newfoundland, were part of a tiny continent that in the Cambrian was in close association with parts of Europe and Africa; subsequently this continent moved away from what are today Europe and Africa and became attached to North America.

Why Wasn't Continental Drift Accepted Right Away?

As can be seen from our brief discussion, Wegener provided a set of compelling evidence that the Earth's continents had moved, yet the idea that the continents are in motion was not accepted by most geologists for a long time: not until almost 50 years after he died. Why not? Part of the reason has to do with the way Wegener presented his information: as a catalog of empirical evidence. Even though much of the evidence he presented in support of continental drift was valid, Wegener had a difficult time convincing geologists because he failed to provide an adequate mechanism to explain how the continents might move. (Wegener did present a mechanism, involving continents moving across the ocean floor as a result of a pole fleeing force generated by the rotation of the Earth; this force is akin to the force that causes the water in a draining sink or bathtub to rotate; geologists rightly rejected this mechanism as having insufficient force to move continents.)

The reason for Wegener's difficulties in convincing people of continental drift are reminiscent of a common theme in science: when you develop a new and exciting idea counter to received wisdom, and continental drift was such an idea, as no less a personage than Charles Darwin remarked that the continents had never changed their position, then you need a convincing mechanism to explain how the new idea would work. Mere facts alone are not enough to convince people to change their viewpoint. We suspect this is true of not only new ideas in science but also new ideas in the art world, in business, and probably in all human endeavors. Without a valid mechanism, people rejected Wegener's theory and instead argued that there had been large, emergent land bridges that had once joined continents; these would allow animals and plants to migrate along their length, and thus fossils from continents separated by oceans could come to resemble one another. It was further argued that these land bridges later sunk into the sea, explaining why they were no longer visible.

There is indeed an interesting parallel with ideas on biological evolution here. For example, as described in Chapter 4, the notion that evolution had happened was an idea introduced long before Darwin published "*On the Origin of Species . . .*";

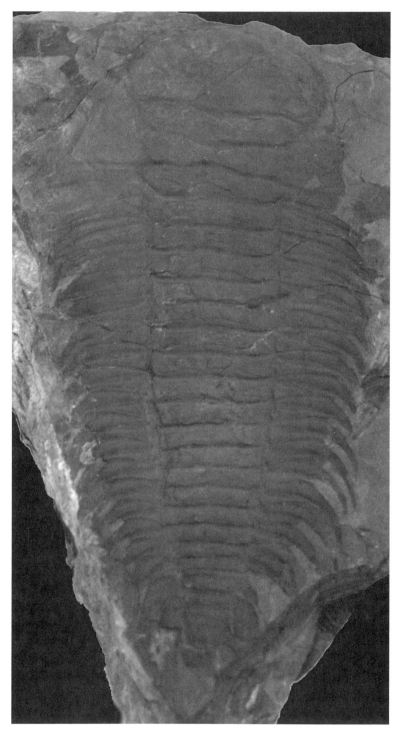

Figure 10.4 *Paradoxides*. This trilobite is from the Cambrian of Newfoundland, Canada. Image by B. S. Lieberman, specimen in the University of Kansas Museum of Invertebrate Paleontology, donated by R. Levi-Setti.

however, the idea was not accepted broadly until after Darwin published his great book in 1859. Part of the explanation for this lies in the fact that Darwin (along with Wallace) was the first to provide a reasonable mechanism for evolutionary change: natural selection. The evidence supporting evolution was not generally accepted until a valid mechanism to explain it was formulated.

Plate Tectonics: Continental Drift in a Different Guise and with a Valid Mechanism

The formulation of plate tectonics and the development of a valid mechanism to explain how the Earth's continents could have moved awaited the detailed examination of the ocean floor, a project undertaken for military reasons. In particular, the United States Navy wanted to know what the seafloor looked like so we would have good places to hide our submarines and also so that we would know the places where the Soviet's would try to hide their submarines. This only goes to show you that sometimes significant scientific advances are accomplished by accident, when people are looking for, or trying to accomplish, something else.

The Ocean Floor: its Topography and Magnetic Character

The maps of the topography of the Earth's oceans were crucial to the development of plate tectonics for several reasons. Of primary importance, they revealed no evidence of sunken land bridges joining continents; thus evaporated one of the arguments used to explain away the fact that organisms like *Mesosaurus* and paradoxidid trilobites could be found on what are now distantly separated continents. Instead, topographic maps revealed a giant ridge system of underwater mountains (Figure 10-5), and a set of deep ocean

Figure 10.5 **Map of the Earth.** Showing the topography of the ocean floor. Image from iStockPhoto.com, © J. Rysavy.

canyons or trenches. Moreover, the sediments covering the ocean floor were comparatively young, in stark contrast to the continents. In particular, no sediments older than the Jurassic were recovered. The significance of the ridges and trenches, and the relatively young rocks of the ocean floor, will be explained shortly.

Some of the most important discoveries ultimately leading to the discovery of plate tectonics involved evidence of the Earth's magnetic field preserved in ancient rocks. Today's Earth has a magnetic field centered at magnetic North, near to, but not precisely at, the geographical North Pole (the reason why magnetic and geographical North do not precisely correspond are beyond the scope of this book); modern lavas (and sedimentary rocks) are magnetized in the same direction as the Earth's magnetic field. This is because tiny magnetic grains inside of the rock become oriented to the Earth's magnetic field. However, older lavas can sometimes be magnetized in the opposite direction, implying that the Earth's magnetic field had flipped and was once centered near the South Pole. The magnetic signature preserved in a lava, and sometimes in a sedimentary rock, not only allows a scientist to determine if the magnetic pole was in the same or opposite direction relative to current magnetic north. The signature can also be used to calculate the latitude of the place where a rock formed, called the paleolatitude. For example, a paleomagnetic analysis conducted on a recently formed rock derived from an active volcano in Hawaii would reveal that the rock formed at the latitude it sits at on Hawaii today: not a surprising result. However, when older rocks are considered, sometimes the paleolatitudes recovered are significantly different from those rocks' present day latitude; typically the older the age of the rock considered the greater the difference between the paleolatitude and the latitude where the rock occurs today. (We should add that of course a geologist who studies the magnetic signature of rocks, a scientist called a paleomagnetician or sometimes just a paleomagician because of the seemingly arcane nature of their subject, must make sure that the rock he or she collects is not just a rock lying loose at the surface; instead, it must be a rock integrated into the surrounding rocks and part of the Earth's crust; further, there are other types of complications. Still, in bare bones, this is what is involved with paleomagnetic analyses.)

The paleomagnetic character of rocks surrounding the Earth's underwater ridge system played an important part in convincing scientists of plate tectonics; on either side of the ridges there are alternating bands or stripes of normal and reversed magnetic polarity and the greater the distance from the ridge the older the age of the ocean floor. It was ultimately recognized that lava emerged from the mountainous ocean ridges (these ridges are akin to underwater volcanoes that circumscribe the globe) and spread outward: a phenomenon called seafloor spreading (Figures 8-19 and 10-6). (There are actually places where the ocean ridges rise above the surface of the sea and become emergent: Iceland represents such an example.) As the lava cooled, it retained the magnetic character of the Earth's field when it formed. The further away from the ridge one went, the older the rock material encountered. Since some of the time when new undersea lava was produced, the Earth's magnetic field may have been in a different direction relative to today's magnetic field, the rock would possess a reversed magnetic polarity. The result would be a pattern of magnetic stripes on either side of the ocean ridges, with some lavas resembling current magnetic polarity, and some with opposite polarity.

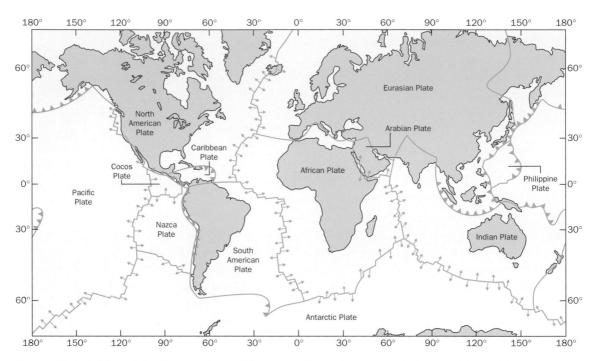

Figure 10.6 A Map of the Earth's plates. The green lines show the relative position of the ocean ridges, with the arrows showing the relative directions of seafloor spreading. Image from R. S. Boardman, A. H. Cheetham, and A. J. Rowell (1987) *Fossil Invertebrates*, Blackwell Publishing; derived from S. Uyeda (1978) *The New View of the Earth*, © W.H. Freeman and Co., San Francisco, CA.

The Earth's Plates in Motion

The continents move around as seafloor spreading occurs: they basically sit passively on the plates that move out slowly but inexorably from ocean ridges. This mechanism does differs substantially from Wegener's explanation of continental drift which suggested that continents plowed through the ocean floor, yet Wegener was right on the mark when he recognized that throughout geological time the continents have moved great distances.

Just as new ocean floor is created at ridges, it is also being destroyed such that the net size of the Earth remains constant. Old ocean floor is destroyed or subducted at the trenches: the deepest part of the seafloor. This ocean floor is heated inside of the Earth and ultimately melted. This is why there is no seafloor older than the Jurassic. All older ocean floor has been subducted. In effect, there is a cycle that operates on a time frame of tens of millions of years with new ocean floor created, then destroyed, and ultimately created again.

Sometimes the movement of plates causes the continents sitting on them to collide with one another. Past collisions between continents are reflected in the position of ancient mountain ranges. For instance, a continental collision between North America, Europe and Africa around 350 million years ago produced the Appalachian mountains, which are still with us today, although they've worn down and eroded away a bit in the intervening interval. The Himalayas are more spectacular than the Appalachians partly because they are younger and have been produced by a more recent collision: the collision between India and Southern Asia.

India once was a continental bloc that sat alongside Madagascar and Africa on Africa's southeastern coast. However, seafloor spreading (and continental rifting, discussed more fully below) that began during the Cretaceous Period separated it from Africa and Madagascar; ultimately India came to lie on a different plate than Africa as it broke away from Africa and moved northward. Eventually, roughly 30 million years ago, it collided with Asia. This continental collision produced the Himalayan mountain range and the Tibetan plateau; actually, the plate India is on continues to move northward, causing the Himalayas and the Tibetan plateau to continue to grow upward (although this growth upwards is partly balanced by erosion of the mountain range). Such continental collisions not only produce mountain ranges but also major earthquakes.

Plate tectonics is such a powerful idea because it explains a great many facts. For instance, the evidence Wegener adduced to support continental drift: the coast lines of South America and Africa fit because 180 million years ago these continents were joined. Eastern Newfoundland and England were once joined which is why the rocks in these places look so similar. Plate tectonics also explains a great many paleontological observations. *Mesosaurus* could occur only in fossilized lakes found in today's Brazil and Africa because these lakes were once in close association with one another. Similarly, the widespread *Glossopteris* plants confined to the southern continents can be explained by the fact that these continents were once together in a landmass scientists still call Gondwana. Further, a Carboniferous glaciation occurred only in the southern Gondwanan continents because this continent once sat at the Earth's South Pole.

The Evolutionary Implications of Plate Tectonics

It can readily be seen how the movement of these plates around the Earth would be important for the history of life. Terrestrial organisms are of course tied to the continents they sit on. Even the bulk of today's marine diversity is largely (though not entirely in the case of mid-ocean dwelling organisms) distributed around the margins of these continents. This was true in the past as well: the bulk of the fossil record of ancient marine life (and all of this record prior to roughly 180 million years ago) comes from rocks that are today found on continents (because many of the continents were once under the sea and represented broad swaths of habitat available to marine organisms). Continents will move with their underlying plates at an average pace of a few centimeters a year. Given enough time, for instance, many millions of years, we can see how plate tectonics will move continents vast distances. This can change the climate a group of animals and plants experiences. For example, Antarctica was not always a frozen continent centered on the South Pole. It once was at more moderate latitudes and experienced less harsh temperatures, when global climate was milder. In these times, a much more diverse and abundant set of life forms lived in Antarctica, including rich plant floras and, back in the Mesozoic, even dinosaurs. Plate tectonic movements, as already described in the case of India, can also cause different continents to separate or join, thereby isolating animals and plants that once lived together, or bringing together animals and plants that once lived apart. Each of these events can often lead to important evolutionary changes.

In short, the existence of plate tectonics implies that there should be a close association between the evolution of the Earth and the evolution of life. Principally, the way plate tectonics is directly tied to evolutionary change is through the process of speciation described earlier in Chapters 5. (It also bears mentioning that there are several indirect connections between plate tectonics and evolution. For instance, changes in the geometry and position of the Earth's continents can have a major affect on global climate [which we'll discuss in the next chapter] and these climatic changes will in turn have important biological and evolutionary effects, with the case of Antarctica described above being one noteworthy example.) To summarize though, one of the primary ways evolution occurs is by speciation: when one lineage divides into two or more new lineages.

Geographical Isolation and Speciation

Biological species (discussed extensively in Chapter 7) are composed of individual organisms and further these organisms are generally divided up into different populations which are congregations of a few to many individuals that live together, for at least part of the year, and interbreed. The number of populations comprising a species varies from one to many. There are a variety of ways that new species form: a process called speciation. The most common way is when one or more populations of the species becomes reproductively isolated from other populations for long periods of time such that they are no longer interbreeding. This type of speciation is called allopatric speciation (Figure 10-7). Given sufficient time, reproductively isolated populations will start to diverge evolutionarily. A new species is formed when the populations can or will no longer interbreed with one another, which happens if they diverge sufficiently.

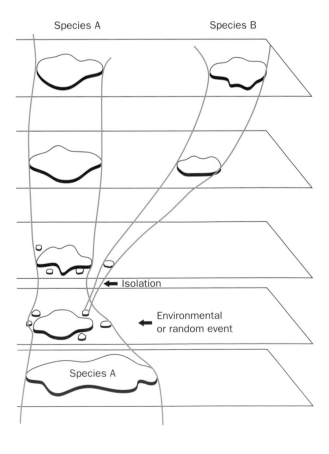

Figure 10.7 Allopatric speciation. An environmental change or perhaps a chance dispersal event leads to one or more populations becoming geographically isolated from the rest of the species. Through time they diverge, and eventually a new species forms. Diagram from M. J. Benton and D. A. T. Harper (2009) *Introduction to Paleobiology and the Fossil Record*, Blackwell Publishing.

There are certain conditions that encourage evolutionary divergence. For instance, if the isolated population(s) contains relatively few individuals, it is more likely to diverge. Also, the manner in which populations become reproductively isolated matters. Usually isolation occurs because some sort of barrier is interposed between the two populations. The smaller or more transitory this barrier is, the easier it will be for individuals from different populations to make it across and thus allow interbreeding. Obviously, the Himalayas are going to be a more effective geographical barrier to the free movement of plants and animals than a small hill. Also, barriers can be geological or climatic. For instance, a nearly flat desert could be an effective barrier to animals because they would die of thirst before they could cross it. Moreover, climatic factors and geology may be interrelated. In particular, one of the primary reasons an animal may not be able to pass over a large mountain range is because it would have to pass through inhospitable climates at high elevations.

Plate Tectonics and Geographical Isolation

Plate tectonics is important from an evolutionary perspective because it can cause geographical barriers to form and these will encourage evolutionary change and speciation. Imagine a single species that is distributed across an open, largely homogeneous continent that lacks prominent internal geographical barriers. Then imagine that the continent begins to split or rift into two pieces (Figure 10-8). This is a process akin to seafloor spreading, which occurs as upwelling lava emerges and spreads out at either side of an ocean ridge: however, in this case instead of the ridge being in the middle of the ocean, it occurs within a continent. The result is a fragmenting continent; the intervening area where rifting occurs becomes lower than the surrounding land and usually is quickly filled with water. If the rift grows large enough it will eventually become an ocean. When this happens populations of many species of land animals will become isolated on either side of the growing rift; these populations will diverge evolutionarily and eventually speciate allopatrically. (Of course, a different situation may play out with marine organisms. A rift inside of a continent, if it leads to the formation of a new waterway, can allow formerly isolated marine species to come into contact with one another. If they are sufficiently diverged they will not interbreed, but other types of events may occur, described more fully below in the discussion of the Great American Interchange.)

Plate tectonic changes will affect not only single species. Instead, they cause geographical barriers to form that isolate many different groups of organisms at roughly the same time, and these can lead to wholesale speciation in literally hundreds or thousands of lineages. In many cases, it seems that without forming geographical barriers there will be little speciation and thus little evolutionary change: if this is true it suggests that plate tectonics is one of the real motive forces of evolution. Data gathered by paleontologist Jim Valentine support this contention. For example, over the last 600 million years, the Earth has oscillated between the condition where all of the Earth's continents were joined together in a single supercontinent (which happened twice, at the end of the Proterozoic and during the Permian-Triassic) and when all of the continents were relatively dispersed (see Figure 6-10 and Plate 10-1), the condition we are in today. Valentine showed that when the continents were less fragmented the total diversity of life on Earth was lower than when the continents were more fragmented. Intuitively this makes sense

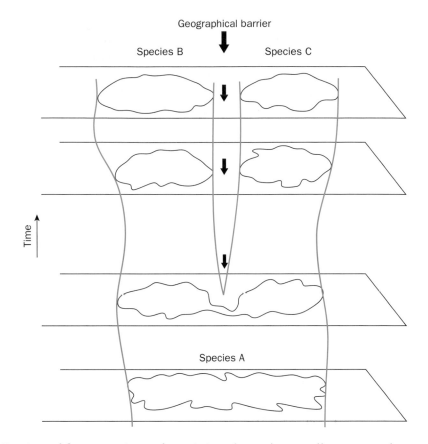

Figure 10.8 Continental fragmentation and speciation. A continent splits apart and a once continuous species is broken up into two separate populations. These become isolated, diverge, and eventually speciate. Diagram from M. J. Benton and D. A. T. Harper (2009) *Introduction to Paleobiology and the Fossil Record*, Blackwell Publishing.

if we think of the important role that geographical isolation plays in generating new species. If the different continents become separated it increases the opportunities for different populations of species to become isolated. By contrast, when all the continents are together there are fewer opportunities for populations to become isolated and thus fewer chances for speciation to occur. The relationship between the configuration of the continents and the overall number of species is evidence that geological processes play an important role in influencing biology, evolution, and the history of life.

Plate Tectonic Changes Can Bring Isolated Biotas Together

Just as plate tectonics can split continents apart, it can also bring them together by initiating continental collisions. Such collisions, at least at first, will allow many different species to nearly simultaneously expand their geographical ranges. For example, when India first collided with Asia in the Cenozoic many species formerly restricted to India moved into Asia; in addition, many species formerly restricted to Asia moved into India. Eventually, however, such continental collisions will cause new geographical barriers to form. Now, the Himalayas, roughly in the region where the continents of India and Asia collided, partly isolates the Indian subcontinent from the rest of Asia. It turns out that India still retains many unique and distinctive species of plants and animals. One of the reasons why India has so many unique types of organisms can be attributed to the fact that the evolutionary origins of these groups extends back to the time when India was a fully independent continent surrounded by oceans: a time when its animals and plants were isolated from those of other continents. Another reason why India has so many distinct species is because it also contains many species of Asian origin that first entered India shortly after India and Asia collided and then subsequently became geographically isolated from the rest of Asia due to the uplift of the Himalayas. Thus, continental collisions can create opportunities both for geographical range expansion and geographical isolation.

Not all continental interactions produce Himalayan geological results; some involve less profound geological changes, yet they can still have major biological consequences. One of the most spectacular examples of this occurred roughly 3 million years ago when North America and South America were joined at the present day site of the Panamanian isthmus. This allowed many groups of organisms to expand their geographical ranges. Before this, and extending back to about 50 million years ago, the South American continent had been isolated from all other continents. This long period of geological isolation allowed the fauna and flora of South America to evolve independently in what the great paleontologist George Simpson famously called "Splendid Isolation:" the net result was a dramatic and distinctive South American Cenozoic fauna. Some of the most spectacular denizens of this fauna were the gigantic, flightless carnivorous birds called "Phorusrhacoids" but we prefer the term "Terror Birds" (Figure 10-9). They were given this name with good reason: they were swift runners, while reaching ten feet in height; they were also armed with massive beaks for tearing and rending prey, and powerful, but short, arms for grasping or chopping unlucky animals that got in their path. They may well have fed on the unique set of mammals that also evolved in South America during its isolation. A great variety of unique mammals including several distinctive marsupials are known as fossils from the Cenozoic rocks of South America and still persist in South America to this day.

Once North and South America collided and became connected formerly isolated land animals (and plants) from each of these continents could now dramatically expand their ranges; many did, with North American species moving south and South American species moving north in an event now called the Great American Interchange (Figure 10-10). Of course not all of the species from the respective continents moved northwards and southwards; many stayed put, and still today, both continents, especially South America, have a unique complement of species. Still, it is accurate to say that after the collision there were many species of North American mammals that invaded South America, while some species of South American mammals invaded North America. Thus, what appears to be a relatively subtle geological event around the Panamanian Isthmus, produced tremendous opportunities for many groups of species to expand their geographical range. Familiar species that invaded North America after the collision include the opossum and the armadillo.

Paleontologists have investigated what happened to mammals during and after the Great American Interchange. What they have found is that a fair number of North American lineages make it into South America, survive, and diversify while the number of South American species that make it into North America to survive and diversify is lower. Originally, this difference between North and South American

Figure 10.9 Terror bird. Skeleton of a giant, carnivorous bird from the Cenozoic of South America. Diagram from R. Cowen (2005) *History of Life*, 4th Edn, Blackwell Publishing; after Sinclair.

mammals was explained by invoking differences in the constitutions and competitive abilities of the organisms in their respective continents. In particular, it was assumed that the mammals of South America were less competitively fit than their northern brethren; therefore, when South American mammals were introduced to North American mammals they would serve as easy prey, perform worse as hunters, and even be less adept at gathering food such that ultimately they would be outcompeted and die off while the fitter North American mammals thrived. Further, it was argued that such a situation could arise because South America was geographically smaller than North America and its fauna had long been isolated; each of these factors would mean that the mammals there had not been exposed to the competitive rigors of the broader world. By contrast, it was suggested that mammals from North America had been exposed to conditions that make them excellent competitors: in particular, North America was a large continent that had more frequently been in contact with other continents (and thus mammals from these continents) including Europe and Asia.

The explanation for the apparent success of the North American mammals possibly at the expense of South America mammals relies on an analogy with the

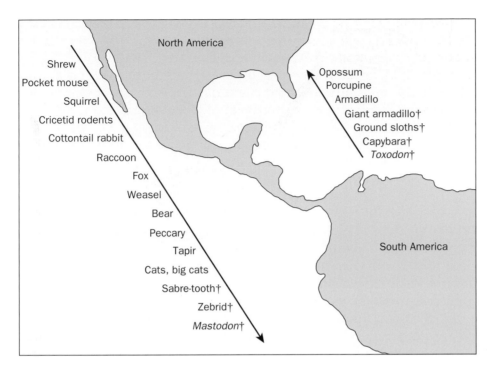

Figure 10.10 The Great American Interchange. Illustrating the mammal species that moved from North to South America and from South to North America. Diagram from C. B. Cox and P. D. Moore (2005) *Biogeography: An Ecological and Evolutionary Approach*, 7th Edn, Blackwell Publishing.

phenomena we introduced in Chapter 6 when we discussed the naiveté of the Falkland Islands fox (recall that Darwin collected live specimens of the fox with his geological hammer). It is true that the distinctive animals today found on small islands fare poorly when brought into contact with humans and animals associated with us like rats, cats, and dogs, and many have succumbed to extinction while others are gravely threatened. Given that animals on small islands seem prone to extinction and are not always such adept competitors, the argument relied on treating South America as a big little (or a little big) island, with an isolated and extinction prone set of mammals.

Although the argument invoking competition is appealing because of its simplicity it may not hold up in the details (and recall from the discussion in Chapter 5 that the fossil record does not support the notion that competition is always an important evolutionary force). The patterns during the Great American Interchange cannot solely be explained by competition because joining North and South America also precipitated a major episode of climate change and global cooling that will be considered in greater detail in the next chapter. Paleontologist and evolutionary biologist Elisabeth Vrba argued that the Great American Interchange cannot be considered apart from the environmental changes occurring at the same time. In particular, she argued that the episode of global cooling that follows the collision between North America and South America would be more detrimental to tropical organisms. Recall the discussion in Chapter 6 of the mass extinction at the end of the Ordovician and the fact that in general tropical organisms tend to be much more diverse than organisms living at higher latitudes. Given that the equator was (and is) centered over South America (Figure 4-4), after global cooling organisms preferring tropical environments will experience extinction as their habitat shrinks and they will also move towards the equator, following their preferred habitat. Given that, the tropical forms in South America will have to stay put; by contrast, the tropical forms present in North America will have to move towards the equator and their preferred environment, which happens to now lie in South America. The direction of movement of the mammals during the Great American Interchange can thus be explained by invoking climate change and its effects on life forms without having to invoke competition. More research will certainly be done on this topic, but Vrba's explanation is an exciting and innovative idea.

Biogeography

The study of where animal and plant groups are found, and the processes that have caused them to occur in these regions, is called biogeography. One of the most important large-scale biogeographical patterns involves the distribution of organisms and was first discovered in the 16th, 17th and 18th centuries when explorers set out from Europe to colonize new regions. They expected that the animals and plants they found would be the same as those that occur in Europe, at least if they went to a place that had a similar climate, but they found quite the contrary. Different regions had very different types of organisms. Prior to the discovery of evolution and plate tectonics this was very hard to explain. Now we know that different regions have unique types of animals and plants because they have a unique geological history when they were separated from other regions by barriers. During this time their organisms evolved independently. For example, Australia has been separated from all of the other continents for about 50 million years and for this reason the animals and plants found there are very different from those found anywhere else. As we go back in time, we can predict which continent will have plants and animals most similar to Australia. Probably the continent that shared the most recent history of connection with Australia is Antarctica, but given that there is little terrestrial animal life on Antarctica, other than penguins, we must turn our attention to South America, the continent next in line in terms of physical connections with Australia: here we do find significant evolutionary commonalities. In particular, quite a few of the mammals of South America are marsupials, and South America and Australia are the only continents with abundant living marsupials (putting aside invasive forms like the opossum that recently moved into North America during the Great American Interchange). A recurrent theme seen in biogeographical studies is that the evolutionary history of a region corresponds to the geological history it experienced.

Concluding Remarks

The important way that plate tectonic changes cause evolution suggests a fundamental tie-in between geology and biology. Indeed, here we can see the significance of the fact that information from biogeography proved instrumental in the development of Darwin's ideas on evolution. Recall from Chapter 4 that one thing that convinced Darwin that evolution had happened were patterns involving the geographical distribution of animals on and around the South American continent. Some of the best examples of such biogeographical patterns came from the birds of the Galapagos Islands. As mentioned in Chapter 4, the closest point on the South American mainland to the Galapagos is part of Ecuador proper, which lies some 500 miles to the East (the Galapagos also happen to be part of Ecuador). The environment found at sea-level in mainland Ecuador is tropical and very different from the environment found in the Galapagos, which is cooler and drier because it is awash in a cool Pacific ocean current. In spite of the profound differences in environment, the closest relatives of many of the birds found in the dry environs of the Galapagos are found in tropical Ecuador. This fact would not make sense if every species had been perfectly created for the environment it lives in, but it does make sense in light of the geological history of the region and evolution. In particular, several species of birds made it from South America to the Galapagos. Those leaving from points closest to the Galapagos were most likely to survive the long ocean voyage and land on the Galapagos. The rest, as they say, is history: in this case, several million years of isolation of the Galapagos birds from the mainland.

This represents an example of what might be thought of as a large-scale connection between geology and evolution, but smaller scale connections also exist. Another interesting biogeographical example, going back to our discussion in Chapter 4, again comes from the Galapagos Islands and this time involves mockingbirds. There are different species of mockingbird found on various major islands of the Galapagos; these species are basically restricted to particular islands and thus do not live together. Not surprisingly, mirroring the pattern already described, the different mockingbird species resemble each other more closely than they resemble the more distantly related mockingbirds of South America: again this makes sense in light of evolution and geology. Again, at a smaller scale, the mockingbirds provide another example of how not every species was perfectly created to fit its environment. This time, the environments of the different Galapagos islands appear to be identical yet they house different mockingbird species. These different mockingbird species diverged from one another because the environmentally similar volcanic islands of the Galapagos remained isolated from one another. We would argue that the connections between geology and biology are thus fundamental; in subsequent chapters we will explore the connections that also exist between geology and climate; further, we will even explore and describe how life in turn has affected climate and geology. These connections show that the Earth, life, and climate systems are all inextricably linked.

Additional Reading

Brooks, D. R., and McLennan, D. A. 2002. *The Nature of Diversity: An Evolutionary Voyage of Discovery*. University of Chicago Press, Chicago, IL; 676 pp.

Cox, A., and Hart, B. R. 1991. *Plate Tectonics: How it Works*. Wiley-Blackwell, Oxford; 416 pp.

Cox, C. B., and Moore, P. D. 2005. *Biogeography: An Ecological and Evolutionary Approach*, 7th edition. Wiley-Blackwell, Oxford; 440 pp.

Hallam, A. 1990. *Great Geological Controversies*, 2nd edition. Oxford University Press, Oxford. 256 pp.

Lieberman, B. S. 2000. *Paleobiogeography*. Kluwer Academic/Plenum Publishing, New York; 208 pp.

Lomolino, M. V., Riddle, B. R. and Brown. J. H. 2005. *Biogeography*, 3rd edition. Sinauer Press. Sunderland, MA; 845 pp.

Mayr, E. 1999 (1942). *Systematics and the Origin of Species from the Viewpoint of a Zoologist,* reprint edition. Harvard University Press, Cambridge, MA; 372 pp.

Simpson, G. G. 1983. *Splendid Isolation: The Curious History of South American Mammals,* reprint edition. Yale University Press, New Haven, CT; 275 pp.

Stehli, F. G., and Webb, S. D. (eds.). 1985. *The Great American Biotic Interchange.* Springer, Berlin; 550 pp.

Valentine, J. W. 1974. *Evolutionary Palaeoecology of the Marine Biosphere.* Prentice Hall, New York; 511 pp.

Vrba, E. S. 1993. Turnover-pulses, the Red Queen, and related topics. *American Journal of Science* **293**: 418–452.

Wiley, E. O. 1981. *Phylogenetics: The Theory and Practice of Phylogenetic Systematics.* John Wiley & Sons, New York; 456 pp.

Chapter 11
Life, Climate, and Geology

Outline

- Introduction
- Some of the Major Factors that Govern the Climate System
- Examples of How Life has Influenced Climate: The Difference Between the Proterozoic and the Permian
- Life Influencing Geology: the Form and Shape of Rivers and the Rocks they Leave Behind
- Plants, Oxygen, and Coal: More Examples of Life Affecting the Atmosphere and Geology
- How Geology Affects Climate: Considering How Plate Tectonic Changes have Contributed to Climate Changes Over the Last 60 Million Years
- Concluding Remarks
- Additional Reading

Introduction

In this chapter we take up and continue our discussion on some of the relationships between the biological, geological, and climatic systems. We may be accustomed to thinking about how climate affects us, and indeed other living organisms: when it becomes cold outside we feel it, and some animals hibernate while others migrate great distances to stay warm; when it is hot we, and other animals, may feel like lazing around. However, the relationship between climate and life is not a one-way street, and one of the fundamental lessons gained from the study of the history of

Prehistoric Life: Evolution and the Fossil Record. 1st edition. By Bruce S. Lieberman and Roger Kaesler. Published 2010 by Blackwell Publishing.

life is that living things can also influence climate. Given the context of life's long and rich history, including its effects on climate, it is much easier to understand how we can be causing changes in our environment today, like global warming. Indeed, other organisms in the past probably caused episodes of global warming and certainly modified the atmosphere. For example, in the next chapter we will describe how the increase of oxygen in the atmosphere was directly related to the activities of living organisms. In this chapter we will focus on some of the effects life has had on climate, principally through its effects on the atmosphere. As part of this discussion, we will also describe some of the major factors that govern and influence the climate system.

We're also accustomed to thinking about how geology affects life; recall in the previous chapter we considered how plate tectonics influences evolution. Again though, it turns out that this relationship is a two-way street, and here we will describe some of the effects that life has had on geology. Finally, we will consider how geology has affected climate. (Recall as from the previous chapter that climate also effects geology, although this topic is beyond the scope of this book).

Some of the Major Factors that Govern the Climate System

Our climate system is awfully complicated. What the precise conditions will be at any one point in time even a few weeks in the future is difficult to understand and predict as any weather forecaster, or anyone who has had to rely on a weather forecast, is bound to know. Thus, our understanding of climate over the scale of thousands and millions of years is bound to be a bit muddled. Still, the complexity of the climate system can be overstated, and likely has been, partly because, dare we say it, it is a "hot button issue." Indeed, we do have some understanding of the broad parameters of climate. For example, if it's spring in the Midwestern United States there are bound to be some strong thunder storms, although we do not know the precise day that they'll arrive. If it's winter in Moscow, Russia, you can bet it will be pretty darn cold, although we cannot predict the precise temperatures, only a range; and if it's August in New York City, forget about it, it probably will be hot. It is in the sense of these generalities that we also have a solid understanding of the climate system, which allows us to recognize some of the major factors that influence climate, and it is worth describing these because they allow us to recognize how and why biology affects climate.

The Albedo Effect

One of the factors that profoundly influences climate is the albedo effect. This is related to what happens to sunlight as it strikes the Earth; much heat and energy is brought to the surface of this planet in the form of sunlight. Certain materials reflect lots of this heat and energy back into space, while other materials are better at retaining that heat and energy (Figure 11-1). Ice, being light in color, is a material that is particularly good at reflecting the Sun's energy back into space. By contrast, darker colored materials, like a parking lot paved with asphalt, tend to absorb and retain heat and energy, making them difficult to walk across in summer time whilst in bare feet. The more ice on the planet, the greater the albedo effect, and the more heat and energy reflected back into space; all things being equal, the more ice on the planet, the cooler the planet becomes, and not simply due to the cooling effects of ice, which is itself important.

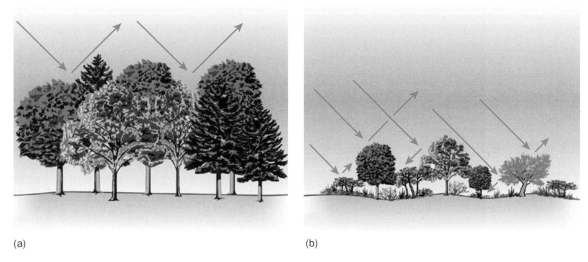

(a)   (b)

Figure 11.1 **The albedo effect**. Certain materials reflect sunlight back into space, cooling the planet. Other materials tend to absorb sunlight, warming the planet. Diagram from C. B. Cox and P. D. Moore (2005) *Biogeography: An Ecological and Evolutionary Approach*, 7th Edn, Blackwell Publishing.

Given how the albedo effect works, one can see how this aspect of the climate system could build upon itself: as the Earth becomes cooler, the amount of ice on the planet will tend to increase, which will further increase the albedo, and in turn the Earth becomes even colder. For example, imagine if the Antarctic ice sheet were to grow: an increase in the volume of ice in Antarctica would certainly act to cool ocean temperatures, in a manner akin to placing more ice in a glass of water; this would also serve to further lower temperatures by increasing the Earth's albedo.

For the same reason, things can also work in reverse, and indeed that is precisely what is happening today. As the Earth gets warmer, more ice on the face of the planet will melt as temperatures rise. The albedo of the Earth then drops, more heat gets absorbed, and it gets even warmer. For this reason, we refer to the albedo effect as an amplifier of the climate system, meaning that it tends to positively reinforce and amplify changes in climate by making a cooling trend become even colder and a warming trend become even hotter.

This is one reason why as global warming continues it will become ever harder to abate its effects. Instead, warming tends to build on itself and it will become ever warmer. The implications are that the planet's climate system may be very sensitive to even small perturbations.

The Greenhouse Effect

Another major factor or effect that governs the climate system is the greenhouse effect. The greenhouse effect was described in Chapter 6. Recall that light, energy, and heat reach our planet's surface from the Sun. Not all of this energy is reflected back into space: some stays and heats the Earth's atmosphere. The phenomenon resembles what happens with a gardener's greenhouse, which can become quite warm on a sunny day, hence the name of the effect. There are certain compounds in the atmosphere that strengthen the greenhouse effect, and these are sometimes called greenhouse gases. One very effective greenhouse gas is carbon dioxide; another is methane. Living organisms produce these compounds, meaning that biology can influence climate. (Living organisms are not the only source of carbon dioxide in the atmosphere; volcanic emissions are a primary source, and other factors like weathering and erosion also substantially influence carbon dioxide concentrations in the atmosphere.) Today, human's are magnifying the concentrations of certain greenhouse gases in the atmosphere. Before, however, we turn to human mediated climate change, we will focus on examples from the fossil record that nicely illustrate the important effects organisms have had on climate, through their influence on the concentrations of greenhouse gases in the atmosphere. These examples point out how current, human mediated climate change is not some unverified phenomenon without past analogues. Organisms have repeatedly modified climate on this planet, suggesting how very conceivable it is that we too, with our complex technologies, could do the same.

Examples of How Life has Influenced Climate: The Difference Between the Proterozoic and the Permian

Twice in the last billion years all of the Earth's continents have been joined together in a single land mass or supercontinent. For instance, there was a supercontinent called Pangaea that existed at the end of the Paleozoic and the beginning of the Mesozoic (Figure 6-10 and Plate 10-1). The existence of Pangaea meant that a Triassic dinosaur could likely have walked from the South Pole to the North Pole without even getting its feet wet. Another supercontinent was in existence at the end of the Proterozoic called Rodinia (Plate 11-1): it began to split apart around 750 Ma. The supercontinents have different names because the different continents that comprised them were in different positions; the different names also reflect the fact that Rodinia broke up first and then Pangaea reassembled much later.

The patterns of ocean circulation and also the climate during the time of Pangaea and Rodinia must have been very different from what exists today. (We discussed some aspects of ocean circulation and climate during the time of Pangaea as part of our discussion of mass extinctions in Chapter 6.) Ocean circulation would have differed because today there are moderately sized oceans fringed by continents whereas in the Proterozoic and the Permian there would have been essentially one great big ocean. Circulation is much poorer, especially between deep water and shallow water, in one large ocean than in many smaller ones. Recall from Chapter 6, when prominent ocean circulation is absent, as in the Proterozoic or the late Paleozoic, the bottom waters tend to become stagnant and deprived of oxygen. Ultimately, this causes carbon dioxide to leave the atmosphere and enter the oceans. Given that CO_2 is an important greenhouse gas and when CO_2 is removed from the atmosphere temperatures tend to drop, aspects of ocean circulation can profoundly influence climate. Recall that these interrelated aspects of ocean circulation and CO_2 concentrations may well have triggered the Permo-Triassic mass extinction.

All things being equal, given the similarities in continental geometry and ocean size we might predict that ocean circulation and its associated effects on climate would be comparable at the end of the Paleozoic, when Pangaea existed, and at the end of the Proterozoic, when Rodinia existed. However, in the Proterozoic there was no mass extinction because plants had not evolved yet, and animals, at least large complex animals, had not evolved yet either (we will describe the evolutionary happenings at the end of the Proterozoic and the start of the Phanerozoic in Chapter 13).

A "Snowball Earth"

Another difference between the end of the Paleozoic and the end of the Proterozoic concerns the climatic changes. As CO_2 levels fell at the end of the Permian, global temperatures fell as well, and in certain places glaciers advanced while polar ice caps grew; however, these happenings were nothing compared to the profound glaciations that occurred at the end of the Proterozoic. Perhaps as many as four times between 800 and 600 million years ago the Earth's continents were covered by glaciers, even at low latitudes near the equator. Glaciation may have been so profound that the entire surface of the Earth was covered by ice resembling a single, massive snowball. This idea has been given the catchy sobriquet "the Snowball Earth hypothesis." Others have argued that it may have been more of a "Slushball" than a "Snowball;" irrespective, it would have been quite cold then, and temperatures on Earth, even in the depths of the most recent Ice Ages, have never since fallen so dramatically. The climate between the several "Snow-" or "Slushball" episodes may have also been exceedingly warm, implying a dynamic oscillation between times of extreme cold and extreme warmth.

There are several important questions related to this phenomenon that are still difficult to answer. One involves how the Earth escaped from the glacial periods, only to later succumb to pervasive glaciation; another is why the profound glacial episodes finally ended after 600 million years ago. The answer to the former question may lie in the fact that as global temperatures dropped oceanic circulation between the deep and shallow oceans would increase, bringing much of the CO_2 being stored in the deep water to the surface, which would increase concentrations of CO_2 in the ocean and atmosphere, thereby increasing the greenhouse effect and raising global temperatures. Another possibility is that there may have been episodes of global volcanism associated with the start of the breakup of Rodinia; these volcanic events would have caused large quantities of carbon dioxide to enter the atmosphere which would increase temperatures. Further, a Snowball condition did not return after 600 Ma because Rodinia had fragmented sufficiently; this would have meant that ocean circulation would have very much changed, because the Earth's surface would have been covered by several moderately sized oceans instead of one large one.

One thing that is well established is that the first relatively large organisms to evolve on Earth appear right at the end of the Proterozoic and notably just as these glaciations are finally ending: the association is unlikely to be a simple coincidence. These organisms are the so-called Ediacarans, which were discussed in Chapter 8, and will be considered more fully in Chapter 13. Environmental conditions would have been trying, to say the least, during times of a "Snowball Earth." The Earth's emergence from the final profound Proterozoic glaciation may likely have facilitated the origins of large life forms on this planet, pointing out how climate plays an important role in influencing biology.

Why no New "Snowball Earth" when Pangaea was Assembled?

As noted, there was another more recent time when all the Earth's continents had become aggregated into a single supercontinent, in the late Paleozoic and early Mesozoic, notably including the Permian, when the supercontinent Pangaea existed. Then, however, the Earth was not covered from equator to pole in a thick blanket of snow or slush. Why not? At this time there is no sure answer. However, one possibility lies in the fact that different life forms populated the planet during these time periods, and this points out how biology can and has impacted climate.

The biological differences between the two time periods involve the evolution of animals and plants. There is some debate about whether the Ediacarans, which appear as fossils at the end of the Proterozoic, represent animals. Recall from Chapter 8 that they were organisms of questionable affinity, possibly animals, that often exhibited a flattened pancake-like appearance that had solved the problems of surface area to volume constraints by growing large while staying flat. There is no debate, however, that the organisms that proliferate at the start of the Phanerozoic during what is called the Cambrian radiation are animals. At the beginning of the Cambrian many of the modern animal phyla, and also some unusual but clearly animal phyla, such as the denizens of the Burgess Shale, evolved and diversified over a few tens of millions of years. Clearly this was a fundamental episode in the history of life, but it also had profound implications for climate because as part of their metabolism animals produce CO_2. When all is said and done, and in the simplest possible terms, animals act as a spur to global warming and increase the greenhouse effect.

Plants evolved more than 40 million years after the end of the Cambrian radiation; their origin clearly had important biological affects including the fact that they became an important source of food for animals. Their origin also had profound geological effects that we consider later in this chapter. However, another notable effect is their impact on the climate system. Plants, as part of their natural metabolism, take in CO_2 from the atmosphere. They convert it along with sunlight and water to oxygen and more plant material. Again, when all is said and done, and in the simplest possible terms, plants act as a check on global warming and serve to decrease the greenhouse effect.

Given the effects animals and plants have on CO_2 concentrations, what are their joint effects on climate in a world populated by both types of organisms? Undoubtedly, their effects are extremely complicated, and there are many feedbacks, but ultimately the presence of these two types of life forms keeps CO_2 concentrations in the atmosphere in a relative balance. This balance is also related to the fact that CO_2 concentrations influence the growth and amount of plant biomass. Each of these factors can be considered in broad brush terms. For instance, when CO_2 concentrations are high, plants grow more rapidly, which ultimately leads to the growth of more plants, which in turn causes CO_2 concentrations to diminish in the atmosphere. As CO_2 concentrations fall, plant activity slows down, which eventually causes plant activity and biomass to diminish; eventually this will allow CO_2 levels to rebound. Also, the more plants around, the more food for animals to eat, which will allow animals to proliferate and multiply. The removal of plants, and the proliferation of animals, will lead to increasing CO_2 levels, which will ultimately trigger the growth of more plants.

These coupled changes in CO_2 concentrations are not exact nor are they instantaneous: they likely take long amounts of time to occur, and never lead to the establishment of a complete equilibrium. Some have argued that the Earth, its life and its climate, and its life forms are intricately connected and self regulating. This perspective has sometimes been referred to as the Gaia view where the Earth and

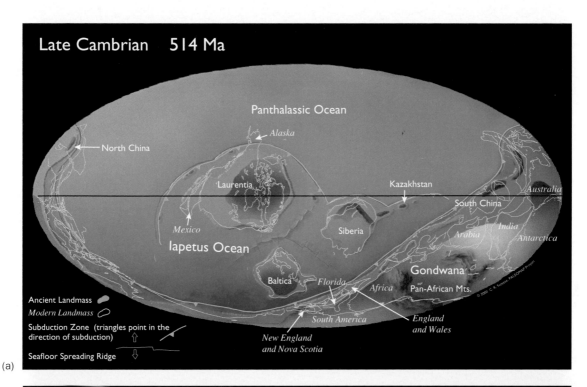

(a)

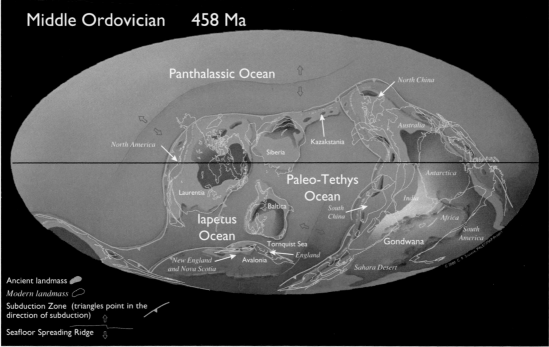

(b)

Plate 10.1 The Earth through time. Paleogeographic reconstructions for (a) the Cambrian, (b) the Ordovician, and (g) the Miocene. Images from C. Scotese, © the Paleomap Project, University of Texas at Arlington. Paleogeographic reconstructions for (c) the Devonian, (d) the Carboniferous, (e) the Cretaceous, and (f) the Eocene. Diagrams from C. B. Cox and P. D. Moore (2005) *Biogeography: An Ecological and Evolutionary Approach*, 7th Edn, Blackwell Publishing.

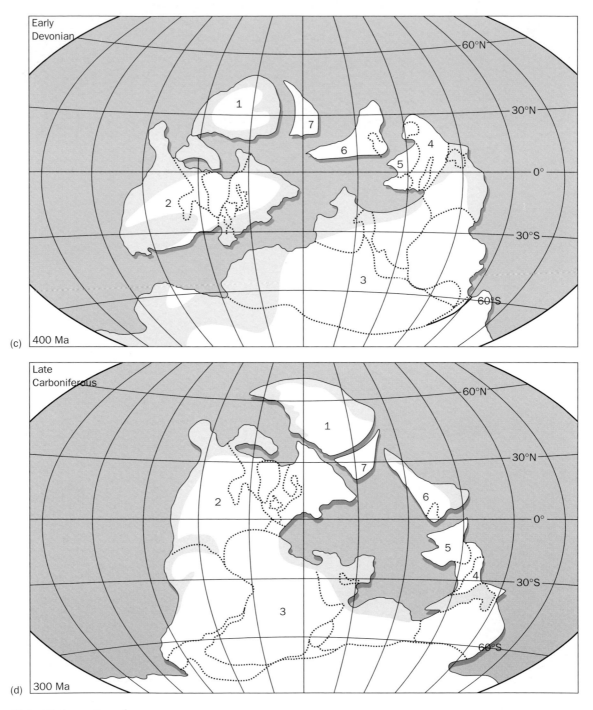

Plate 10.1 *continued*

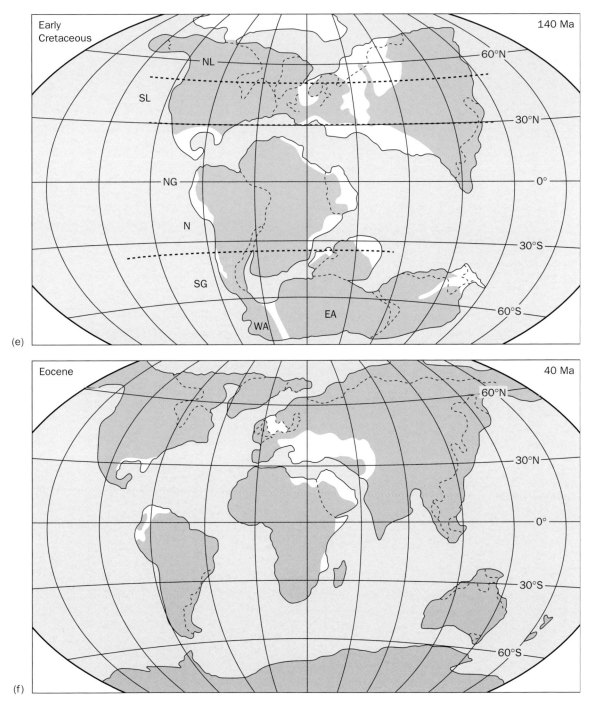

Plate 10.1 *continued*

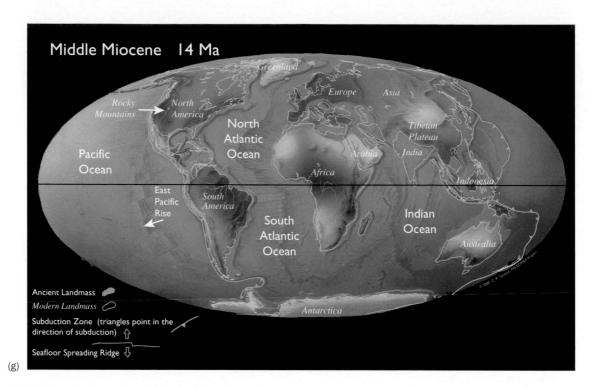

(g)

Plate 10.1 *continued*

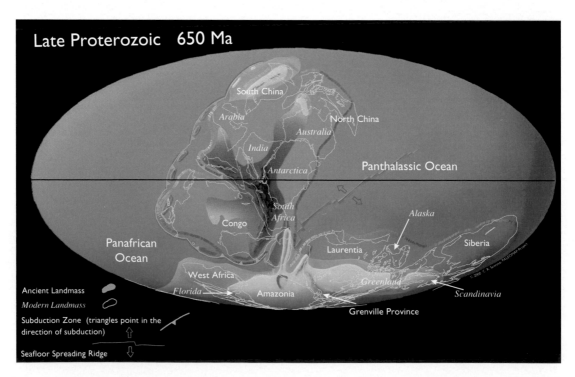

Plate 11.1 Earth in the late Proterozoic. Paleogeographic reconstruction. Image from C. Scotese, © the Paleomap Project, University of Texas at Arlington.

its ecological system exist as a kind of a self-regulating organism. We do not fully endorse this view because the connections between life and the Earth are not purposeful, nor is there precise "self" regulation occurring. However, aspects of the Gaia view do have merit in one respect because it is true that the different aspects of the Earth system, including climate and life, affect one another and are inextricably linked: each has affected the other; whether or not one chooses to call such a system Gaia might represent a matter of personal opinion.

One net effect of the evolution of plants and animals has been that the concentration of CO_2 in the atmosphere (and the oceans) is somewhat buffered and indeed has been for hundreds of millions of years. This buffering means that the climate system should be more stable when plants and animals coexist, while the climate would have been less stable when they had not yet evolved: notably, plants and animals did not coexist during the Proterozoic when there were massive swings in climate. Clearly, there were differences in ocean circulation patterns between the time when Rodinia existed and the time when Pangaea existed, and these would have attendant effects on climate; however, likely an important reason why the climate was so different during these two time periods seems to be fact that when Rodinia was assembled no large animals and plants existed; the biologically imposed buffers of the climate system had not yet evolved. By the time Pangaea formed, animals and plants were part of the thin, living veneer of our planet's surface, and their joint presence served to control climate and prevent major climatic swings from occurring. In short, life impacts the atmosphere, and led to the profound differences between Proterozoic and Phanerozoic climate: biology may partly explain why there was no Snowball Earth during the late Paleozoic and early Mesozoic.

Earlier we described how the existence of the albedo effect implies that the climate system should be particularly susceptible to perturbation. By contrast, the interrelationships between plants and animals and their joint effects on the atmosphere imply that the climate system should be partly buffered and resistant to change. Of course, there are also complex feedbacks between the albedo effect and the balancing influence of plants and animals that we have not considered, but we can think of plants and animals as having effects that serve to abate dramatic temperature increases or decreases.

How do These Aspects of the Climate System Relate to the Effects Humans are Having Today?

With our discussions of the albedo effect we have emphasized that there may be elements of intrinsic instability in the global climate system. However, we have also emphasized that given the relationship between plants and animals there is also a tendency for the climate system to be buffered. Each of these is worth considering in light of the effects humans are having on climate today. One of the major effects humans are having on the climate system is that we are increasing the concentration of CO_2 in the atmosphere through various activities, including the burning of fossil fuels: this has increased the greenhouse effect and led to rising temperatures (the state of current temperatures and their future is a topic we will take up in detail in the last chapter of the book).

Given the feedbacks we've already described, one might predict that all things being equal, the rising CO_2 levels should engender increasing plant activity which will partly act to counterbalance this; thus, we have less to fear from the effects of our own species on Earth's climate. Unfortunately, these hopes are dashed because of another major negative impact our species is having on the planet: everyday thousands, maybe even tens of thousands of acres of land are cleared due to suburbanization and deforestation. Humans are clearing tremendous quantities of plants from the face of the Earth and thereby eliminating what would have been an important repository for the growing levels of atmospheric CO_2 we are causing. We are thus removing some of the important forces that serve to counterbalance temperature change and buffer the climate system. This will make the albedo effect, which destabilizes climate and amplifies temperature change in any one direction, become more important. The net implications are that human-induced global warming will be exacerbated and it will become even warmer than it might have been: we may be returning to a situation as in the Proterozoic when greater, more profound, and thus more destructive swings in climate can occur. Thus, it is critically important that we not only reverse increasing CO_2 concentrations in the atmosphere, but that we also halt the dramatic deforestation and suburbanization we are causing: indeed, ameliorating the latter will help keep in check the negative effects that burning fossil fuels has.

Methane, the Other Greenhouse Gas, or Did Dino Farts Warm up the Cretaceous World?

CO_2 is not the only greenhouse gas; another very important one is methane. Its concentration in the atmosphere is nowhere near as high as that of CO_2, but pound for pound it is even more effective at producing a greenhouse effect than CO_2. Some have even argued that because of this we should not bother regulating CO_2 concentrations in the atmosphere and instead focus on methane. We disagree and believe we should work to mitigate anything that can have a potentially serious negative effect on climate. CO_2 is an important greenhouse gas and therefore we should be very serious about preventing the increase of its concentration in the atmosphere; by the same token, given that methane can have such major effects, efforts to control its concentration should also be put in place.

Just as CO_2 concentrations are influenced in an important manner by biological activity in general, and human activities in particular, methane is also produced by biological activity (though inorganic processes produce it as well). Here is another interesting example of how life effects climate. Today, an important atmospheric source of methane happens to be cow flatulence or farts (Figure 11-2).

The reason cows produce so much methane is that their main food source, grass, is difficult to digest, and they have large, multichambered stomachs to accomplish this: a byproduct of the difficulty of digesting grass, and the long time required for digestion, is excessive and malodorous flatulence. The arrival of large numbers of domestic cattle on the scene is a relatively new phenomenon, only true of roughly the last few thousand years; the population of cattle has grown tremendous recently, to feed the dramatically expanding human population. The massive growth of the cattle population has also contributed in an important way to global warming, although not as much as our increasing use of fossil fuels has. In any event, the role of cow flatulence in producing enhanced concentrations of atmospheric methane and subsequent warming again shows the important way that living things can modify climate.

Figure 11.2 Cow flatulence and global warming. Diagram from iStockPhoto.com © B. Foutch.

An interesting question has become whether changes in methane concentrations played any role in causing past episodes of climate change. There is some indication that an episode of major global warming roughly 55 million years ago at the so-called Paleocene–Eocene boundary was caused by an increase in atmospheric methane; this increase in atmospheric methane in turn may have been the result of methane rich "ice" on the seafloor (such patches of methane on the seafloor also exist today) becoming mobilized into ocean water and eventually the atmosphere. Notably this time of major global warming also corresponds to a significant extinction event both in mammals and other types of animals.

Methane also may have played a role in the extremely warm conditions our planet experienced during the Cretaceous, although this link is more tenuous. The Cretaceous happens to have been one of the warmest periods of the Phanerozoic, a fact that will also be discussed later in this chapter; it also notably was the time when dinosaurs were most diverse and abundant. Some have speculated that the

Figure 11.3 Why was it so warm in the Cretaceous?. Could dinosaur flatulence be to blame? Based on diagrams modified from iStockPhoto.com and © B. Foutch, D. Murray, and A. Tooley.

reason it was so warm in the Cretaceous was the great number of plant eating dinosaurs farting up a storm (Figure 11-3). The argument is based on the supposition that large plant-eating dinosaurs may have employed a digestive system much like modern cattle. As a byproduct, these plant-eating dinosaurs would have produced an extensive amount of methane, arising as a consequence of dinosaur flatulence.

   Notably temperatures also decline after the dinosaurs die off at the end of the Cretaceous. The truth is, the idea has little scientific support thus far, and it is difficult or impossible to test: we shudder to think how a scientist searches for (or finds) evidence of 75 million year old farts. Still, it serves as a humorous little anecdote to our exposition of a general theme in the history of life: life has had a major impact on the composition of the Earth's atmosphere and the state of Earth's climate.

Life Influencing Geology: the Form and Shape of Rivers and the Rocks they Leave Behind

In the previous chapter we focused on the important role that geological changes play in influencing biology through their effects on evolution. There is also

Figure 11.4 A modern desert. What terrestrial environments looked like before plants evolved. Image from iStockPhoto.com © J. Rihak.

substantial evidence that life influences geology. One excellent set of evidence comes from terrestrial habitats and the rocks that preserve ancient terrestrial habitats. Before the start of the Silurian period, roughly 440 million years ago, there were no large animals or plants on land. There were certainly bacteria, and simple algae, but little else; a traveler to Earth from outer space would have been confronted with barren, rocky vistas: no plants, no grass, no trees, not to mention no insects and no vertebrates. Maybe the closest situation in the modern world is certain extremely barren deserts (Figure 11-4).

Geological Effects of the Origins of Plants

The Silurian was the time when the first land plants demonstrably evolved (there is sketchy evidence for an origin back in the late Ordovician). Although they started out small, less than a yard high, these plants represented a fundamental breakthrough. Plants obtain energy from the Sun using photosynthesis; before the Silurian, there were certainly single- and multicelled photosynthetic organisms (algae) in the oceans and very likely single-celled photosynthetic organisms on land. But at the start of the Silurian there was a significant transition as multicelled, and large, photosynthetic organisms began to colonize the land in great numbers. Today plants sit at the base of the terrestrial foodchain and essentially all other terrestrial animals depend on plants, or they depend on animals that depend on plants; when

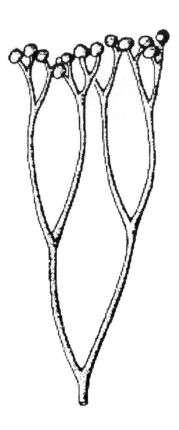

Figure 11.5 An early land plant.
A specimen from the Silurian of
Wales, diagram from R. Cowen
(2005) *History of Life*, 4th Edn,
Blackwell Publishing; after Edwards.

plants evolved it opened up a whole new set of evolutionary opportunities for other types of organisms.

The earliest land plants belonged to groups now totally extinct (Figure 11-5). They lacked roots and leaves and basically consisted of a stem and nothing more. Eventually leaves and roots evolved and by the end of the Devonian there were large trees more than 100 feet tall that made up lush forests (Figure 11-6). These largest Devonian trees are extinct relatives of ferns and mosses. It might be hard to believe, but extinct relatives of modern mosses once grew to huge sizes. The relatively quick and pervasive spread of plants to the far corners of the world initiated dramatic changes. One of these changes involved the influence on the climate system already described: without a concomitant expansion in the variety of animals that eat plants, CO_2 levels in the atmosphere and global temperatures would fall. A second set of profound changes involved geological effects, with a major change in the type of sediments, and thus rocks, that could form at the Earth's surface.

The effects of vegetation on sediments and rocks can be studied in the modern world. Probably all of us have seen, either through pictures or in person, rampant deforestation: clearcuts of formerly forested lands in places like British Columbia, Canada, Washington State, near Seattle, or in the Amazon rain forest of Brazil (Figure 11-7). Plants are significant for geology because they bind soils and prevent erosion. When trees and plants are eliminated, soils are quickly eroded away by either wind or rain; the eroding soils are sediments. In the case of deforestation in British Columbia and the Amazon, this sediment clogs up the rivers before it is carried far out to sea. An excellent example of a related phenomenon is the Dust

Figure 11.6 **The first tree.** A specimen from the Devonian. Diagram from R. Cowen (2005) *History of Life*, 4th Edn, Blackwell Publishing; after Beck.

Figure 11.7 **A clearcut in a forest.** From British Columbia, Canada. Image from iStockPhoto.com © F. Leung.

Figure 11.8 The dust bowl. Kansas in the 1930s. Image by F. Conrad from KansasMemory.org, Kansas State Historical Society, Copy and Reuse Restrictions Apply.

Bowl of the 1930s (Figure 11-8) in the central United States, which was caused by over farming. During the Dust Bowl, soils were carried great distances by the wind.

Before the Silurian, continental landscapes would have resembled a situation similar to the Midwest during the Dust Bowl or a forest after a clearcut. There were no plants that bound the soil; continents were barren and there would have been much more erosion of the continents with great volumes of sediments carried from the land to the ocean by wind and rain. As a brief digression, this would have had biological effects as well, because there are certain types of marine animals that depend on clear, sediment-free waters to live; an example are the corals. Before the Silurian, corals would have found many nearshore, shallow-water environments inhospitable, because in the absence of vegetation large volumes of sediment would have eroded from the continents and clouded marine environments.

The prominent geological effects of vegetation include the shapes of rivers and the types of sediments and rocks left behind by these rivers. We can find analogues for what pre-Silurian terrestrial environments would have looked like by visiting modern environments that lack significant vegetation: for example, deserts in places like Death Valley, California. Rivers are ephemeral here, but they do form after major thunderstorms, and they have a geometry resembling braided hair, which has led geologists to call them braided rivers. Such braided rivers are also found at high altitudes or in the very cold tundra (Figure 11-9). What causes them to have such a braided appearance is that, in the absence of any vegetation to stabilize their banks, their sides quickly erode and the river follows the straightest course down the mountain or through the desert.

Figure 11.9 A braided river. From the Northwest Territories, Canada. Image by B. S. Lieberman.

Rivers of very different appearance are found in vegetated regions, especially at lower elevations. Here they follow a winding or meandering course and are thus called meandering rivers (Figure 11-10). Excellent examples of these include the Missouri River and the Mississippi River in the United States, the Thames River in England, and parts of the Amazon River in Brazil. Plants stabilize the banks of these rivers and prevent erosion.

Just as ancient marine environments are preserved in the geological record, so too are ancient river systems. The rocks that geologists find can tell them what kind of river system produced them. In particular, because the braided rivers are made up of a complex network of channels that run together, the sediments and thus rocks that these form are typically made up of large sized grains of sand or even larger sized particles that make up rocks called conglomerates (Figure 2-1). By contrast, in meandering rivers there is an oscillation between relatively coarse sediments and fine muds (Figure 2-1), the latter produced when the river periodically floods its banks.

When geologists find rocks of pre-Silurian river systems, they are always of the type associated with a braided river system. By contrast, those from the Silurian and subsequent time periods can be associated with rocks typical of either braided or meandering rivers. When the former are found, it indicates deposition at high altitudes, in a desert, or in a very cold environment, with high rates of erosion. Rocks associated with meandering rivers are not present in the rock record before land plants evolved in the Silurian. Only braided rivers existed before the Silurian, and only the characteristic deposits they left behind can be found in pre-Silurian rocks.

This not only provides a clear example of how life affects geology; it might even help guide us as we search for evidence of past life on other planets, including Mars. In particular, geologists, or the robots they design, will now have a clear litmus test as they search for evidence of past life on other planets. For instance, finding sedimentary rocks on other planets that had been produced by meandering river

Figure 11.10 A meandering river. From Indonesia. Image from iStockPhoto.com © M. Terao.

systems deposits would provide a clear indication that something like vegetation once existed on that planet, even if today the planet's surface was barren. If only braided river deposits were found, by contrast, it would indicate that evidence for Earth-like vegetation on a planet was absent.

Plants, Oxygen, and Coal: More Examples of Life Affecting the Atmosphere and Geology

We will consider the rise of oxygen, its causes, and its consequences, more fully in the next chapter. However, given the discussion of plants here, it is worthwhile to briefly consider how terrestrial plants have modified the composition of oxygen in

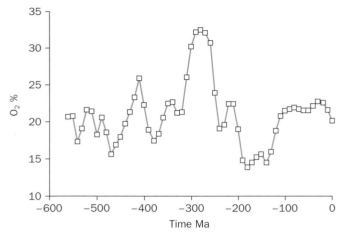

Figure 11.11 Phanerozoic O_2 levels. Diagram from R. Berner, Yale University.

the atmosphere over the last roughly 450 million years. Early in the Phanerozoic, before the origin and proliferation of land plants, oxygen levels are estimated (it is hard to know how accurate these estimates are and since it is not possible to calculate oxygen levels directly modeling techniques must be used) to have been about 50 percent of the present atmospheric level (today oxygen comprises about 20 per cent of our atmosphere). Thus, during the Ordovician the total make-up of the atmosphere was about 10 percent oxygen, implying that humans would have found movement on land laborious without bottled oxygen to assist them. Oxygen levels stayed relatively constant until the Carboniferous period, when they rose dramatically (Figure 11-11). Atmospheric oxygen concentrations during this time period may have reached 40 percent, twice as high as today's levels. Such values are rather provocative because when atmospheric levels of O_2 reach 40 percent there is a strong probability of spontaneous combustion: oxygen is a very volatile compound and can easily explode. The implications are that during the Carboniferous static electricity or an open flame could have had potentially disastrous consequences: can anyone say "firestorm?" It just so happens that the times of very high oxygen levels correspond to the widespread development of huge tropical forests on our planet; oxygen concentrations were high in the Carboniferous because there were so many organisms actively photosynthesizing.

The Time of Coal

The world's most prolific coal deposits also come from the Carboniferous: as we mentioned in Chapter 2 that name in fact can be translated from the Latin as "coal" or "carbon rich." Coal is a rock produced when dead plant matter is buried and then compressed (Figure 2-4); before plants evolved, there basically were no coals. The Carboniferous was a time when plants were tremendously productive and they proliferated in giant swamps (Figure 11-12); indeed, this level of plant productivity may never have been equaled again.

Volumetrically coals comprise an important part of the entire sedimentary rock record: a part generated by biological activity. Other sedimentary rocks, including limestones (Figure 2-2), may be composed almost entirely of biological material, or their formation is initiated by chemical reactions caused by organisms. Each

Figure 11.12 A Carboniferous coal swamp. Image courtesy of Library Special Collections, American Museum of Natural History.

of these represents clear instances when biology has fundamentally influenced geology. Today, the United States (and many other countries) uses coal to generate most of its electricity; coal contains carbon that had been removed or sequestered from the Earth's atmosphere, sometimes hundreds of millions of years ago. By igniting coal we generate energy but we also as a byproduct return that long sequestered carbon back to the atmosphere in the form of CO_2 (and also soot). This is how power generation through coal acts directly to cause global warming.

How Geology Affects Climate: Considering How Plate Tectonic Changes have Contributed to Climate Changes Over the Last 60 Million Years

Earlier in this chapter we introduced the amusing, albeit largely untestable idea, that dinosaur farts may have contributed to warm conditions in the Cretaceous world; as was already described, the modern world is much colder than the Cretaceous. Perhaps a part of that temperature differential was the absence of dinosaur flatulence in the Cenozoic world, but the major part of the explanation lies in large-scale changes in the positions of the Earth's continents following the Cretaceous, initiated by plate tectonics. Thus, just as we have already considered the connections between biology and climate, and biology and geology, we round out this chapter with some concrete examples of how geology influences climate.

The paleontologist and geologist Al Fischer used the terms "greenhouse" and "icehouse" to describe time periods of relatively warm and relatively cold climate. Cretaceous climate epitomized greenhouse conditions (towards which we appear to be moving today, with potentially disastrous results): there were no polar ice caps and few if any glaciers even at high elevations. Because there was little water trapped as sea ice, sea-level was much higher and parts of the continents that are today exposed were covered by water (Figure 6-7 and Plate 10-1).

Over time, from the Cretaceous to the Cenozoic, the Earth shifted from this greenhouse state into what is termed an icehouse setting. Here we outline the series of contingent historical events related to plate tectonics that occurred during this greenhouse to icehouse transition. In actuality, the Earth's setting today could be very much characterized as an icehouse condition: on average it is much colder now than it was throughout most of the Phanerozoic. Polar ice caps are present today, as are alpine glaciers; with all that water trapped as ice, sea-level is much lower than it was in the Cretaceous. If nothing else, this means that there is much more land available for terrestrial animals, and plants, and us, to live (and also in our case to grow crops), although back in the Cretaceous exposed high-latitude land was much more hospitable to terrestrial life than such land is today. Part of the concern about human-induced global warming is that as temperatures rise and ice melts there will be less room available for humans to live (and farm): as we described in Chapter 6, this is the type of situation that increases the risk of extinction for any species.

The Formation of a Current Around Antarctica

During the Cretaceous and early Cenozoic there were no land masses directly at the poles (see Figure 6-7 and Plate 10-1). This has an effect on climate because a land mass cools (and heats) more rapidly than ocean water. Consider the fact that temperature on land is usually more equable in cities near ocean water than in cities in the middle of continents: throughout the year temperatures vary much more in Kansas than they do in Seattle or Washington, DC. Even in the Cenozoic, before there was a separated land mass directly over the South Pole, the world was a much warmer place, with crocodiles, turtles, and tropical plants living as far north as the Arctic circle. However, over the course of several millions of years, Antarctica became separated from South America and it settled over the South Pole. Further, probably by the Eocene, an ocean current became established around this isolated Antarctica (Figure 11-13). This initiated the Antarctic deep freeze. The landmass sitting at the pole, in the dark for many months out of every given year, in conjunction with a current circulating around a continent which was quickly becoming a large ice cube, encouraged cooling not just within Antarctica but globally. The once diverse biota that Antarctica supported, including marsupials and many plant species, died off (Figure 11-14).

Today, Antarctica is almost a completely dead continent, at least in terms of the animals and plants that live on top of the ice: not surprising considering that in most places it is covered with a sheet of ice that is up to three miles thick. The growth of the prodigious volume of ice over Antarctica not only acted like a large ice cube to cool ocean temperatures, but it also served to increase the albedo effect, reflecting more energy back into space (Figure 11-1), leading to further global cooling.

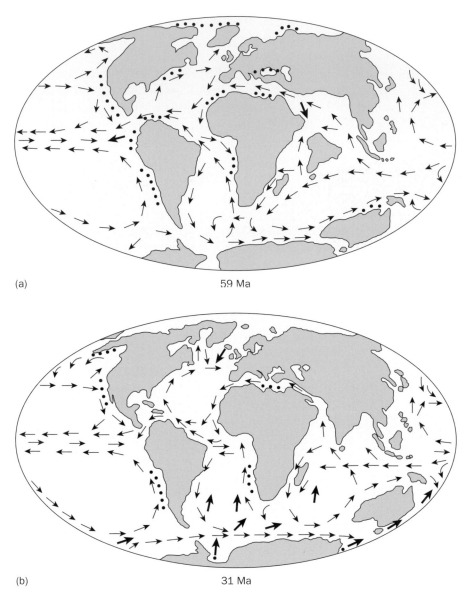

(a)   59 Ma

(b)   31 Ma

Figure 11.13 Ocean circulation and the establishment of the Antarctic current. Currents before (a) and after (b) Antarctica became an isolated continent sitting at the pole. Diagram from C. B. Cox and P. D. Moore (2005) *Biogeography: An Ecological and Evolutionary Approach*, 7th Edn, Blackwell Publishing.

The Messinian Salinity Crisis

The next geological event that led the Earth further on the path towards an icehouse state involved tectonic movements between Europe and Africa. During the Cenozoic, and even today, the African plate has been moving slowly but inexorably northwards, setting it on a course of active collision with Europe (see Plate 10.1). These plate movements have led to the formation of the Alpine mountain range in France, Switzerland, and Italy, and also the active volcanism that occurs in parts of Italy, including the famous Mount Vesuvius.

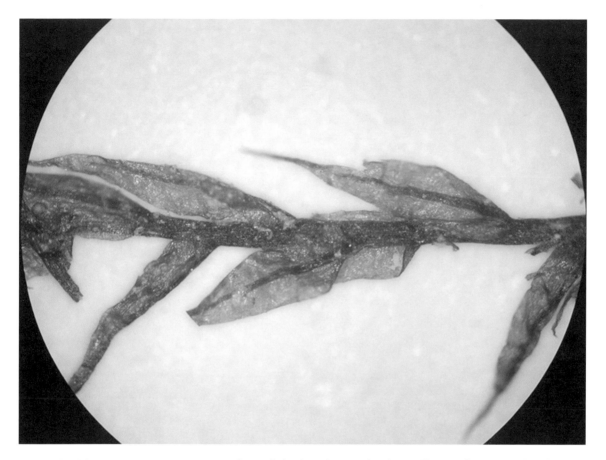

Figure 11.14 Miocene Antarctic moss. One of the last Antarctic plants. Image from D. Marchant, Boston University.

The northward movement of Africa ultimately meant that there was only a narrow opening that joined the modern Mediterranean with the Atlantic Ocean: the pillars of Heracles to the ancient Greeks, and the straits of Gibraltar to modern geographers. Only a subtle shift in the relative position of Europe and Africa would close the gap between these two continents and close the western portal between the Mediterranean and the Atlantic; when this happens, a giant pond will form. The Mediterranean first became an isolated pond sometime late in the Miocene Epoch, between about 11 and 5 million years ago; when that happened the Mediterranean Sea dried up very quickly, in perhaps as little as 1,000 years. As the waters of the Mediterranean dried up, they left behind a substantial volume of salt, much like what is left behind when a smaller body of water dries up, for example, the Great Salt Lake in Utah. Just as a subtle movement between the two continents would be enough to close the Straits of Gibraltar (Figure 11-15), a similarly subtle movement would be enough to open the Straits. And when that happens, the waterfall created would make Niagara Falls look puny by comparison (Figure 11-16). Geologists have examined the volume of salt that lies beneath the modern Mediterranean; its thickness is significant and indicates that the sea must have dried out, only to later fill again, as many as 20 times. This event is called the Messinian Salinity crisis, in

Figure 11.15 Straits of Gibraltar from space. Image from NASA.

Figure 11.16 Niagara falls. Ontario, Canada. Image from iStockPhoto.com © F. Leung.

reference to the Mediterranean (or Messinian) Sea, and the large amount of salt removed from the oceans and deposited on the sea floor.

The repeated drying and refilling of the Mediterranean would act to further exacerbate global cooling by removing significant volumes of salt from the Earth's oceans. The less salt ocean water contains, the easier it freezes; this is why a cup of salt water takes longer to freeze than a cup of fresh water. During the Messinian Salinity crisis a large amount of salt was removed from the Earth's oceans and lower salt concentrations in the Earth's oceans would make it easier to form pack ice. This in turn would amplify the amount of incident radiation from the Sun reflected back to space due to the albedo effect and further accelerate global cooling. It appears that global temperatures took quite a plunge sometime around 5 million years ago, during or shortly after the Salinity crisis.

The Collision Between North and South America

The final geological event that plunged our planet into the icehouse condition it is in today was the joining of North and South America: this event was described in the previous chapter as part of the Great American Interchange and occurred around 3.5 million years ago when the Isthmus of Panama docked with northern South America. The primary way the closure of the Isthmus of Panama impacted climate was by perturbing ocean circulation. Before these two continents collided, there was a warm tropical current joining the Atlantic and Pacific Oceans (Figure 11-17). This current kept the oceans relatively warm. When the Isthmus closed, the moderating effects of these tropical waters were eliminated (Figure 6-11). Eventually this triggered the formation of ice in the Atlantic and Arctic Oceans, greatly expanding ice volume and increasing the planet's albedo.

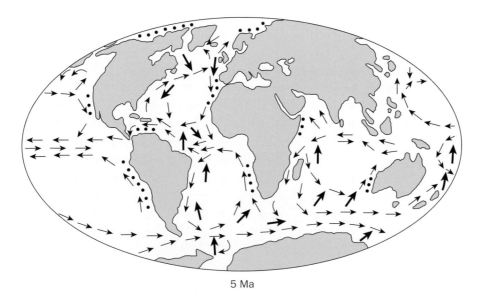

5 Ma

Figure 11.17 Ocean circulation just before the collision of North and South America. Diagram from C. B. Cox and P. D. Moore (2005) *Biogeography: An Ecological and Evolutionary Approach*, 7th Edn, Blackwell Publishing.

Milankovitch Cycles

Although the Earth is in an icehouse condition today, there are still times when it is relatively cooler and other times when it is relatively warmer. In fact, in the last 3.5 million years about 20 times the Earth's climate has oscillated back and forth between Ice Ages and times of relative warmth, called interglacials or glacial minima. During Ice Ages, the oceans would have been blocked with pack ice as far south as New York City; inland, ice sheets at times reached points as far South as Kansas (Figure 11-18). During times of glacial minima, hippopotamuses would have frolicked in the Thames River, near the site of present day London.

The reason that climate has oscillated so much over the last 3.5 million years can be attributed to astronomical cycles called Milankovitch cycles, named in honor of the person who first discovered them. Milankovitch cycles describe the fact that

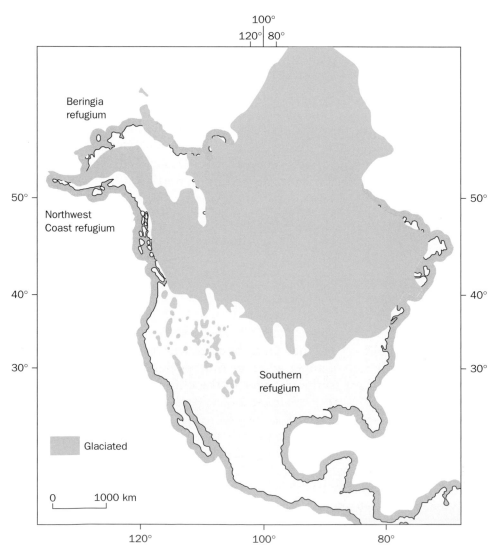

Figure 11.18 Maximal extent of ice sheets in North America. During the last glacial advance, roughly 15,000 years ago. Diagram from C. B. Cox and P. D. Moore (2005) *Biogeography: An Ecological and Evolutionary Approach*, 7th Edn, Blackwell Publishing.

over intervals of tens of thousands of years, the distance between the Earth and the Sun, the angle of the Earth relative to the Sun, and the way the Earth's axis wobbles, all vary. Subtle changes in these parameters determine how much heat energy reaches the Earth from the Sun. Climate is relatively colder when the northern hemisphere is more angled away from the Sun and also when the Earth is further from the Sun. It turns out that these cycles have been operating for literally billions of years, but at other times in Earth history, especially when the Earth was in a greenhouse state, they caused more subtle shifts in climate. Now that the Earth is in an icehouse state, the differences in the energy reaching the Earth are sufficient to move the Earth into and out of an Ice Age. Therefore, given the current position of the continents, there can be some predictability about the relative future appearance of Ice Ages and times of glacial minima: we happen to be about 10,000 years out from the next Ice Age, assuming humans have not sufficiently perturbed climate by then to have changed the very nature of the current climate system itself. Today scientists can predict at the scale of thousands and tens of thousands of years when Ice Ages are coming, thanks to our understanding of Milankovitch cycles. However, to a greater extent, the most important climatic changes over the last 60 million years occurred because of changes caused by plate tectonics. These were historical or contingent events that happened over tens of millions of years and made the climate system what it is today.

Concluding Remarks

One of the fundamental lessons gained from the study of geology and paleontology is that the Earth, Life, and Climate systems are all fundamentally interconnected. Organisms have a long tradition of modifying the landscape and climate of our planet; the same is true of humans today, but on a vastly greater scale. Of particular concern is that we are not only modifying climate but also removing some of the bulwarks that happen to keep larger climate swings in check. Not only do we know that climate changes have happened in the past, but they also tend to happen swiftly, at least in geological terms; the swifter and greater the changes, the greater the ensuing biological effects. If change is too rapid and profound, organisms may no longer be able to migrate to their preferred environment, particularly if their preferred environment no longer exists. If they do not adapt to changing conditions, and when changes are fast and extreme that becomes less likely, the only alternative will be extinction. We should expect that many species will not be able to weather the storm of impending climate change: whether our species will is an open question.

Fortunately, the solutions to the changes in the climate system we are causing are far from intractable and are in fact readily at hand. All we need do is no longer continue to swell atmospheric concentrations of CO_2 and methane, and discontinue rampant deforestation and suburbanization. Perhaps easier said than done, you may say, but given knowledge of the solution the future actually should not look all that grim. Moreover, given the gravity of the potential problem, it is also encouraging that each individual can make personal and lifestyle decisions that directly contribute to ameliorating this problem.

Additional Reading

Agusti, J., Oms, O., and Meulenkamp, J. E. (eds). 2006. Late Miocene to Early Pliocene Environment and Climate Change in the Mediterranean Area. *Palaeogeography, Palaeoclimatology, Palaeoecology* **238**: Special issue.

Algeo, T. J., Berner, R. A., Maynard, J. P., and Scheckler, S. E. 1995. Late Devonian oceanic anoxic events and biotic crises: "Rooted" in the evolution of vascular land plants? *GSA Today* 5(45): 64–66.

Beerling, D. 2007. *The Emerald Planet: How Plants Changed Earth's History*. Oxford University Press, Oxford; 304 pp.

Burroughs, W. J. 2005. *Climate Change in Prehistory: The End of the Reign of Chaos*. Cambridge University Press, Cambridge; 368 pp.

Frakes, L. A., Francis, J. E., and Syktus, J. I. 1992. *Climate Modes of the Phanerozoic*. Cambridge University Press, Cambridge; 286 pp.

Gore, A. 2006. *An Inconvenient Truth*. Rodale Books, Emmaus, PA; 336 pp.

Hsu, K. 1987. *The Mediterranean was a Desert: A Voyage of the Glomar Challenger*. Princeton University Press, Princeton, NJ; 197 pp.

Knoll, A. H., Bambach, R. K., Canfield, D. E., and Grotzinger. J. P. 1996. Comparative Earth history and late Permian mass extinction. *Science* **273**: 452–457.

Kolbert, E. 2007. *Field Notes from a Catastrophe*. Bloomsbury Publishing. London; 240 pp.

Lovelock, J. 2000. *Gaia: A New Look at Life on Earth*. Oxford University Press, Oxford; 168 pp.

Redfern, R. 2001. *Origins: The Evolution of Continents, Oceans and Life*. University of Oklahoma Press, Norman, OK; 360 pp.

Walker, G. 2004. *Snowball Earth*. Bloomsbury Publishing, London; 288 pp.

Willis, K. J., and J. C. McElwain. 2002. *The Evolution of Plants*. Oxford University Press, Oxford. 392 pp.

Chapter 12

Patterns and Processes of Precambrian Evolution

Outline

- Introduction
- The Earliest Evidence for Life in the Geological Record
- The Time of Pond Scum and the Rise of Oxygen
- For Billions of Years Organisms Have Been Modifying the Atmosphere and Their Environment
- More Effects of Rising Oxygen Levels
- The Evolution of the Eukaryotic Cell
- Concluding Remarks
- Additional Reading

Introduction

One of the lessons evident from the study of the history of life is that there are large-scale processes that influence living things and the world around them. Recall how the relationship between surface area and volume, as described in Chapter 8, determined the size and shape of organisms and also the distribution of craters on planets in our solar system. One way that we search for overarching processes in paleontology and evolutionary biology is by studying different groups of animals, plants, or other life forms and searching for commonalities that they share. Really the scientific method in paleontology involves studying the history of different things and then seeing if there are any similarities in the patterns and details of that history. By way of an example, consider the individual events, the history if you will, of any one person's life: you are born; you go to elementary school; you go to college; you get a job; and so on. That is just one person's history of course and it is hard to extract any general principles from that. Now, however, imagine we trace the history of several people. We may find there are general processes or principles that recur in every human individual's life such that we don't have to view any life in isolation. Maybe junior high school, or college, is a time when we all experience similar things, some of them trying. Thus, from single individual histories we can reconstruct processes that repeatedly occur, and we can extract generalities. This is what the scientific approach in paleontology is all about, extracting the generalities from a set of historical patterns. Through the study of several repeated events that occur throughout the history of life we can extract generalities and processes.

This chapter focuses on roughly the first 2 billion years in the history of life: don't panic though, we'll be brief and focus primarily on the major patterns which comprise the pivotal episodes. As such we will cover the first evidence for life on Earth, the proliferation of bacteria and the effects this had on the oceans and atmosphere, and finally we'll conclude with a discussion on the origins of the eukaryotic cell. We'll also discuss the scientific processes and principles these topics elucidate. When thinking of the general scientific approach used by paleontologists you might recognize one complication with events like the origins of life, or the origins of the eukaryotic cell: as far as we know, each of these events has happened only once (barring some strange scenario played out on the television show "*X-files*"). Thus, our task of looking for generalities in these pivotal events in the history of life is not as easy as comparing the lives of many different human individuals.

Thankfully though, the task of looking for generalities is not insurmountable. The reason is the existence of deep time: the fossil record on Earth extends back nearly 4 billion years. When that much time is present it is likely that similar or analogous events recur frequently. For example, it is true that the eukaryotic cell evolved only once, but the mechanisms responsible for this evolution also played a role in other key episodes in the history of life. It is also true that the evolution of life ultimately triggered certain profound changes in the atmosphere and in the oceans. However, organisms throughout the history of life, even those alive today, have influenced the composition of the atmosphere and the oceans; the effects produced at these other times, even today, are different from the earliest effects life had. Still, they are ultimately analogous to the manner in which life first exerted its influence on the atmosphere and the oceans. The scientific method in paleontology is partly possible because of the tremendous stretch of geological time; even if similar events occurred only once every few hundred million years, that's still pretty often.

The Earliest Evidence for Life in the Geological Record

Of the three types of rocks described in Chapter 2, only sedimentary rocks are likely to preserve traces of ancient life. That is because these rocks were not originally generated as super hot magmas (like igneous rocks), nor were they cooked to great temperatures or squeezed under great pressures (like metamorphic rocks); each of these circumstances would conspire to bake away any traces of ancient life. Unfortunately, sedimentary rocks from early in Earth history are relatively rare, for a few reasons. First, the longer a rock sits around (really the older it is), the longer it is exposed to the elements and the greater the chance that it will be eroded away. Think of it like a video game where someone aims at a target; the longer the target is around, the more likely it will be hit. Another process that destroys ancient sedimentary rocks is metamorphism due to heat and pressure that cooks the rock and alters its original composition, including whatever fossils it might have contained. As with erosion, the longer a rock sits around, the more likely it is to become metamorphosed. Finally, older rocks are more likely to be buried by younger rocks and thus become largely inaccessible to paleontologists searching for fossils in them.

Given this, it is amazing that any sedimentary rocks are exposed at the surface that exist from early in Earth's history but thankfully there are some: the record of sedimentary rocks extends back to roughly 3.8 billion years ago. What is even more amazing, however, is that these 3.8 billion year old sedimentary rocks contain evidence of ancient life.

The Oldest Fossils on Earth

Usually, when we think of traces of ancient life, we think of fossils (the various types of fossils are described in Chapters 1 and 2). The earliest traces of life, however, are not akin to fossils as they are traditionally defined; for example, a body fossil like a trilobite (Figure 12-1). Instead, the evidence for ancient life fits the broader definition of fossils given in Chapters 1 and 2: fossils represent any trace of ancient life. This includes not only a fossil trilobite but the crawling trail a trilobite left as it moved along the seafloor hundreds of millions of years ago (Figure 12-2).

Figure 12.1 Trilobites. From the Ordovician of Ohio. Image by J. Counts, specimens in the University of Kansas Museum of Invertebrate Paleontology.

Figure 12.2 Trilobite trackway. Diagram from E. N. K. Clarkson (1998) *Invertebrate Palaeontology and Evolution*, 4th Edn, Blackwell Publishing.

Again though, we do not find the actual trackways or physical traces of ancient life in 3.8 billion year old rocks. Instead, in these rocks scientists have found chemical signatures of ancient life that also qualify as fossils. There are not physical remains of the organisms present in these sedimentary rocks, that is true, nor are there physical imprints left over from the movement of organisms in ancient mud or sand before it became a rock; however, there are certain types of chemical compounds or signatures that are only left by organisms and which otherwise would not occur in ancient environments. If scientists find these in a rock they know an organism was around, even if they do not find the actual organism that produced them. These types of chemical signatures have been found in 3.8 billion year old sedimentary rocks.

One way scientists can use chemical signatures to infer the presence of past life forms is through the study of certain types of carbon atoms. Remember, in Chapter 1 we discussed one type of carbon atom, carbon-14, that can be used to calculate the radiometric dates or ages of certain fossils; there are also carbon-13 and carbon-12 atoms. These atoms differ from one another in the number of neutrons in their nucleus. Animals, plants, and other organisms need carbon to live, and they preferentially use carbon-12 as opposed to carbon-13; the upshot is that as organisms take in carbon, the carbon-12 in their environment gets depleted more quickly than carbon-13 does. If a scientist studies the chemistry of a sedimentary rock and determines a higher proportion of carbon-13 relative to carbon-12 than would typically be predicted, in the absence of organic activity, it means something was living in the area when the rock was deposited. Such a biased proportion of the ratio of carbon-13 to carbon-12 atoms is found in the earliest sedimentary rocks. The oldest rocks that could contain evidence of life do so, indicating that life evolved on our planet at least as far back as 3.8 billion years ago.

The Significance of these Early Chemical Fossils

What is the significance of a chemical fossil in a 3.8 billion year old rock? Recall, as described in Chapter 1, the oldest rocks, which appear to be metamorphic rocks, are 3.9 billion years old. Further, the Earth probably formed around 4.5 billion years ago, based on the ages of meteorites from our solar system, but it was likely fully or partly molten until perhaps as "recently" as 4 billion years ago. Before 4 billion years ago, life as we know it probably could not have existed on Earth. Thus, chemical fossils in 3.8 billion year old rocks indicate that life had evolved soon after much of the Earth's surface was no longer molten. This indicates that life itself may not be difficult to evolve and the traditional view that life is hard to evolve or "unlikely" may well be wrong too. Given the right conditions the origins of life should be expected, meaning that life may exist (or may have existed) on many other planets, a topic that we will take up in greater detail in Chapter 17. For now though we note that it is still a big jump to say that intelligent life has evolved elsewhere, some might even question whether it has evolved on Earth, but we find the prospect of other life in the universe fascinating.

The Oldest Body Fossils: Simple Bacteria

Although the oldest sedimentary rocks have not yet revealed body fossils, the evidence for life on Earth in the form of body fossils is still extremely old: rocks as old as 3.5 billion years have yielded such fossils. These ancient body fossils are simple and microscopic, representing single-celled organisms of various types (Figure 12-3; some may be related to the cyanobacteria and these are sometimes called blue-green algae). Cyanobacteria obtain energy from the Sun via the process of photosynthesis: the same type of photosynthesis that occurs in living plants. As a byproduct of photosynthesis these bacteria, like plants today, produced oxygen, perhaps as early 3.5 billion years ago. Incredibly, nearly the very same types of single-celled bacteria persist today and can be found in freshwater ponds and streams throughout the world (Figure 12-4).

Stromatolites

When we go as far back as 3.5 billion years ago, sometimes these cyanobacteria formed large mats or carpets, in various places covering many square meters of the

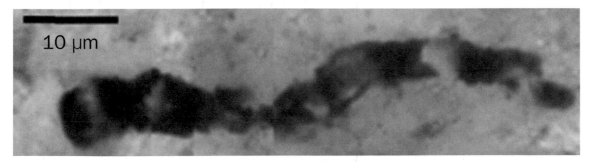

10 μm

Figure 12.3 One of the oldest body fossils. This single celled bacterium is from 3.485 billion year old rocks in northwestern Australia. Image from J. W. Schopf, University of California, Los Angeles.

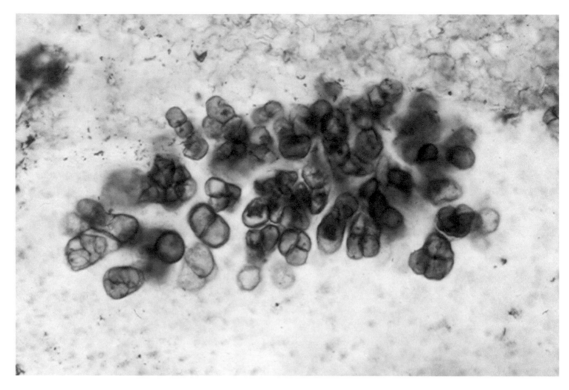

Figure 12.4 Fossil cyanobacteria. These fossilized single celled organisms occur in 750–800 million year old rocks in Greenland and are indistinguishable from modern cyanobacteria. Image from A. Knoll, Harvard University.

Figure 12.5 Stromatolites. These 1.5 billion year old structures from Siberia were built by the activities of cyanobacteria. Image from A. Knoll, Harvard University.

shallow-water seafloor. Over time, these mats would build up in layers separated by a thin veneer of sediment, and could form huge columns called stromatolites. Furthermore, sometimes these columns were grouped together to form large reef-like structures (Figure 12-5). Stromatolites are sedimentary structures, but they were often made through the action of cyanobacteria.

Although cyanobacteria can still be found in abundance today, only rarely can places be found where they agglomerate into large mats and form stromatolites. As in the distant past, such places need to be in water shallow enough for sunlight to penetrate, allowing photosynthesis to occur. However, whereas in the distant past stromatolites were found in all types of shallow-water ocean environments, today (Figure 12-6) they are primarily found in environments that are very salty, in fact, in environments that are too salty for animals like worms, snails and fish to live in. Their restriction to such environments today is simple: if conditions were less salty, large animals could and would live there and in turn would feed on cyanobacterial mats and thus eat away the stromatolites. Thus, the presence of stromatolites in a modern, or fossil, environment, is a good indication that in that particular place and at that particular time there were few if any large animals around: if they were around they would have devoured any stromatolites.

It is reasonable to conclude that when the first stromatolites appeared 3.5 billion years ago there would have been no life forms grazing on these mats at this time and thus no animals around. Stromatolite fossils are abundant throughout the rest of the Archean and the Proterozoic, and only start to disappear from the fossil record in the Phanerozoic, after the first large animals arrived on the scene in abundance.

The Time of Pond Scum and the Rise of Oxygen

It seems almost incomprehensible, but that was it: for the next 1.5 billion years basically the only organisms we see in the fossil record are bacterial forms, especially cyanobacteria, and the only large structures were stromatolites; as the paleontologist J. W. (Bill) Schopf so famously remarked, at least the first 1.5 billion years of the history of life was dominated by pond scum. Given that our own species has been on this planet for a few hundreds of thousands of years this is a fairly sobering thought. However, for all that time, and for all their passiveness, these simple cyanobacteria were doing something that ultimately impacted us in a profound way; they were performing photosynthesis, which has the general chemical equation:

$$CO_2 + 2H_2O \rightarrow \text{energy or sugar} + O_2 + H_2O$$

and thereby pumping oxygen into the atmosphere. Prior to the appearance of these photosynthetic organisms there was little or no oxygen in the Earth's atmosphere, but after the cyanobacteria and other photosynthetic forms evolved, oxygen levels in the atmosphere slowly began to rise. This was a crucial event in the history of life because we after all need oxygen to live and breathe, as do most animals. We will not ask for a moment of silence for these pond scum, but we can all appreciate what their activities during this monumental stretch of time meant for us. (The oxygen we breathe today is primarily provided by plants, but if not for the initial increase in oxygen it is doubtful that the early animals that ultimately gave rise to us could have ever evolved in the first place.)

Until about 2.3 billion years ago oxygen levels were only about 1–2 percent of what they are in the present day. One way we can infer this is from the distribution

Figure 12.6 **Modern stromatolites**. From Shark's Bay, Western Australia. Image from iStockPhoto.com © A. Hafemann.

of minerals in the rock record. There are certain minerals that when exposed to oxygen dissolve rapidly. Two such minerals are uraninite, which is a mineral that contains uranium, and pyrite (Figure 12-7), which is made up of iron and sulfur. It is often called fool's gold, because although golden in color you'd have to be a fool to think it actually was gold.

You do not see these minerals forming in modern beach or river environments today; they only form in places were oxygen concentrations are less than 1–2 percent of their present levels. However, before about 2.3 billion years ago, these minerals were abundant in all types of rocks including those that were deposited in your typical shallow ocean environment: akin to what you would find

Figure 12.7 Pyrite. Fool's gold. Image from iStockPhoto.com © D. Wilkie.

today near a nice stretch of beach. Since 2.3 billion years ago, sediments from typical shallow-water environments containing pyrite and uraninite are rare; thus, by this time oxygen concentrations had risen to a threshold level, thanks to the activities of simple cyanobacteria.

Banded Iron Formations

Other types of deposits in the fossil record also reflect a rise in oxygen. The most spectacular are the banded iron formations (Figure 12-8). These types of rocks typically occur in ancient marine deposits dating from between 3.5 billion and 1.9 billion years ago, and they reached their zenith around 2.0 billion years ago. Banded iron formations are composed of thin beds that are iron rich, alternating with iron-poor beds. Each individual bed may only be a few centimeters thick, but they appear together in prodigious numbers of couplets that sometimes reach thicknesses of many kilometers. Basically most of the economically valuable sources of iron on Earth are developed as banded iron formations. There are especially large developments of banded iron formations near the surface in and around Michigan, and this explains at least the former concentrations of the auto and steel industries in that region; again, as we shall see, part of the former industrial might of the United States was derived from the activities of simple bacteria that lived billions of years ago (is another moment of silence for tiny, long dead bacteria in order?).

Figure 12.8 Banded Iron Formation. Showing alternate light and dark bands. Image from D. Lowe, Stanford University.

How such banded iron formations were created has always been a bit of a puzzle because in modern oceans there is very little chemically free iron: for obvious reasons iron in the oceans is a necessary ingredient if banded iron formations are to form. The fact that banded iron formations were deposited suggests that billions of years ago there was literally thousands of tons of free iron in the oceans. One well supported theory suggests that most banded iron formations formed as a result of Archean and early Proterozoic photosynthetic bacteria; these ocean-dwelling bacteria produced oxygen that trapped the iron that was present in seawater and it was precipitated out. Then, for some reason bacterial activity decreased and low oxygen conditions prevailed; then the iron-poor bands of the banded iron formations were produced. This sequence would have happened literally hundreds of thousands of times.

It is straightforward to understand how photosynthetic bacteria would add oxygen to the water which would cause iron to precipitate out of the water. It is perhaps on the face of it harder to understand what may have been responsible for the conditions that caused the iron poor bands to accumulate. The solution has not been fully worked out yet, but could lie in the nature of oxygen itself, which is actually a highly toxic substance. This may seem paradoxical in the case of humans for don't we after all need oxygen to live? Even though we depend on oxygen to breathe, many of the negative effects that aging has on the human body are related to the effects of free oxygen atoms in our body (whereas the oxygen we breathe is made up of two joined oxygen atoms). Intuitively many of us know this, as this is why certain commercial products, for example, makeups, tout their role as antioxidants that are supposed to "reverse" the aging process. Various diseases like cancers may also be partly triggered by the destructive movement of free oxygen atoms through our cells. This is why it is recommended that we consume vitamins and other foods rich in antioxidants to reduce our risk of certain types of cancers.

Not only do oxygen atoms play a role in some human diseases, and earn makeup companies a lot of money, they also are toxic for many forms of life including certain types of bacteria. The toxicity of oxygen to many bacterial forms may be tied up with the creation of both iron-rich and iron-poor layers in banded iron formations. It is possible that the photosynthetic bacteria may have poisoned themselves or their cousins by the very oxygen they were making as a byproduct of the way they got their energy. In particular, as bacterial numbers climbed, oxygen levels nearby in the water would have also increased. During this interval the iron-rich layers would have been deposited on the ocean floor. Ultimately oxygen levels could have reached a threshold where they were not only high enough to bond with free iron in the surrounding water, but also sufficient to be toxic for the bacterial colonies that were producing them. As these toxic effects ramified, the photosynthetic bacteria would be largely wiped out. Eventually, with fewer photosynthetic bacteria around, oxygen levels would drop. During this interval little free-iron would be removed from the oceans and the iron-poor layers of the banded iron formations would be deposited on the ocean floor. Once oxygen levels dropped to low enough levels, in the absence of bacterial photosynthesis, conditions would ameliorate; eventually the numbers of photosynthetic bacteria could increase again; as these bacteria multiplied oxygen levels would increase, and once more iron would be precipitated out of seawater. This cycle would have repeated literally thousands of times, leading to tremendous buildups of thin iron-rich layers alternating with thin iron-poor layers.

For Billions of Years Organisms Have Been Modifying the Atmosphere and Their Environment

As mentioned above, after about 2 billion years ago, banded iron formations become far less abundant. The explanation may be that the photosynthetic bacteria, through all the oxygen that they generated, ultimately siphoned all of the free iron out of the oceans and into the geological record. This too actually would be a pivotal point in the history of life. Before this time, any of the oxygen generated by photosynthetic bacteria would not remain for long in the oceans; instead it would have been mostly bound up with free iron and deposited as banded iron formations at the bottom of the oceans: little oxygen could reach the atmosphere. Therefore, until abundant free iron (and also perhaps compounds containing sulfur) was removed from the oceans, oxygen concentrations in the atmosphere would stay relatively low. However, once the extensive free iron had been removed from the oceans, oxygen would no longer be quickly removed from the oceans and instead could start building up both in the oceans and the atmosphere.

It is clear that during this time major changes were occurring both in the chemistry of the oceans and also the atmosphere, and ultimately these changes were all triggered by the activities of organisms. Scientists are still working out the details of what precisely happened with oxygen levels and when it happened. However, in broad strokes, these events serve as an important parable for us today. Recognize that very simple organisms, such as bacteria, have profoundly modified the environment around them. Indeed, some of these modifications, like increasing oxygen levels, were to their profound detriment. Further, these detrimental effects were simply a byproduct of the activities they needed to perform in order to live: photosynthesis. Bacteria are not only relatively simple but of course they lack technology and advanced civilizations, yet they ultimately poisoned or polluted themselves. If simple organisms like these can modify their environment, thereby poisoning themselves, realize what complex organisms like humans can do with our advanced technology. This puts the debate about human-induced global warming, and whether we can negatively modify our own environment, into a new light.

More Effects of Rising Oxygen Levels

The history of life suggests that organisms have sometimes unintentionally modified their environments, with negative consequences, a lesson for ourselves today. Sometimes, however, these modifications can ultimately trigger changes that lead to new evolutionary opportunities. We are not suggesting that this is the case with the pollution and other atmospheric effects like global warming that we are causing today, but the effects of early photosynthetic bacteria, which led to increasing oxygen levels in the atmosphere, is one such example. Increasing oxygen levels may have been harmful to some types of bacteria, but this also eventually facilitated the origins of large animals like ourselves as since we discussed earlier all large animals today need oxygen to persist and breathe. Clearly, increasing atmospheric oxygen levels was not a goal that photosynthetic bacteria had in mind; bacteria do not have minds, and even if they did, it would not have occurred to them that in order to one day make the environment on Earth just right for humans they should produce oxygen. Still, in the long term their activities did facilitate subsequent evolutionary events which we justifiably see as important today. Another significant effect of rising oxygen levels was that it led to the formation of the ozone layer: again, this proved pivotally important for the subsequent pathway evolution took.

The Ozone Layer

Since the Earth formed, it has been bombarded by ultraviolet radiation from various sources in space, especially our own Sun. Today, less than 2 percent of all of the ultraviolet (UV) radiation that strikes the Earth's upper atmosphere reaches the surface; this is because our planet's surface is "shielded" by the ozone layer. This is significant because UV radiation, which comes in the form of high-energy, short-wavelength light, is highly toxic to almost all living organisms (animals and plants in particular): in humans (and other animals) it can cause burning of the skin and mutations that lead to various types of skin cancer. Such burns and cancers result because when high-energy UV radiation strikes skin cells it can literally burn them; a particularly severe burn, or multiple burns, can damage the DNA inside the skin cells; DNA that is damaged in certain ways will began to replicate in a faulty manner; this produces the cascade of effects doctors refer to as cancer.

Ozone is the chemical compound O_3 and therefore cannot form without oxygen present; further, without oxygen high up in the atmosphere there would be no ozone layer. The manner the ozone layer forms is relatively simple chemically and is worth elucidating: O_3 forms when the molecule O_2, the oxygen we breathe (two atoms of oxygen joined together), is struck by UV radiation. The high energy implicit in UV radiation causes some of these O_2 molecules to split up into two separate oxygens. Each one of these atoms is highly reactive and will combine with any other available O_2 around. When this happens, the compound O_3 is formed. Before oxygen levels in the atmosphere grew, initially as a result of the activities of photosynthetic bacteria including cyanobacteria, there would have been no ozone layer shielding the Earth's surface. Any organisms sitting on the Earth's surface, or floating in shallow water, would have been proverbial sitting ducks, waiting to be fried by harmful UV radiation. Thanks to rising levels of oxygen in the atmosphere, by about 2 billion years ago the initial ozone layer had formed, making the environment at the Earth's surface (land and sea) much less harsh.

When we survey the modern biological kingdom we find that cyanobacteria are the most tolerant of high levels of UV radiation. Given the role the early cyanobacteria likely played in generating the first significant quantities of oxygen in Earth's atmosphere this makes sense. They had to survive before the ozone layer formed, when the Earth's surface was exposed to significant levels of UV radiation. Today large green plants produce the bulk of the oxygen that keeps the ozone layer intact, but the initial credit for forming this crucial barrier that shields us from harmful UV radiation must be given to the cyanobacteria. Ultimately this facilitated the evolution of many other types of organisms, including plants and animals, among the latter of course humans, that happen not to be able to tolerate high levels of UV radiation.

It is somewhat ironic, and unfortunate, that humans have produced chemical compounds that have caused the ozone layer to break down. Thankfully, a treaty signed at the end of the 20th century meant that these compounds would no longer be produced. Although these compounds will linger in the atmosphere for a while, making sun bathing a more dangerous prospect than usual, eventually the levels of atmospheric ozone will start to climb again. A sign that although humans can inadvertently negatively effect our atmosphere, we can also take the steps needed to rectify our damage. This signifies that humans have the potential to reduce the negative effects we are having on our environment; our environmental problems need not be viewed as insurmountable.

The Evolution of the Eukaryotic Cell

With oxygen levels rising and the ozone layer at least partly in place, it meant that life could evolve in a less harsh and rigorous environment. Interestingly, around 2 billion years ago, or relatively soon thereafter, an extremely important evolutionary transition does occur. This is not to say that nothing important happened prior to 2 billion years ago: many important bacterial lineages were proliferating, but the first 1.5–2 billion years on this Earth was the time of pond scum nonetheless. The major evolutionary transition we refer to around 2 billion years ago was the origin of the eukaryotic cell. This transition may have happened as far back as 2 billion years ago (it could have even happened earlier, but we do not yet have evidence from the fossil record to demonstrate this) or it could have happened as "recently" as 1.7 billion years ago. The ambiguity in timing is attributable to the fact that molecular markers specific to eukaryotes occur in 2 billion year old rocks but body fossils that can definitively be identified as eukaryotes date from 1.7 billion years ago.

Before the origins of eukaryotes, all the organisms on Earth were single-celled bacteria of various forms that had a prokaryotic type of cell structure. By contrast, the type of cells that make up our bodies and also those of all plants, animals, and fungi have a very different and dramatically more complex structure termed eukaryotic. Our bodies are made up of literally billions of eukaryotic cells. The first eukaryotes to evolve, and indeed many types of eukaryotes alive today, were stand-alone, single-celled organisms; all plants, animals, and fungi are derived from these earliest eukaryotes.

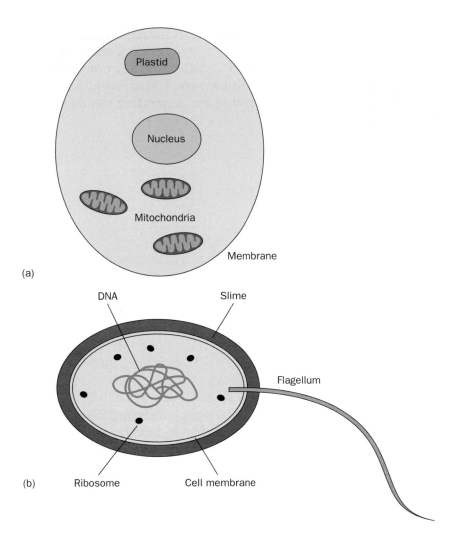

Figure 12.9 Different cell types. (a) Eukaryotic and (b) prokaryotic, showing some of the structures present in each. Diagram from R. Cowen (2005) *History of Life*, 4th Edn, Blackwell Publishing.

There are major differences between the structure and form of prokaryotic cells and eukaryotic cells (Figure 12-9). For example, all prokaryotic cells are very small. The eukaryotic cells that comprise our body are of course small too, but they are on average more than ten times the size of prokaryotic cells. Also, the position of the DNA or genetic material is very different in prokaryotic and eukaryotic cells. In prokaryotic cells the DNA is loosely bonded inside the cell. By contrast, in eukaryotic cells the DNA is bound into strands on chromosomes and these are housed inside of a nucleus, a central, cellular structure. If you were to examine the cell of a single celled eukaryote or one of the cells of your body under a microscope the nucleus would appear as a rounded, dark structure; you would see no such body if you examined a prokaryotic cell under a microscope. Eukaryotic cells also contain many other types of small bodies, and none of these are present in prokaryotic cells. For example, some single-celled eukaryotes have small, oval shaped, green structures called chloroplasts.

Eukaryotic cells in plants also have these; by contrast, animals do not have chloroplasts. Chloroplasts perform a very handy function, they conduct photosynthesis, a process already described, whereby they convert sunlight and carbon dioxide to energy and oxygen. Eukaryotic cells also house small bodies called mitochondria. These are responsible for generating the energy that is used by the cell to move, feed, etc.

There are many other types of small bodies inside eukaryotic cells that for our purposes are not essential to discuss. However, these, like the aforementioned nucleus, chloroplast, and mitochondria, each performs a specialized function such that within the eukaryotic cell there is a division of labor. In a way, each of the small bodies inside a eukaryotic cell acts like the organs, the heart, lungs, and liver, inside the body of an animal. For this reason, these tiny cellular structures are called organelles.

Finally, another important difference between cells of eukaryotic and prokaryotic type is their mode of reproduction. Prokaryotic cells reproduce asexually, by splitting off a direct copy of themselves. Eukaryotes can reproduce asexually or they can reproduce sexually. This has important evolutionary implications that we will discuss shortly.

From this discussion, it is clear that there are many differences between cells of prokaryotic type and eukaryotic cells and the latter are far more complex and sophisticated. Because of these dramatic differences in size and complexity, one of the big puzzles in evolutionary biology has been how did the eukaryotic cell evolve from prokaryotic ancestors. The first person to definitively answer this puzzle was the evolutionary biologist Lynn Margulis. She recognized that to tease apart this question we need to focus not on the differences between these two cell types, but rather their similarities. It turns out that the similarities between prokaryotic cells and eukaryotic cells are profound, although they are not necessarily what one would expect. For example, some of the tiny organelles inside of eukaryotic cells closely resemble stand-alone prokaryotic bacteria. The chloroplast organelles present in some eukaryotes (Figure 12-10) (including the single-celled *Euglena* and plants) look remarkably like the photosynthetic cyanobacteria (Figure 12-4).

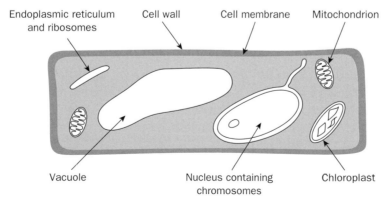

Figure 12.10 Eukaryotic cell. Emphasizing some of the different organelles. Diagram from M. J. Benton and D. A. T. Harper (2009) *Introduction to Paleobiology and the Fossil Record*, Blackwell Publishing.

There are other types of bacteria with prokaryotic cell types that closely resemble mitochondria found in eukaryotic cells. Another manner in which organelles resemble prokaryotic bacteria is that the organelles are surrounded by a membrane, just as prokaryotic (and eukaryotic) cells are.

The similarities between the organelles inside of eukaryotic cells and prokaryotic bacteria are not confined to simple physical similarities. There are also similarities in the distribution and makeup of the DNA in the different cell types. As we described above, in eukaryotes most of the DNA is housed on chromosomes that sit within the nucleus, but that is only part of the story. Some of the different organelles like the mitochondria and the chloroplast (when present) also possess their own genetic material. In these organelles the DNA is dispersed and not on chromosomes, very much akin to the distribution of the DNA in stand-alone prokaryotic bacteria. Part of the process of reproduction in eukaryotic (or prokaryotic cells) involves making copies of the DNA they contain. Depending on whether reproduction is sexual or asexual, there are differences in the copying process and the byproducts, but the net result is that some form of copy of the DNA is made. During eukaryotic reproduction not only does the genetic material in the nucleus within chromosomes make copies of itself, but also the genetic material in the mitochondria and the chloroplast makes copies of itself. The similarity in the general appearance of organelles and prokaryotic bacteria, the fact that they are both surrounded by a similar type of membrane, and their shared possession of DNA suggests that there is some kinship between individual prokaryotic bacteria and the organelles inside the eukaryotic cell.

The composition of the DNA possessed by organelles provided the final clue that was used to sort out the origins of the eukaryotic cell. As described in Chapter 7, scientists can compare the sequences of the DNA molecules from different organisms to determine how they are related evolutionarily. For instance, when scientists study the genetic code of a mouse and a rat their code shares more significant commonalities with one another than they share with let's say a cat. This is an important bit of evidence that rats and mice are more closely related to one another evolutionarily than they are to cats, much to the cat's relief. Scientists are able to routinely sample the DNA not only inside the nucleus of a eukaryotic cell, but also the DNA inside a cell's mitochondria; they also routinely sample the DNA of prokaryotic bacteria. What they have found is that there is a profound difference between the DNA in the nucleus of a cell and the DNA inside that same cell's mitochondria. What is even more fascinating is that the DNA inside the mitochondria shares the most significant commonalities with the DNA inside those prokaryotes that most closely resemble mitochondria. This provides genetic evidence that the mitochondrion and a prokaryotic bacteria are more closely related to one another evolutionarily than they are to the nucleus of the eukaryotic cell that a mitochondrion sits inside of. A similar pattern was found with chloroplast DNA: the DNA of the chloroplast inside a eukaryotic cell shares more significant commonalities with the DNA of a cyanobacterium, that it physically resembles, than with the DNA inside the eukaryotic cell in which it is housed.

The Origins of Cooperation

The radical idea put forward to explain these various clues was that somehow several different prokaryotes came together and started to live inside a single cell. Eventually all of these once separate cells subsumed their individual needs to that of the whole; some continued to make copies of their DNA at reproduction, others

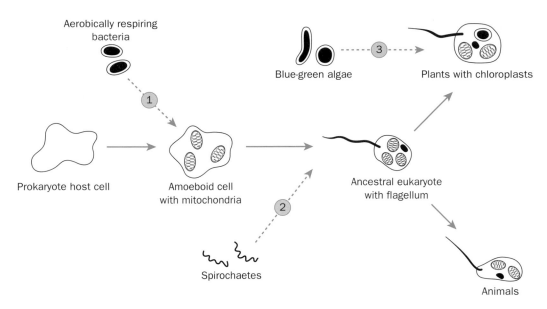

Figure 12.11 **The origins of the eukaryotic cell.** Showing how serial endosymbiosis may have occurred. Diagram from M. J. Benton and D. A. T. Harper (2009) *Introduction to Paleobiology and the Fossil Record*, Blackwell Publishing.

lost their DNA altogether. Thus was the first eukaryotic cell born. Lynn Margulis termed this idea the serial endosymbiont theory which in layman's terms just means that there was a sequence of events where several bacteria came together to live and work in a cooperative manner (Figure 12-11). The evidence for serial endosymbiosis is overwhelming, but scientists are still not clear precisely how it may have transpired. One possible explanation lies in the fact that bacteria eat other bacteria by engulfing them. Perhaps one bacterium engulfed another one but for some reason did not digest it. Several different bacteria would have been engulfed by this single cell; eventually, all of these formerly separated "prokaryotic" bacteria started working together and cooperating in the interests of the whole. For example, the mitochondrion would have started producing the energy necessary for the cell to live. The chloroplast would have provided nutrition via photosynthesis, etc.

The traditional view is to think of competition as the primary force that drives evolution; indeed, such a view is implicit in Darwin's formulation of natural selection, described in Chapter 5. We also provided some evidence though in Chapters 5, 6, and 7 to suggest that competition is not the sole force that drives evolution, nor is it necessarily the most important evolutionary process. The origins of the eukaryotic cell reiterate this, suggesting that some of the key episodes in the history of life are more about cooperation than competition. In the case of the eukaryotic cell, different lineages came together and sublimated their own interests to that of the whole. Of course one might ask "Did not these formerly separated lineages gain some sort of competitive edge by cooperating?" and indeed the answer must be yes, but to totally ignore the aspects of cooperation also involved with this event would be a serious distortion. Some other key episodes in the history of life also involve cooperation, not simply competition. For example, the origins of multicellularity, and associated with it the origins of animals, is an event that was as much about cooperation as it was about competition. We shall describe this event more fully in Chapter 15.

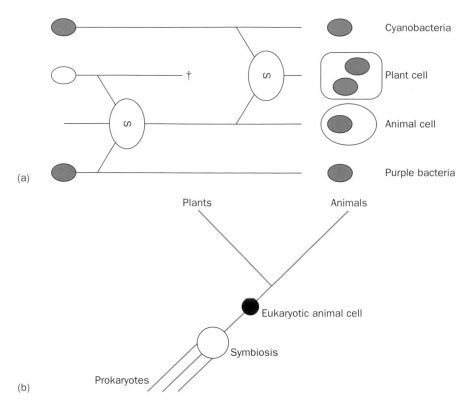

Figure 12.12 **The origins of the eukaryotic cell** *continued*. (a) Some of the different lineages that may have combined to produce the plant and animal eukaryotic cell, where the "S" refers to symbiosis and the "†" denotes extinction. (b) The event in the context of a cladogram, where the symbioses pre-date the origin of plants and animals. Diagrams from R. Cowen (2005) *History of Life*, 4th Edn, Blackwell Publishing.

Evolutionary Significance of the Origins of the Eukaryotic Cell

One reason why the origins of the eukaryotic cell is such a fundamental episode in the history of life is because it happened in a manner so different from the way evolution typically occurs. Recognize from what we described in Chapters 4, 5, and 7 that usually evolution involves different lineages diverging from one another, producing an evolutionary tree involving separated branches (or lineages); once lineages are separated they typically never come back together and coalesce. With the origins of the eukaryotic cell there were several different lineages that had started diverging, perhaps more than 3.5 billion years ago; after perhaps 1.5–2 billion years of separate evolutionary history some of these lineages ultimately came back together to form a new type of organism (Figure 12-12).

Another reason why the origins of the eukaryotic cell was such a fundamental episode in the history of life is that this cell type had a much more complex architecture and could perform many more ecological roles than a single prokaryotic cell. We recognize that typically all of the organisms held to be complex, including the multicellular animals, plants, and fungi are descended from this original single-celled eukaryote. There are also many unicellular eukaryotes included in this group that are still alive today. The proliferation of these complex forms after the origins of the eukaryotic cell is not to say that prokaryotes are somehow less fit evolutionarily than eukaryotes. Indeed, as described in Chapter 5 it could be argued that even with all of the different types of more complex eukaryotes that have evolved, prokaryotic bacteria, among the first life forms to evolve on Earth, still rule the world today. The reasons for the proliferation of all of these complex forms of eukaryotes likely does not signify that they were more evolutionarily fit or "better." Instead, it is explicable because the origins of the eukaryotic cell are associated with the origins of the sexual mode of reproduction. This is another reason why the origins of the eukaryotic cell should be viewed as a fundamental episode in the history of life.

The Origins of Sex and the Increasing Potential for Evolutionary Variation

The origin of sex was important, from an evolutionary perspective, because sexual reproduction produces variation and variation is the material that fuels more evolution. It is instructive to compare sexual reproduction with asexual reproduction. Asexual reproduction involves making a direct copy from the parent cell; sometimes there are errors or mutations in this copying process that can introduce limited variability. During sexual reproduction there can be mutation but there is also recombination, which introduces much greater variability. Recombination involves taking part of the mother's genes and part of the father's genes and mixing them in new combinations; this is why we do not look exactly like our parents. As we described in Chapters 4 and 5, one fundamental prerequisite for evolution is genetic variation. Without genetic variation, life cannot evolve. Stated in other words, if everything is exactly the same genetically, then there is no real potential for change. By contrast, the greater the genetic variation, the greater the potential for evolutionary change. After the evolution of the eukaryotic cell, eventually there was a great proliferation of complex organisms, including animals and plants, and sexual reproduction was the catalyst.

One thing that is important to recognize is that the eukaryotic cell did not originate in order to produce more variation, that was just a byproduct of the event. Again though, this event had huge consequences for the history of life. If it had not happened, our own species, and indeed no animals and plants, would have evolved.

Concluding Remarks

There are several fundamental lessons from the first 2 billion years of life on this planet. First, the origin of life itself seems a relatively simple transition, at least when measured by the time it took to happen; there is chemical evidence that life evolved relatively quickly, with life forms appearing on Earth shortly after the planet was no longer molten; contrast that with the origins of the eukaryotic cell that may have required nearly 2 billion years of evolution. If life evolves relatively easily, it suggests that we may not be alone in the universe. Another important lesson is that right from the start life began modifying the Earth's atmosphere: oxygen and the ozone layer were originally derived from the activities of simple bacteria. If such simple bacteria could so profoundly influence our planet's atmosphere, imagine what we can do with our complex civilizations and technology. Not only did these bacteria modify their environment, but the chemical byproducts they produced may have been detrimental to their own way of life; again, this can help inform the effects of our own species today. Biological modification of the atmosphere is not some hypothesis with little support; it is well corroborated, and the lessons for us are clear: the ensuing damage we do to our atmosphere through pollution and climate change have direct analogues in the past. We ignore them at our own peril.

Further, when we look back at the first two billion years of the history of life, much of that history can be characterized as a monotonous time of pond scum. The end of that time period is punctuated by the origins of the eukaryotic cell, which emerges as another pivotal event in the history of life. The origins of this more complex cell triggered all sorts of subsequent evolution, engendered primarily by the associated origin of sexual reproduction which increased the potential for evolutionary variation. This event is fascinating, not just because it signaled the proliferation of more complex organisms including ultimately plants and animals, but also because it involved a little recognized but important evolutionary mechanism: cooperation. In particular, lineages that had been separated for upwards of 1.5 billion years came together and sublimated their interests to that of the whole.

Additional Reading

Bengtson, S. (ed.) 1994. *Early Life on Earth*. Columbia University Press, New York; 630 pp.

Bengtsson, L. O., and Hammer, C. U. 2001. *Geosphere–Biosphere Interactions and Climate*. Cambridge University Press, New York; 304 pp.

Cockell, C., and Blaustein, A. R. (eds). 2001. *Ecosystems, Evolution and Ultraviolet Radiation*. Springer, New York; 221 pp.

Huggett, R. J. 2006. *The Natural History of Earth: Debating Long-term Change in the Geosphere, Biosphere, and Ecosphere*. Routledge, London; 203 pp.

Knoll, A. H. 2003. *Life on a Young Planet*. Princeton University Press. Princeton, NJ; 277 pp.

Margulis, L. 2003. *Symbiotic Planet: A New Look at Evolution*. Basic Books, New York; 160 pp.

Margulis, L., and Dolan, M. F. 2002. *Early Life*. Jones and Bartlett, New York; 224 pp.

Margulis, L., and Sagan, D. F. 1997. *Microcosmos: Four Billion Years of Microbial Evolution*. University of California Press, Berkeley, CA; 304 pp.

Schopf, J. W., and Klein, C. (eds.). 1992. *The Proterozoic Biosphere*. Cambridge University Press, New York; 1374 pp.

Schopf, J. W. 2001. *Cradle of Life: The Discovery of Earth's Earliest Fossils*. Princeton University Press, Princeton, NJ; 336 pp.

Schopf, J. W. 2002. *Life's Origin: The Beginnings of Biological Evolution*. University of California Press. Berkeley, California; 224 pp.

Staley, J. T., and Reysenbach, A.-L. 2001. *Biodiversity of Microbial Life: Foundation of Earth's Biosphere*. John Wiley & Sons, New York; 592 pp.

Stanley, S. M. 2004. *Earth System History*. W. H. Freeman, San Francisco, CA; 608 pp.

Chapter 13

The Cambrian Radiation and Beyond: Understanding Biology's Big Bang

Outline

- Introduction
- Life Before the Cambrian Radiation
- The Burgess Shale and the Cambrian Radiation
- The Ordovician Radiation and Concluding Remarks
- Additional Reading

Introduction

The Cambrian radiation marks one of the key episodes in the history of life. At the grand scale, this is a time of major transition in the fossil record between roughly 540–510 million years ago, including the bulk of the early part of the entire Cambrian period. In pre-Cambrian rocks that date from prior to the radiation, fossils of large organisms are rare and the evolutionary affinities of many of these fossils are enigmatic: some argue these fossils are the remains of animals; others suggest they represent something else. After the start of the Cambrian, however, fossils clearly indicate the presence of several distinct types of relatively large animals including various types of arthropods (the group including trilobites, crustaceans, and relatives of horseshoe crabs), chordates and vertebrates, molluscs (the group including snails and clams), cnidarians (the group including jellyfish and anemones), brachiopods, and echinoderms (the group including crinoids, sea

Prehistoric Life: Evolution and the Fossil Record. 1st edition. By Bruce S. Lieberman and Roger Kaesler. Published 2010 by Blackwell Publishing.

cucumbers, and starfish). Further, fairly soon after the start of the Cambrian, these groups became relatively abundant, common, and diverse.

Another hallmark feature of the Cambrian radiation is that the organisms appearing and diversifying were largely, though not entirely, things with skeletons, such as the hard rigid, external skeleton of a trilobite and clam, or the hard, rigid, internal skeleton of a vertebrate (although the earliest vertebrates lacked these). The nature of this evolutionary radiation, and its significance, has puzzled paleontologists and evolutionary biologists at least since the time of Darwin. Here we will focus on the run-up to the Cambrian radiation, including paleontology and evolution during the very end of the Proterozoic, while discussing the Ediacaran organisms that we have already mentioned. We will also describe the Cambrian radiation itself. Further, we will describe the various environmental and biological changes that may have triggered the radiation. Finally, we will conclude with a discussion of the major evolutionary event that follows the Cambrian radiation: the Ordovician radiation.

Life Before the Cambrian Radiation

The oceans were not dead zones before the Cambrian, far from it; recall that life has been present for most of the history of this planet. However, until the very end of the Proterozoic, almost all life was small and single celled (unicellular). Further, a key evolutionary event that happened during the Proterozoic, the origin of the eukaryotic cell, was described in Chapter 12. Because of this, eukaryotic organisms were certainly present in Proterozoic oceans and many of these were algal forms. That is to say, they obtained their energy from the Sun via photosynthesis. Many of these algae were unicellular and microscopic, which makes them harder to find in, and retrieve from, these ancient rocks, although they can be found. There were also some larger organisms that evolved during the Proterozoic; for instance, there were multicellular algae, akin to certain types of modern seaweeds (Figure 13-1). Tiny algae (albeit different from these Proterozoic forms) are still a very important part of life in the oceans. Indeed, they comprise the base of the foodchain that many types of animals feed on. However, there was nothing like the large things alive today that prey upon these algae, or prey upon the things that prey upon these algae. That was to change following some profound environmental and geological changes that occurred during the last half of the Proterozoic.

Global Change at the End of the Proterozoic

Some of the changes that occurred during this time have already been described in Chapter 11. Recall that before about 750 million years ago all of the Earth's continents were assembled into one supercontinent, Rodinia, that formed roughly a billion years ago. The positions of the Earth's continents were very different then (Plate 11-1). However, around 750 million years ago, give or take a few tens of millions of years, this supercontinent started to fragment into smaller parts, and the outlines of some of these were more similar to the continents we are familiar with today. Over many tens of millions of years, and in a complicated and still not fully understood sequence, these different parts of Rodinia sometimes collided with each other and sometimes moved off on their own. A different, ephemeral supercontinent may have even formed for a short time, yet

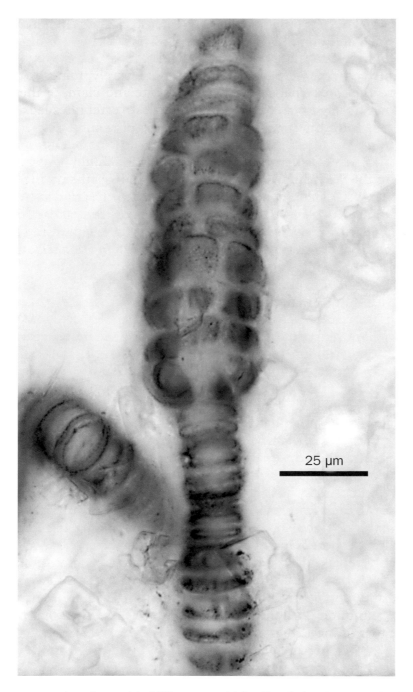

Figure 13.1 **Proterozoic alga.** A roughly billion year old fossil red alga from arctic Canada. Image from N. Butterfield, University of Cambridge.

by the Cambrian the world consisted of several smaller and more dispersed continental blocs (Plate 10-1).

We also described in Chapter 11 how at roughly the same time as these significant geological changes there were a series of major environmental changes that occurred. In particular, perhaps as many as four times between roughly 800 and 600 million years ago the Earth was plunged into profound, glacial cold:

the so-called "Snowball Earth hypothesis." Oscillating with these extreme times of cold there were also long bouts of steamy warmth. Finally, the concentration of oxygen was increasing in the atmosphere. Each of these changes would have been important because animals typically find extremes of temperature rather inhospitable; any amelioration of such climatic swings would make animal life "easier" in a sense, and perhaps free up evolutionary opportunity. Further, animals depend on oxygen, and at low oxygen concentrations large animals, which are multicellular, cannot persist. A rise in oxygen concentrations would increase the potential body size that animals could achieve, and again, represent an environmentally mitigating event. Finally, fragmenting the continents would geographically isolate different organisms on different continents. Geographical isolation is one of the primary mechanisms of evolution: it is the motor that causes speciation to occur. In short, environmental conditions became milder (from most animals' perspectives) around this time, and the engine driving evolution was revving. Perhaps not coincidentally, after both the surface and atmosphere of the planet underwent these changes, there were profound changes in the biosphere. These environmental changes may have been the primary impetus that caused the radiation to happen when it did. Another possibility is that there were inherent biological changes in the organisms themselves that caused the radiation to happen when it did, a topic we take up more fully in our discussion of the Burgess Shale.

The Ediacarans

Roughly 600 million years ago the Ediacarans appeared on the scene. These have already been described in Chapter 8, and, as mentioned, there has been some debate about whether these represent animals, a separate evolutionary origin of multicellularity, or some combination of the two (Figure 8-9). Consensus does seem to be emerging that at least some of the Ediacarans belong to the phylum Cnidaria, or are at least closely related to this phylum. Cnidarians are described in detail in Chapter 14, and include sea pens, anemones, and jellyfish. The Ediacarans persisted over about 70 million years of the late Proterozoic; through time the attendant size and range of form of these types of organisms increased as new forms evolved. By the very end of the Proterozoic there were quite a range of different Ediacarans, running the gamut in form from starfish-like (Figure 13-2a) to arthropod-like (Figure 13-2b) to worm-like (Figure 8-9) (though whether they can actually be assigned to any of these groups is still not known). However, none of these organisms possessed rigid, mineralized skeletons. Further, nearly all of the Ediacarans succumbed at the very end of the Proterozoic. This may have been the first "mass extinction" that impacted multicellular life, although it is not counted as one of the big five mass extinctions. (Nor perhaps should it be, as the total amount of diversity eliminated at the time was not very high, although on a percentage basis it was devastating for the Ediacarans: only a very few cryptic, Ediacaran-like specimens are known from Cambrian rocks.) The cause of the extinction of most of the Ediacarans is not yet known, but it did roughly correspond to a time when a fairly large asteroid collided with planet Earth.

　Irrespective of various arguments about the nature of the Ediacarans, it probably is safe to conclude that at least *some* of them were cnidarians, or close evolutionary relatives of cnidarians. The evidence for other major animal groups in the Proterozoic is disputed, with a few important exceptions. Sponges, which belong to the phylum Porifera, are one exception. Tiny pieces of sponge skeletons are known from the

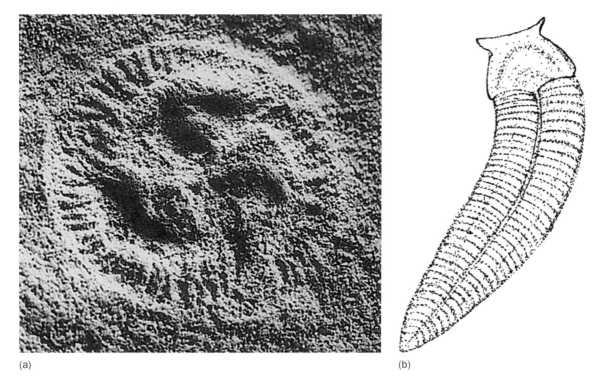

(a)   (b)

Figure 13.2 Ediacarans. Examples of these late Proterozoic fossils from Australia. The ones shown vaguely resemble: (a) a starfish, image from R. Cowen (2005) *History of Life*, 4th Edn, Blackwell Publishing—by M. Glaessner; and (b) a trilobite, diagram from E. N. K. Clarkson (1998) *Invertebrate Palaeontology and Evolution*, 4th Edn, Blackwell Publishing—redrawn from Glaessner and Wade.

late Proterozoic interval containing the Ediacarans. Another important exception involves one of the most spectacular fossil finds of the last 30 years.

Proterozoic Animal Embryos and Other Evidence for Animals Before the Cambrian Radiation

We have already described some of the reasons why the fossil record is incomplete. For these reasons, many might be surprised to know that fossilized animal embryos are known, but indeed they are, and some digression here is merited. In particular, one spectacular example is the record of embryos from the dinosaur *Oviraptor*; these are preserved inside Cretaceous eggs recovered in Mongolia's Gobi Desert. Incidentally, these eggs vindicated the dinosaur *Oviraptor*. Its name means "egg thief", and when the first fossils of the carnivorous dinosaur *Oviraptor* (Figure 13-3) were found, they were near eggs presumed to belong to a different dinosaur, *Protoceratops*, whose fossils were also found in the vicinity. Since *Protoceratops* (Figure 9-9) was a herbivore it was assumed that *Oviraptor* was feasting on *Protoceratops'* eggs, and thus it was given its name. Most eggs were empty, but a few weren't, and because of these many years later paleontologists determined that the eggs in fact contained developing embryos of *Oviraptor* (Figure 13-4). *Oviraptor* was no thief, and instead it turns out it was simply guarding its own eggs, not stealing them from *Protoceratops*.

Figure 13.3 *Oviraptor*. A Cretaceous dinosaur from Mongolia. Diagram from R. Cowen (2005) *History of Life*, 4th Edn, Blackwell Publishing; © B. Giuliani, Dover Publications, New York.

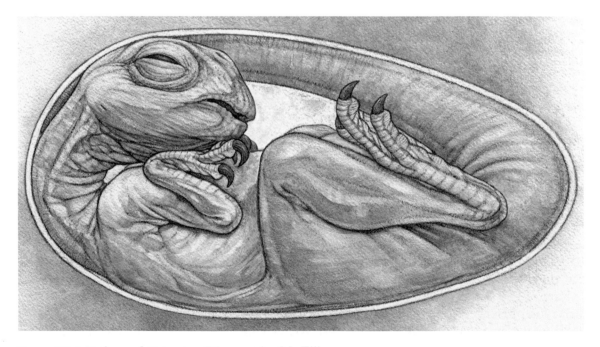

Figure 13.4 Embryo of *Oviraptor*. Diagram by M. Ellison.

In any event, even given the many reasons why the fossil record is incomplete, perhaps one might have predicted the eventual discovery of Cretaceous dinosaur embryos. However, the existence of fossilized Proterozoic embryos seemed almost beyond the realm of the possible, but back in 1998 the impossible became possible, thanks to a study by paleontologists Shuhai Xiao, Y. Zhang, and Andy Knoll. These paleontologists uncovered exquisitely preserved, microscopic fossil embryos,

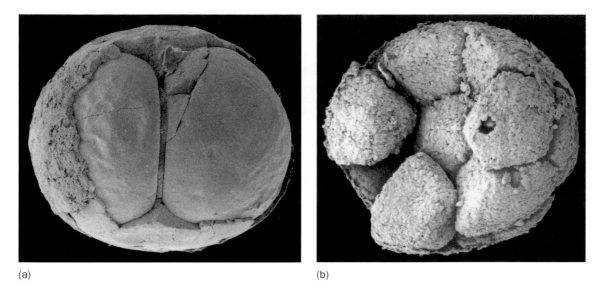

(a)   (b)

Figure 13.5 **Late Proterozoic animal embryos.** From Doushantuo, China, showing embryos at two different early cell stages shortly after fertilization. Image from S. Xiao, Virginia Polytechnic Institute.

documenting various stages in an animal's life shortly after fertilization occurred (Figure 13-5); and we can indeed say "an animal's life" because the embryos share certain similarities with various types of living animal embryos; (although they do not precisely match those referable to any one group, cnidarians are a possibility). These embryos hail from China and date from roughly 570 million years ago. It seems that either these embryos were from some Ediacaran (and these would thus have been animals), or there were various animals lurking in the shadows of the Ediacaran's world, metaphorically waiting to take over after the Ediacarans went extinct at the end of the Proterozoic (much like the case with mammals and dinosaurs at the end of the Cretaceous). Subsequently, other animal embryos have also been recovered from younger, Cambrian aged rocks in China, perhaps suggesting that if paleontologists look hard enough there may be other embryos out there, waiting to be discovered.

Part of the puzzle paleontologists are now facing is why are there remains of embryos preserved when remains of adults are not present, for at the deposit yielding embryos no fossils of adults have been recovered. However, it seems to be related to the way fossilization happens, and in particular, fossilization processes preserving tiny embryos tend to be biased against preserving larger things.

Recent work by one of us (BSL) in collaboration with geologist Joe Meert also suggests that the first trilobites had evolved during the end of the Proterozoic, even though the earliest trilobite fossils did not appear until the start of the Cambrian a few tens of millions of years later. Given that trilobites are arthropods, and considering there are many modern animal phyla separating arthropods from cnidarians and sponges (Figure 13-6), this suggests that many other phyla must have also evolved by the end of the Proterozoic. Results from a study of Cambrian fossil jellyfish by evolutionary biologist Paulyn Cartwright and colleagues also support this view. They described the oldest definitive true jellyfish; these are remarkably similar to modern jellyfish (Figure 13-7). This suggests a lengthy

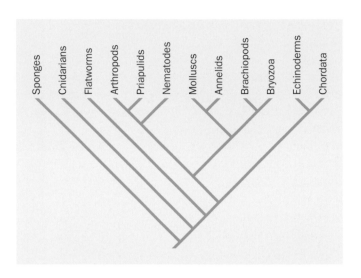

Figure 13.6 Animal phylogeny. A cladogram showing the relationships of the major animal phyla. Diagram by F. Abe, University of Kansas.

Figure 13.7 Cambrian jellyfish. Specimen from Utah, in the collections of the University of Utah, courtesy of S. Halgedahl and R. Jarrard. Image by B. S. Lieberman and J. Hendricks.

period of jellyfish evolution, currently hidden from view, in the Proterozoic fossil record.

Taken in total, the results from cnidarians, trilobites and fossilized animal embryos are part of the growing body of evidence suggesting that sponges, cnidarians, arthropods, and many other types of animals were present, but exceedingly rare (again depending on the status of the Ediacarans) before the start of the Cambrian Period and the inception of the Cambrian radiation. In short, animals were certainly around and had at least diversified somewhat before the Cambrian radiation started. Sometimes in a fit of paleontological sensationalism the Cambrian radiation has been called the "Cambrian explosion;" one can safely use this catchy title, but they should recognize that given the evolutionary evidence for animals down in the Proterozoic, it appears that this "explosion" may have resulted in a Big Bang, but there was a long fuse (a few tens of millions of years) associated with it.

Is the Cambrian Radiation Real, or is it an Artifact of a Poor Fossil Record?

Recall that one of the important aspects of the Cambrian radiation was that it marked the first time when animals with hard skeletons became abundant in the fossil record. Given this, it has been suggested that perhaps the radiation does not represent a time of profound evolutionary change, but instead the time when many different groups acquired hard skeletons and suddenly became far more likely to be fossilized. This is an idea that goes back at least to Darwin and he discussed it in his "*On the Origin of Species. . . .*" If this view is accurate, it suggests that maybe the radiation is not so much an evolutionary event, but instead reflects a change in life's fossilization potential. Maybe animals had evolved tens of millions of years before the Cambrian radiation, but they lacked skeletons and were soft and squishy, or maybe they were exceedingly small? Either one of these conditions would make an organism less likely to become a fossil and be discovered by a paleontologist. As mentioned, the accumulating evidence suggests that there appears to be some stretch of time when the animals we associate with the Phanerozoic had evolved, but were not manifest in the fossil record.

Why they were not manifest though is still a difficult question. The explanation that early animals were large but soft and squishy may not work, for instance, because at the same time that the diversity of fossils from different animal phyla increased in the Cambrian radiation, there was a concomitant increase in the number and type of tracks and trails that fossil organisms left behind. There is not a major rise in the number of tracks and trails before this time in the Proterozoic. Presumably if large but soft and squishy animals were alive during the Proterozoic, they would have left a rich record of fossil trackways, but the diversity and variety of these tracks and trails is very low right until the end of the Proterozoic, and only starts to climb during the radiation. (Some of these Cambrian trackways were left by animals like trilobites that had a hard skeleton, but others were left by animals that were worm-like and lacked a skeleton.) It is still of course possible that a variety of animals were present during the Proterozoic, but perhaps they were small and this small size (possibly combined with an absent skeleton) made them less likely to fossilize. (This may parallel the situation during the Cretaceous when mammals were small and relatively rare as compared with dinosaurs, and thus far less likely to fossilize.) Another possibility is that animals were present during the late Proterozoic, but they were rather rare and confined to environments that usually do not become fossil deposits.

The Burgess Shale and the Cambrian Radiation

One of the most famous fossil localities in the world dates from shortly after the Cambrian radiation, and it offers significant insight for our understanding of this key episode in the history of life. This fossil locality, known as the Burgess Shale, is important because it showcases the rich variety of animal life present shortly after the radiation. In particular, not only are the typically fossilized remains of Phanerozoic animals—their hard skeletons—preserved at this locality, but along with these skeletons, soft tissues, for instance, guts, muscles, etc., remain; even animals that completely lacked skeletons are known from the Burgess Shale. If we look around at the oceans today, only about 30 percent of all marine species have hard skeletons; the rest consist solely of soft tissues and thus are not likely to be fossilized, yet they surely represent an important aspect of living diversity and ecology. The Burgess Shale provides us with insight about what had evolved shortly after the Cambrian radiation not only for the standard 30 percent of typically fossilized life forms, but also for the broader panoply of life extant at that time.

How are Soft Tissues Fossilized?

Preserved remains of soft and squishy body parts, or soft and squishy animals, are rare in fossil deposits (as we described in Chapter 2), and thus how and why they get preserved is interesting. In order to become a fossil an organism has to do a few things. First, it has to die; then, its body has to be buried in sediment. That hard mineralized skeletons are the things that typically get fossilized is not surprising as these are more difficult to break down physically; these often also contain less organic matter, so scavengers and other organisms are less likely to attack them, chew them, and break them apart. After an organism dies, it begins to decay. There is the physical decay already described, and then there is the biological decay initiated by munching scavenging animals. Finally, bacteria cause a dead organism to decay. Imagine, but do not attempt (for your own sake) the following simple experiment: think what would happen to the smell of your kitchen if you left a raw piece of steak out. Assuming no small mammals (like rats or mice, arguments for keeping your home neat) or small arthropods (like roaches, additional arguments for keeping your home neat) the primary engine of decay will be the action of bacteria.

The result of both physical and biological decay is that soft flesh is quickly removed from a carcass, and is very unlikely to enter the fossil record. The hard mineralized external skeletons of, for instance, molluscs, arthropods such as trilobites, and echinoderms, and the hard mineralized internal skeletons and teeth of vertebrates are much more resistant to decay and these are the most common fossils. (However, certain scavengers do target bones because of the marrow they contain, and further, strong physical forces like moving currents will break down skeletons and teeth; burial will shield a skeleton from these types of forces, but it typically will not arrest bacterial decay.) Still, there are times when soft tissues of animals are preserved (Figure 13-8).

Deposits such as these usually form when a carcass is buried very rapidly (shielding it from physical forces and scavengers) and put in conditions that inhibit tissue decay by bacteria: for instance, low-oxygen environments. (There are cases, however, when certain types of bacteria actually help the anatomies of soft tissues to become fossilized.) Fossil deposits with this type of preservation are variously referred to as "soft-bodied" or lagerstätten; they represent unique windows on the past.

Figure 13.8 Early Cambrian trilobite. From Pennsylvania showing soft parts preserved, in particular, the antennae. Specimen in the collections of the Yale University Peabody Museum of Natural History. Image by B. S. Lieberman.

Background on the Burgess Shale

At no other time in the history of life before or after the Cambrian radiation do so many different types of major groups of animals or "phyla" appear in the fossil record. Over the roughly 30 million years of the Cambrian radiation, from start to finish, more than 20 phyla of animals appeared in the fossil record; only a single animal phylum (of those phyla likely to fossilize) appeared in the fossil record subsequently. The Burgess Shale is a window on the diversity of animals shortly after this radiation. It is situated in Middle Cambrian rocks from British Columbia, Canada, in Yoho National Park (Figure 13-9) and fossils of many different types of animals have been found there.

The Burgess Shale was first made famous by the paleontologist Charles Walcott in the early 1900s, but its significance and scientific importance was revivified and expanded by paleontologists Harry Whittington, Derek Briggs, Simon Conway

Figure 13.9 Burgess Shale quarry. Yoho National Park, British Columbia, Canada. Image by A. Stigall.

Morris, Richard Fortey and others in the 1970s and 1980s. Shortly thereafter, the paleontologist and evolutionary biologist Stephen Jay Gould published the book *Wonderful Life* and documented the relevance of the work on the Burgess Shale.

Paleontologists such as Walcott, Whittington, Briggs, Conway Morris, and Fortey uncovered a rich variety of soft-bodied animals from the Burgess Shale. A prime example of these is *Canadaspis*, one of the world's oldest crustaceans (Figure 13-10). Many other Burgess Shale fossils are not so easily classified with modern organisms, a point they recognized and Gould stressed. By way of example, in terms of overall species diversity, arthropods rule the roost today, and deposits like the Burgess Shale tell us that this was also true back in the Cambrian. Arthropods today include crustaceans (crabs, lobsters, and pill bugs), millipedes and centipedes, insects, and chelicerates (spiders and horseshoe crabs), and trilobites are an important extinct arthropod group; all possess a rigid and segmented external skeleton, jointed limbs, and a compound eye. However, many of the Burgess Shale arthropods differ from modern arthropods in important respects. In particular, modern arthropods (and trilobites; Figure 13-11) possess characteristic numbers of segments and legs in different parts of the body, especially the head. So, if you were to closely examine a crustacean you will find it will always have five segments in the head and these bear five pairs of limbs differentiated as two pairs of antennae, and three pairs of appendages used to process food and stuff it into the mouth. There is also a characteristic limb pattern that is always found in chelicerates, a characteristic pattern always found in insects, etc. It so happens that there is no intergradation

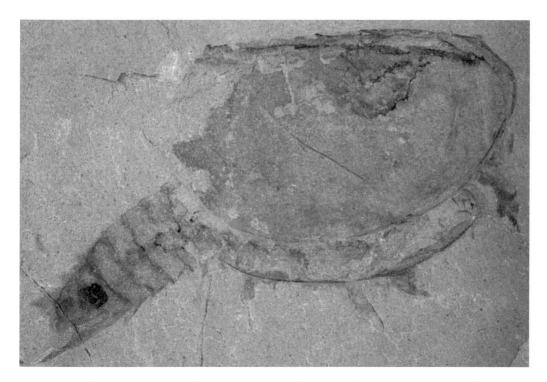

Figure 13.10 Cambrian crustacean. *Canadaspis* from a Burgess Shale type deposit in Nevada. Image by B. S. Lieberman, specimen in the collections of the University of Kansas Museum of Invertebrate Paleontology.

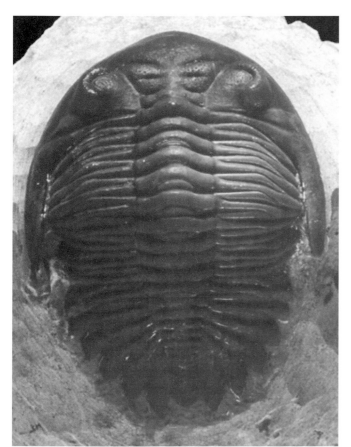

Figure 13.11 Trilobite. From the Devonian of Morocco. Image by A. Modell.

between living groups in these traits, and this actually holds true since their earliest appearance in the fossil record, which in many cases dates to the Cambrian.

The situation is different though during the Cambrian, because back in the Cambrian there were also many other types of arthropods around. This is one reason why the Burgess Shale and other Cambrian deposits with soft-bodied fossils are important: they document the rich variety of now extinct arthropod body plans.

When paleontologists have examined the patterns of segmentation in the head and the body of these Cambrian arthropod fossils they have identified segmentation types and numbers distinct and different from those present today (Figure 13-12). Further, there are fossils from the Burgess Shale that are certainly arthropod-like,

Figure 13.12 *Marrella*. A distinctive Cambrian arthropod from the Burgess Shale. Image from E. N. K. Clarkson (1998) *Invertebrate Palaeontology and Evolution*, 4th Edn, Blackwell Publishing.

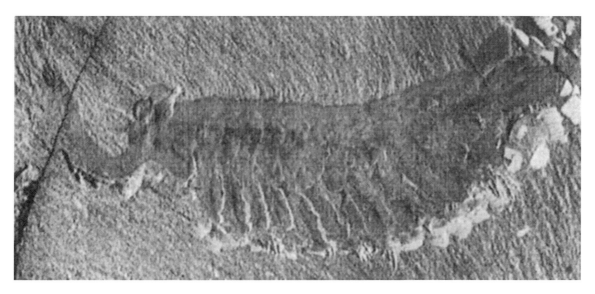

Figure 13.13 *Opabinia*. A distinctive Cambrian arthropod-like animal from the Burgess Shale. Image from E. N. K. Clarkson (1998) *Invertebrate Palaeontology and Evolution*, 4th Edn, Blackwell Publishing.

but they differ in subtle to important ways from modern arthropod groups and trilobites, including in their absence of jointed limbs on most of the body column. For instance, the Cambrian arthropod-like forms *Opabinia* and *Anomalocaris* were endowed with flaps instead of limbs (we will describe each of these beasts more fully below) (Figures 13-13 and 13-14). Consider today there are three distinct arthropod body plans; throw in the richly diverse trilobites and that makes four. However, back in the Cambrian, there may have been upwards of ten distinct types of arthropod and arthropod-like groups.

The Burgess Shale, and other Cambrian deposits with soft-bodied fossils, also preserve other animals that represent now wholly extinct phyla. Although the Burgess Shale quarry is just known from a single small outcrop on the side of a mountain in British Columbia, Canada, subsequent to its discovery other similar deposits have been found throughout the world including Greenland, Utah, Pennsylvania, Australia, and most importantly, the Chengjiang Biota from China. In particular, studies on the Chengjiang Biota by Chen, Hou, Shu, Aldridge, Edgecombe, the Siveter brothers, and other paleontologists brought the Chinese deposits to light. When we consider the Burgess Shale in isolation, or all of the Cambrian deposits with soft-bodied fossils in their entirety, it is a surprising but necessary conclusion first forcefully argued by Gould that there were as many or more distinct types of animals present back in the Cambrian, shortly after animals evolved, than exist today. This is on first blush rather surprising, as it might have been anticipated that the number of distinct types of organisms should have slowly and gradually increased through time and could be characterized as a cone expanding upwards and outwards (Figure 13-15). Instead, the lesson from the Burgess Shale and other Cambrian deposits like the Chengjiang is that there was a radiation, perhaps even an explosion of body types, shortly after the start of the Cambrian that has never since been equaled. The overall number of distinct, major body types has stayed relatively constant since the Cambrian, and could be

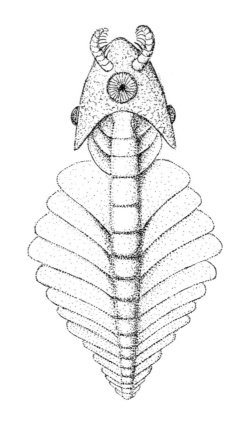

Figure 13.14 *Anomalocaris*. Another distinctive Cambrian arthropod-like animal known from the Burgess Shale and other Cambrian fossil deposits. Diagram from E. N. K. Clarkson (1998) *Invertebrate Palaeontology and Evolution*, 4th Edn, Blackwell Publishing; based on Briggs et al.

Figure 13.15 **Cone**. Expanding upwards and outwards. Image from iStockPhoto.com.

Figure 13.16 Evergreen trees. Banff National Park, Canada. Image by B. S. Lieberman.

characterized as a cylinder, or perhaps the number of distinct body types has even decreased and fits the shape of a Christmas tree (Figure 13-16).

The key evolutionary question that Stephen Jay Gould first brought into focus is the issue of why there were so many weird types of animals early on in the history of animal life, shortly after the animals evolved and diversified, whereas today there are relatively few basic types.

The Difference Between Diversity and Disparity

Gould used the term "disparity" to refer to distinct types of organisms (and focus on the existence of so many distinct types of animals, the greater disparity of animal life, back in the Cambrian). Gould argued that when it came to the number of distinct species or total animal "diversity," there are many more species alive today relative to the Cambrian world. For instance, today more than a million species of animals have been described; back in the Cambrian only a few thousand likely existed. However, today species in the modern world vary around just a few stereotyped, basic designs. Consider that today literally hundreds of thousands of species of beetles exist. Except to someone familiar with beetles, which we respectfully submit we are not, they may not differ by that much from one another. By contrast, in the Burgess Shale there were animals that to us seem massively peculiar (partly that represents a bit of an anthropomorphic bias, which we shall describe more fully below). Consider *Opabinia* (Figure 13-13): it had five eyes on its head, a long vacuum cleaner shaped "nose" with sharp grasping teeth at the end, and is rather distinct from anything alive today. In fact, paleontologists now consider *Opabinia* on the early evolutionary line that leads to modern arthropods.

Another rather bizarre Cambrian soft-bodied animal on the early evolutionary line leading to arthropods was *Anomalocaris* (Figure 13-14). This animal was the giant predator of the Cambrian seas. Many of the fossils found in the Burgess Shale are just a few centimeters long, but *Anomalocaris* may have reached more than a meter in length. Probably the two long appendages hanging beneath the front of the animal would have brought prey to its round mouth (Figure 13-17). In close up, this round mouth is armed with sharp teeth along its interior. *Opabinia* and

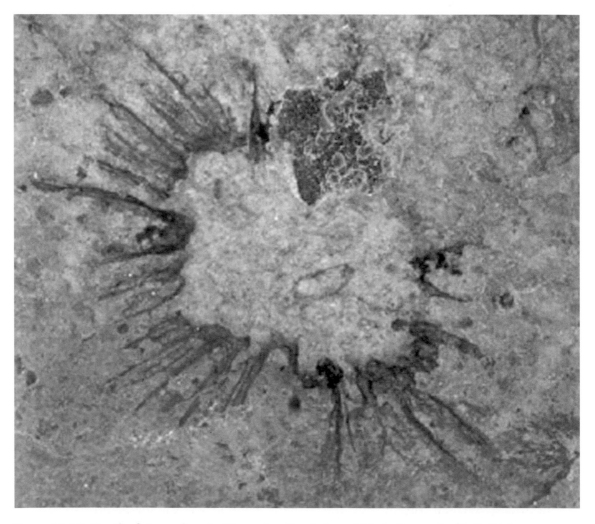

Figure 13.17 Mouth of *Anomalocaris*. From the Cambrian of Utah. Image by B. S. Lieberman, specimen in the collections of the University of Kansas Museum of Invertebrate Paleontology.

Figure 13.18 Appendage of *Anomalocaris*. From the Cambrian of Utah. Image by J. Hendricks, specimen in the collections of the University of Kansas Museum of Invertebrate Paleontology.

Anomalocaris were just two of the many distinctive and highly "disparate" animals to have evolved during the Cambrian.

As an interesting aside, the mouth and appendages of *Anomalocaris* were first found as isolated fossils. The appendages were originally considered to be part of a strange shrimp (Figure 13-18) (the name "*Anomalocaris*" translates as "unusual shrimp"); the mouth was described as an unusual kind of jellyfish (a cnidarian) (Figure 13-17). Harry Whittington and Derek Briggs were the first to find the complete animal and they recognized that these different structures all belonged to the same beast. It so happens that there are trilobites known from various Cambrian deposits, including those from the Middle Cambrian, that have semicircular bite marks that match up very nicely with the mouth of *Anomalocaris* (Figure 13-19).

Why Hasn't Disparity Increased Through Time?

A key evolutionary question relating to the nature of the Cambrian and the significance of the Burgess Shale is what is the reason why disparity has not increased appreciably through time (or why may it have even declined since the Cambrian), even as diversity has climbed dramatically. Gould argued that there seems to have been two primary reasons. First, something about animal evolution was less constrained then, such that early on evolutionary change occurred more easily, whereas today substantial evolutionary change is more constrained and thus more difficult. We can think of the way an animal looks as being controlled by a genetic and developmental system. There is some genetic blueprint which specifies how an animal grows from a fertilized egg to an adult, such that a lion, for instance,

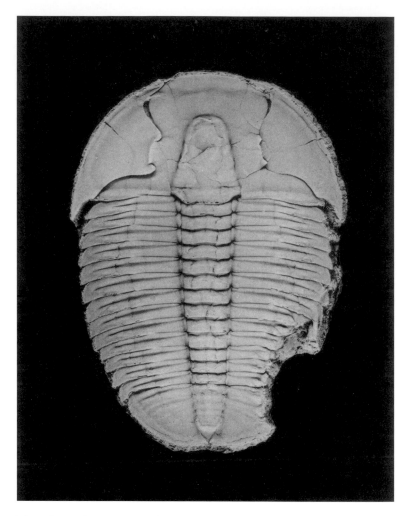

Figure 13.19 Trilobite bitten by an *Anomalocaris*. From the Cambrian of Utah. Image by L. Babcock, Ohio State University.

always looks like a lion. Maybe early on in the history of animals this blueprint was a bit more fuzzy? The lack of specificity may have allowed greater changes to occur in the Cambrian relative to later in the history of animals. This is an interesting idea, but it is difficult to test since it is impossible to sequence the genomes of animals extinct for 500 million years.

Gould recognized that another reason why disparity has not continued to climb or has even fallen since the Cambrian relates to extinction. A large number of the different types of animals present during the Cambrian in the Burgess Shale and the Chengjiang Biota had to have gone extinct: the evolutionary tree of animal life at the grand scale has been extensively pruned (Figure 13-20). Because soft-bodied deposits that preserve fossils like those found in the Burgess Shale are rare, we do not know precisely when this pruning happened, but some of it seems to have occurred before the middle part of the Ordovician period. Some of the pruning may have occurred later as relatives of the characteristic Burgess Shale animals *Anomalocaris* and *Marrella* survived at least to the Devonian.

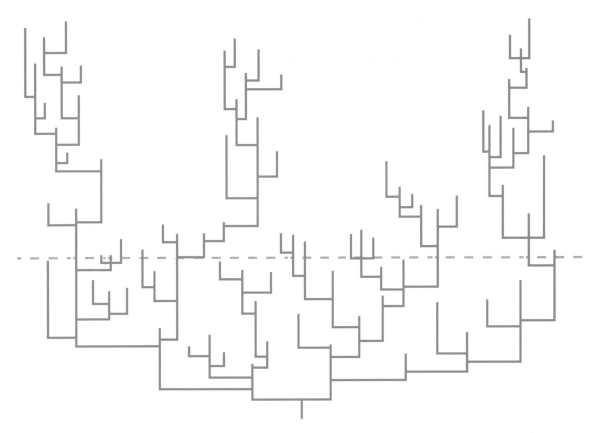

Figure 13.20 **Pruned evolutionary tree**. An evolutionary tree showing an extinction event, the dashed line, that eliminated many lineages. Diagram by F. Abe, University of Kansas.

The Importance of Extinction: Contingency Revisited

The phenomenon of contingency in the history of life was already described in Chapter 6. This idea, developed by Stephen Jay Gould in his book *Wonderful Life*, is quite apposite to the Burgess Shale. In particular, many of the soft-bodied animals found in deposits like the Burgess Shale went extinct; however, some of the groups that were around at this time survived, for instance, the crustaceans as well as the groups that gave rise to the spiders and the insects. Gould argued that the reasons for the differences in survival pattern are unclear and there is no scientific justification for saying that the forms that went extinct were somehow less well adapted than those that survived. Further, he posited that if anything some of the animals we think of us as "weird" were more abundant in the Cambrian than the animals like *Canadaspis* that, as a crustacean, we would not call weird (or as weird) because it ended up surviving. Gould argued that survival in the case of the Burgess Shale animals, and in the case of life in general, is largely a matter of chance. As we discussed in Chapters 5 and 6, long-term evolutionary survival has little to do with organisms being more competitively fit and more complex or progressive. It would be impossible to scientifically argue that an *Anomalocaris* is less fit or less complex than a crustacean. Part of the lesson of the Burgess Shale to Gould, and we feel this is a legitimate lesson, is that the history of life is often just a crapshoot. Some animals make it, but most don't.

Further, our view of *Opabinia* and *Anomalocaris* as "bizarre" or "weird" is largely a matter of our anthropomorphic bias, and an implicit and invalid assumption that life today is the only possible course that evolution could have followed. If, by contrast, different types of animals had happened to survive whatever extinction event, or events, that eliminated most of the representatives of the Cambrian soft-bodied fauna we would have a very different type of modern biota. Gould further suggested it was worth imagining what it would be like if the crustaceans went extinct and *Opabinia* and *Anomalocaris* survived: then the fossil crustaceans would be the things that looked weird to us.

Contingency and Chordates in the Cambrian

Although we have been primarily describing arthropods and the other more distinctive animals from Cambrian soft-bodied faunas, we now turn our attention to a different group, the chordates. We will take up a detailed discussion of chordate and vertebrate evolution in Chapter 16. However, chordates are known from the Burgess Shale and the Chengjiang Biota and interestingly, in the former, they are extremely rare (a single specimen of the early chordate *Pikaia* (Figure 13-21) exists). Gould, in *Wonderful Life*, used the record of chordates in Cambrian soft-bodied faunas, and in particular, their paucity, to illustrate an important point.

The assumption is often made, wrongly, that the evolution of conscious thought, and perhaps even humans, or at least human-like forms, is highly likely, even preordained, because of the natural tendency for life to evolve greater intelligence;

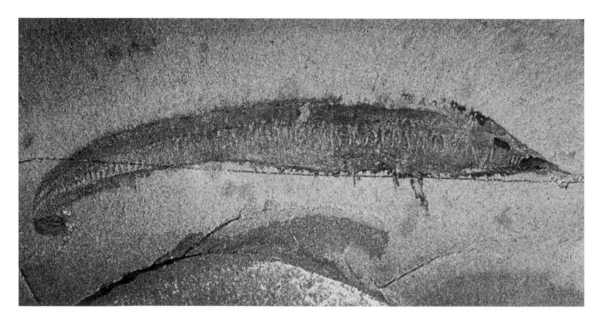

Figure 13.21 *Pikaia*. The early chordate from the Burgess Shale. Image from S. Conway Morris, University of Cambridge.

this assumption is based on the supposition that through time intelligence must produce some competitive advantage. We will discuss the evolution of the vertebrate brain in greater detail in Chapter 15 but the truth is, for better or worse, this isn't the case. Gould raised the important point that if one could go back in time to the Cambrian there would be no rational grounds for predicting that one day the chordates would evolve conscious thought and include the species (us humans) that has had a prominent effect on the Earth. Chordates are rare in the Cambrian, especially in the Burgess Shale, whereas it is the arthropod-like things, in particular, *Marrella* (Figure 13-12), that is known from tens of thousands of specimens and appears to dominate, if you will. Now partly the reason there are so many *Marrellas* known from the Burgess Shale may relate to the unique conditions that allowed this deposit to form and preserve soft tissues. However, Gould's point that back in the Cambrian chordates were not a great success story is well taken. Any set of betting aliens visiting our planet in the Cambrian would have put their money on *Marrella* as the future success, not *Pikaia*, yet because the history of life is governed by contingency, all bets are off and predictability is out the window. Gould argued that if *Pikaia* and its other chordate relatives perished in the extinction event(s) that eliminated many of the things present in the Burgess Shale we of course would not be alive today. By contrast, something like *Marrella* or *Anomalocaris* might be, and further, conscious thought might never have evolved on this planet. In short, Gould not only used the Burgess Shale to focus on the nature of how genetics and development may cause evolution to become more constrained through time, but he also used it to show how simple processes like adaptation and competition might not explain all, or even much, of evolution.

The Aftermath of the Radiation, and What the Radiation can Tell us about Evolution

The conclusion of the Cambrian radiation is usually placed around 510 million years ago. It is followed by the close of the Cambrian period, around 490 million years ago. The Cambrian was essentially the last time that many distinct major animal groups—phyla—evolved. However, of course this does not mean that evolution ground to a halt after the Cambrian period. Although evolution became more about variations on a theme, there were certainly some exciting evolutionary events to follow, and one of these, the Ordovician radiation, we will also discuss in this chapter. The significance of the Cambrian radiation lies in the fact that it marks the first appearance of abundant animal remains in the fossil record and the first time of significant animal evolution. Some of the animal groups that first appeared during the Cambrian probably evolved a few tens of millions of years earlier, in the Proterozoic. One of the things that paleontologists are doing now is scouring the latest Proterozoic and earliest Cambrian for various types of fossils, including embryos, to try to tease apart the exact timing and nature of these evolutionary events.

For now, it looks like what is evolutionarily unique about the Cambrian radiation devolves to two things. First, this was a time of relatively rapid evolution. For instance, rates of speciation were high during the early part of the Cambrian in some groups, like trilobites, although interestingly rates of evolution are even higher much later in the Phanerozoic, for instance, when the mammals start to diversify after the dinosaurs went extinct. (This is another interesting parallel between the events of the Proterozoic–Cambrian and the Cretaceous–Tertiary.) Why might there have been such rapid evolution? It could be because of the geological events described earlier, especially the breakup of a supercontinent, which would have amplified opportunities for geographical isolation and thus speciation. The Cambrian radiation was also a time of greater evolutionary flexibility: the amount of evolutionary change that occurred at individual speciation events may have been greater during the Cambrian then later on. The reason for this may relate to the issue that Gould identified: there could have been something about the genetic makeup of organisms, and how organisms developed from a fertilized egg to an adult, that was fuzzier and less strongly specified back in the Cambrian, compared with today.

The Ordovician Radiation and Concluding Remarks

The Ordovician radiation is an important evolutionary event that follows the Cambrian radiation. It represented a sustained period of evolution over tens of millions of years during the lengthy Ordovician Period. It involved not an increase in the disparity of animals present, as was the case with the Cambrian radiation, but instead in their overall species diversity and also in their ecological habit. The ways that different Cambrian organisms made a living were relatively simple, to the extent that there were not many organisms that were actively swimming (although the aforementioned *Anomalocaris* is a notable exception), and there were not many animals burrowing deep into the sediment. Instead, most organisms simply scurried along the seafloor or were living at the very level of the seafloor and fixed. Of course bottom dwellers and scurriers still persisted in the Ordovician, but during the Ordovician a much greater variety of different types of predators and prey with more different ways to make a living appeared (Figure 13-22). For instance, during the Ordovician animals evolved the ability to burrow at various different depths into the seafloor. Also, animals rooted to the seafloor had evolved in the Cambrian, but during the Ordovician many more types of these evolved; further these fed at a variety of different heights above the seafloor. In addition, during the Ordovician a much greater number of actively swimming forms evolved, including both predators and prey. Finally, animal life started to make the move into freshwater. This Ordovician expansion in ecological complexity was associated with a concomitant rise in overall species diversity, even though only one new phylum, the Bryozoa, makes its appearance during the Ordovician.

As with the Cambrian, what may have caused this proliferation has long been a puzzle to paleontologists. Originally, it was thought that perhaps the appearance of new types of animal ecologies facilitated the appearance of other types of animal ecologies. In a sense, this is akin to the idea that evolution builds momentum and the origin of certain types of animals (for instance, a new swimming animal, or something that lived at a new depth above the seafloor) enabled the evolution of other types of animals (for instance, predators to eat that swimming animal, or something else that could live at a different depth above the seafloor). Recent research findings suggest that this view of evolution building on itself may be too simplistic for the case of the Ordovician (and probably in general too). In particular, an exciting recent study by Schmitz, Harper and colleagues has shown that this time of ecological expansion and proliferation occurred during a time that large asteroids were colliding with the planet. This might seem paradoxical at first, given that asteroid impacts are frequently associated with times of mass extinction on this planet, as discussed in Chapter 6. However, maybe times of elevated extinction

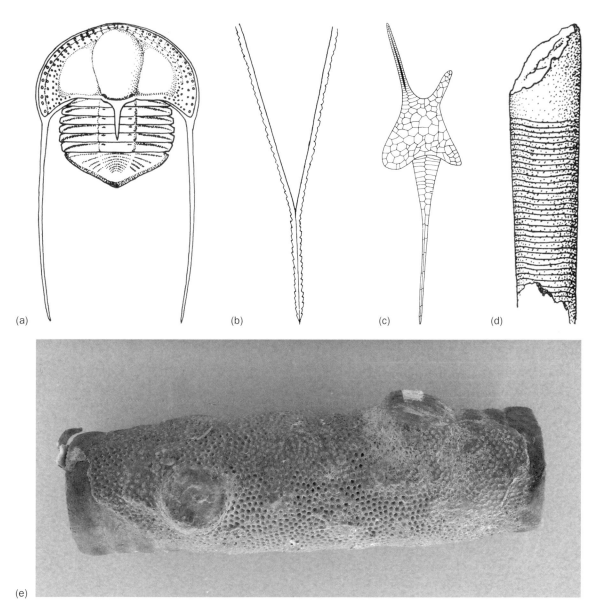

Figure 13.22 Some Ordovician animals. (a) A trinucleid trilobite, (b) a graptolite,
(c) a homalozoan, (d) a nautiloid cephalopod, and (e) a bryozoan encrusting a crinoid stem
(note this fossil is from the Carboniferous but these groups originated or underwent major
evolutionary radiations during the Ordovician). (a–d) from E. N. K. Clarkson (1998)
Invertebrate Palaeontology and Evolution, 4th Edn, Blackwell Publishing: (a) redrawn from
Treatise on Invertebrate Paleontology; (c) redrawn from Jefferies. Image in (e) by B. S. Lieberman,
specimen in the University of Kansas Museum of Invertebrate Paleontology.

might also allow subsequent evolution: again, consider the case of the dinosaurs and
mammals at the end of the Cretaceous. This is a fascinating area of research that
may revolutionize our understanding of the Ordovician radiation. Also, in parallel
with our discussion of the Cambrian radiation and various mass extinctions, it
points out how often it is a change in the physical environment (in the case of the
Ordovician radiation a change caused by something from outer space) that most
stimulates evolution.

Additional Reading

Benton, M. J., and Harper, D. A. T. 2008. *Introduction to Paleobiology and the Fossil Record*, 2nd edition. Wiley-Blackwell, Oxford; 576 pp.

Briggs, D. E. G., and Crowther, R. A. 2001. *Palaeobiology II*. Wiley-Blackwell, Oxford; 600 pp.

Briggs, D. E. G., Erwin, D. H., and Collier, F. J. 1994. *The Fossils of the Burgess Shale*. Smithsonian Institution Press, Washington, DC; 238 pp.

Edgecombe, G. D. 1998. *Arthropod Fossils and Phylogeny*. Columbia University Press, New York; 347 pp.

Fedonkin, M. A., Gehling, J. G., Grey, K., Narbonne, G. M., and Vickers-Rich, P. 2008. *The Rise of Animals: Evolution and Diversification of the Kingdom Animalia*. Johns Hopkins University Press, Baltimore, MD; 344 pp.

Fortey, R. A. 2001. *Trilobite: Eyewitness to Evolution*. Vintage Press, London; 320 pp.

Gould, S. J. 1989. *Wonderful Life*. W. W. Norton, New York; 352 pp.

Hou, X.-G., Aldridge, R. J., Bergstrom, J., Siveter, D. J., Siveter, D. J., and Feng, X.-H. 2004. *The Cambrian Fossils of Chengjiang, China: The Flowering of Early Animal Life*. Wiley-Blackwell, Oxford; 248 pp.

Knoll, A. H. 2003. *Life on a Young Planet*. Princeton University Press. Princeton, NJ; 277 pp.

Lieberman, B. S. 2001. A probabilistic analysis of rates of speciation during the Cambrian radiation. *Proceedings of the Royal Society, Biological Sciences* **268**: 1707–1714.

Lieberman, B. S. 2003. Taking the pulse of the Cambrian radiation. *Journal of Integrative and Comparative Biology* **43**: 229–237.

Schmitz, B., Harper, D. A. T., Peucker-Ehrenbrink, B., Stouge, S., Alwmark, C., Cronholm, A., Bergström, S. M., Tassinari, M., and Xiaofeng, W. 2008. Asteroid breakup linked to the Great Ordovician biodiversification event. *Nature Geoscience* **1**: 49–53.

Selden, P. A., and Nudds, J. R. 2005. *Evolution of Fossil Ecosystems*. University of Chicago Press, Chicago, IL; 192 pp.

Vickers-Rich, P., and Komarower, P. 2007. *The Rise and Fall of the Ediacaran Biota*. Special Publication 286, Geological Society Publishing House, Bath; 470 pp.

Whittington, H. B. 1985. *The Burgess Shale*. Yale University Press, New Haven, CT; 168 pp.

Chapter 14

The Evolution and Extinction of Reefs Through Time: From the Precambrian to the Current Biodiversity Crisis

Outline

- Introduction
- Reef-Forming Organisms Today
- Cnidarians and Outer Space
- How Modern Cnidarian Corals Feed
- Reefs Through Geological Time

- Corals and the Biodiversity Crisis
- Lessons from Human Effects on Modern Reefs
- Concluding Remarks
- Additional Reading

Introduction

In Chapter 6 we described the big five mass extinctions, their commonalities, and also what they could tell us about the current, human-induced biodiversity crisis. Recall that all of these past crises, and even the current one, were caused by habitat degradation: this is the fundamental mechanism of extinction in general and mass extinction in particular. By way of review, mass extinctions were times when many ecologically and taxonomically diverse groups of organisms kicked the proverbial bucket. For instance, the nonflying dinosaurs buy it at the end of the Cretaceous, along with the ammonites. At the end of the Ordovician, many groups of trilobites bade the planet a fond farewell; sadly the trilobites finally succumbed, along with many other life forms, during the cataclysm at the end of the Permian. The characteristics of the species that survived any one of these episodes, and also the

Prehistoric Life: Evolution and the Fossil Record. 1st edition. By Bruce S. Lieberman and Roger Kaesler. Published 2010 by Blackwell Publishing.

characteristics of those species that didn't, provide paleontologists with information about the particular causes of habitat degradation during any one of these mass extinction events. However, it is also true that there are certain types of organisms, broadly categorized ecologically, that do ultimately persist in the face of mass extinctions. Insight into what happened to these types of organisms across mass extinction boundaries, and also understanding how they changed through time, can help us understand the nature of mass extinctions and the history of life: it will be the focus of this chapter. In particular, we will consider one of the best examples of these: reef-forming organisms. Reefs form in shallow ocean water, typically as the mineralized skeletons of organisms (Figure 14-1) (though technically they are any large structure in oceans that collects or baffles sediment); they are large structures occupied by the organisms that form them, and they also provide refuges and hiding places for a host of other types of organisms, both predators and prey. For this reason, they represent places where large numbers of species are concentrated in the oceans of our planet, and are best thought of as biodiversity hotspots. Much of

Figure 14.1 **Modern coral reef.** From Northern Territory, Australia. Image from A. Crowther, University of Kansas.

today's diversity in the sea lives in, on, or around reefs. Eliminate this diversity, and life in the oceans takes a major hit.

As we shall describe in this chapter, it turns out that at any one time during the history of life different organisms have played the role of the dominant reef-forming organisms. At each of the mass extinctions, typically the dominant reef-forming organism was eliminated and thus displaced. Ultimately though, a new type of organism evolved to fill the vacated role. We will focus on some of the different major organisms that have built reefs, as we march through time in this chapter and briefly survey reef ecology and evolution from the Precambrian forward; we will also focus on the current status of today's dominant reef-forming organisms. These happen to be gravely threatened by habitat degradation caused by human activities.

Reef-Forming Organisms Today

Reefs today are primarily made by small organisms called cnidarians (Figure 14-2); these are soft polyps surrounded by a nonliving layer of calcium carbonate or

(a)

Figure 14.2 Close-ups of living cnidarians. Showing at different degrees of magnification from lowest (a) to highest (d) the large number of tiny polyps scattered across the surface of cnidarians. Specimens in (a)–(c) are corals from Northern Territory, Australia, images from A. Crowther, University of Kansas; polyps in (d) are on a hydrozoan from Japan, image from A. Migotto, Universidade de Sao Paulo with assistance of P. Cartwright, University of Kansas and A. Collins, Smithsonian Institution.

(b)

(c)

Figure 14.2 *continued*

(d)

Figure 14.2 *continued*

CaCO$_3$ that makes up the reef. As the cnidarian polyps grow they secrete more and more CaCO$_3$ such that the reef builds up through time, but as a result only the thin layer at the top of the reef is actually living. Volumetrically, the bulk of cnidarian reefs are formed by colonies of many simple, quasi-independent polyps; these polyps are joined together. Each polyp has some independence, but the entire colony also can behave as a unit. For example, each individual polyp can gather food, but this food is shared by the whole colony. Further, some of the polyps on the colony are responsible for reproduction and synchronize their spawning so that the entire colony reproduces effectively as a single organism. Different types of cnidarians have formed reefs through time, and the phylum Cnidaria not only includes corals, but also jellyfish (Figure 14-3).

Cnidarians and Outer Space

It is an interesting bit of trivia that in fact the largest "built" structure on Earth was not made by humans but instead by cnidarians. We refer to the Great Barrier Reef, situated off the eastern coast of Australia (Figure 14-4), which is visible from the Moon without a telescope. Cnidarians not only provide vistas from the Moon; they also give us some insight into celestial mechanics and the evolution of our solar system, especially the coupled Earth–Moon system. Modern coral CaCO$_3$ skeletons happen to grow in little increments each day, such that they form daily growth lines; these growth lines are also organized into larger seasonal and annual bands, because at different times of the year corals grow more or less rapidly, depending on their environment. Thus, it is no surprise that if you examine modern corals they have 365 annual growth lines. The surprise comes when you look at long extinct

Figure 14.3 Jellyfish. Modern specimen from Japan. Image by A. Migotto, Universidade de Sao Paulo with assistance from P. Cartwright, University of Kansas and A. Collins, Smithsonian Institution.

Paleozoic corals. The cnidarians that currently dominate the reef-forming environment evolved some 240 million years ago, but the record of reef-forming cnidarians extends well back into the Paleozoic. One type of Paleozoic reef-forming cnidarian were the rugosans (Figure 14-5), and these all vanished at the end of the Permian. They were somewhat different from the reef-forming cnidarians alive today. However, one thing they do share with modern reef-forming organisms is they also accreted daily growth bands aggregated into larger seasonal and annual bundles.

When you count the number of daily bands in a year in a 400 million year old cnidarian coral the result is a bit of a surprise: there happens to be about 400 lines/year. Why were there more days in the year back in the Devonian? A year is calculated as the number of days it takes for the Earth to revolve around the Sun. Perhaps the reason why there were more days back in the Devonian, relative to today, might be explained by the fact that the Earth is slowly falling towards the

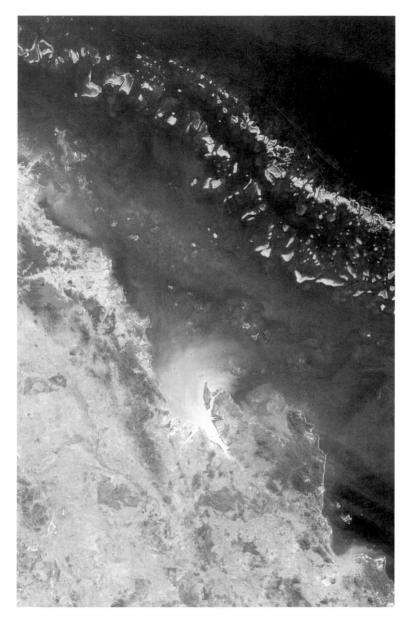

Figure 14.4 **Great Barrier Reef from outer space**. Showing the northeast coast of Australia. The patches to the top and right are reef while those to the bottom and left are land. Image from NASA.

Sun? That seems not to be the case, for our solar system would not be stable nor could it persist for long periods of time if the relative distances between the various planets and our Sun changed significantly. Instead, the reason for the diminishing number of days in a year seems to lie in the interactions between the Earth and the Moon. The Earth not only revolves around the Sun but it is also rotating on its axis, while the Moon revolves around the Earth. Currently it takes 24 hours for the Earth to rotate on its axis: hence the length of a day. As the Earth rotates on its axis, and the Moon revolves around it, the Moon's gravity gently tugs on our planet; indeed,

Figure 14.5 Rugose coral. A specimen
from the Silurian. Diagram
E. N. K. Clarkson (1998) *Invertebrate
Palaeontology and Evolution*, 4th Edn,
Blackwell Publishing; redrawn from
Milne-Edwards and Haime.

this is what causes the difference between high and low tide. In the short term, this
gravitational tug doesn't do much, other than cause the oceans to subtly swash
around. However, over the scale of hundreds of millions of years this gravitational
tug slowly but inexorably decreases the rate of rotation of our planet. As the rotation
of the Earth slows, fewer days can be packed into a year. The net effect is that today
the Earth has a 24 hour day, but back in the Devonian a day only lasted about
22 hours. (We're not sure if this is better or worse from the human perspective, but
it's certainly a little bit different.) Interestingly, physicists could estimate the size of
the gravitational effect of the Moon on the Earth's rotation but they lacked empirical
confirmation: empirical confirmation about the dynamics of the Earth and Moon,
and indeed our solar system, came from the study of Devonian corals. Who might
have guessed that thanks to corals now we know that a day used to be shorter
400 million years ago?

How Modern Cnidarian Corals Feed

Corals happen to feed in two distinct ways. One way they feed is that the polyps
that make up a coral will fire stinging cells when small, microscopic organisms
approach their tentacles. The stinging cells kill or injure their prey, which can then
be drawn into the mouth, eaten and digested. These stinging cells are just smaller,
weaker versions of the stinging cells that humans have come to fear in jellyfish and
that allow the Portuguese Man-O-War to catch fish as large as tuna. Although this is
certainly a dynamic and exciting way of feeding, as far as corals are concerned it's
not the only way, nor is it the most important way, of feeding.

Photosynthetic, Symbiotic Algae

Many modern corals also have microscopic algae that live inside their tissues. These algae are photosynthetic and extract energy from the Sun; they have a symbiotic relationship with corals, meaning each benefits the other in some way. For instance, as the algae take in energy from the Sun and carbon dioxide from the ocean, they produce sugars (carbohydrates) that are used by the coral for food; the corals in turn provide shelter for the algae. These "algae-provided" carbs are where corals get most of their nutrition from and it actually allows them to live in places otherwise lacking in food and nutrients. Also though, because their symbiotic algae require sunlight to carry out photosynthesis and they depend on these algae for a substantial component of their food, it means that corals are constrained to live in shallow water: too deep and the water filters out the necessary rays and energy from sunlight. This is why in deeper water the number of coral species declines precipitously. The energy the algae provide is also instrumental because it helps corals to grow their $CaCO_3$ skeletons more rapidly. Maintaining high growth rates of the coral skeleton is important because it allows them to better fight disease and repair damage caused by storms.

Ultimately, the photosynthetic, symbiotic algae are pivotal in the equation of where corals live. Reefs thrive in clear tropical waters. These are deceptively rich in marine life. It is true that all kinds of brilliantly hued fish swim around the reef, while many other kinds of animals live on and around it too. However, tropical waters are quite low in productivity, explaining why the water is so clear. We might think that cloudy ocean water is in some ways unclean, and that may be the case, but oftentimes water can appear cloudy because it has large numbers of microorganisms living in it; this is the food that animals need to eat and bigger animals in turn eat the animals that eat the microorganisms, etc., etc. Really, the fact that tropical waters are clear is deceptive, because it means they are largely devoid of the microorganisms that animals need to eat. In fact, were it not for coral reefs the tropics would be the blue deserts of the oceans. Their presence attracts a rich diversity of species that could not exist there otherwise, such as animals that live in cracks and crevices of the reef and also feed on the reef, or on the organisms that live on and around the reef. Coral reef success in the tropics is based not on nutrient-rich waters, but on the photosynthetic algae they contain.

Corals are Very Sensitive Environmentally

Environmental demands also determine the geographical range of modern coral reefs (Figure 14-6). Warm, constant temperatures (27–29° C in winter, less than 3° C warmer than this in summer) of the tropical seas are required. Further, poor water visibility excludes corals from eastern South American and west African coasts where large rivers (e.g., the Amazon, Orinoco, and Niger) discharge lots of sediment into waters. Salinities of open oceans (water having a salt concentration of 36 parts per thousand) are also necessary. If there is too much freshwater in the vicinity, for instance if a river discharges nearby and salt levels drop, or if a bay is very isolated and there is extensive evaporation and salt levels rise, corals cannot survive. The constant, stress-free environmental requirements of reef communities in general make them extremely sensitive to any environmental perturbations.

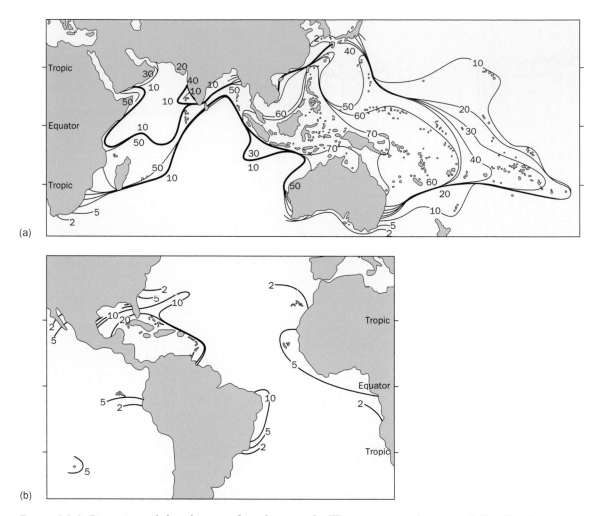

Figure 14.6 Diversity and distribution of modern corals. The contours show coral distribution and the numbers next to the contours represent the maximum number of genera present in (a) the Indian and western Pacific Oceans and (b) in the Atlantic Ocean. Diagram from C. B. Cox and P. D. Moore (2005) *Biogeography: An Ecological and Evolutionary Approach*, 7th Edn, Blackwell Publishing; after Veron.

Reefs Through Geological Time

Reefs have a long history on this planet, and evidence suggests that basically the same environmental conditions that characterize reefs today also characterized the reefs of the past. However, as we already mentioned, the organisms responsible for reef construction have changed dramatically over time, and further these changes happened in fits and starts. Recall Chapter 12 when we discussed cyanobacteria and the large mats they built called stromatolites. Sometimes these stromatolites formed large and impressive structures (see Figure 12-5). Stromatolites were the first reefs, and date back to 3 billion years ago. Thus, cyanobacteria were the very first reef-forming organisms, and the dominant ones for much of the history of life. Stromatolites' hegemony as the dominant reef-forming organisms was not toppled until the end of the Precambrian. This is not to say that no Phanerozoic stromatolite reefs exist, but these certainly became much rarer in the fossil record post-540 million years ago.

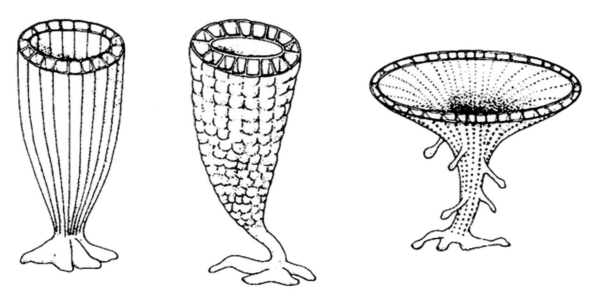

Figure 14.7 Archaeocyathids. Extinct Cambrian reef-building organisms. Diagrams from R. Cowen (2005) *History of Life*, 4th Edn, Blackwell Publishing; derived from R. S. Boardman, A. H. Cheetham, and A. J. Rowell (1987) *Fossil Invertebrates*, Blackwell Publishing.

The Earliest Animal Reefs

The first reef-building animals (recall that cyanobacteria are single-celled organisms and not animals) did not evolve until the Cambrian radiation (an event discussed in detail in Chapter 13). During this initial profusion of animals in the fossil record the reef-forming archaeocyathids also evolved (Figure 14-7). They were fixed and could perhaps best be described as cone-like or irregularly shaped; they are distantly related to modern sponges. An archaeocyathid reef forms as different individual archaeocyathids stack one on top of another and accumulate through time. Some of these reefs grow quite large and today in places like modern day Russia are preserved as dome shaped mounds a few hundred feet high. Alas, their time on Earth was relatively limited and they disappeared from the fossil record before the end of the Cambrian period.

Sponge Dominated Reefs

Following somewhat on the heels of this extinction, and by the mid-Ordovician (about 480 Ma), a different and more complex reef-building community had evolved. This included small algae that make hard skeletons but were different from stromatolites, although animals were still ascendant, in particular, stromatoporoid sponges (Figure 14-8), which are very different from most modern sponges (Figure 14-9), though still of course sponges at base. The fossil reefs these sponges left behind can still be found in Silurian rocks; for example, in Sweden they formed huge masses 20 meters high and 200 meters in diameter. By the mid-Paleozoic, they had waned significantly and no longer comprised and produced the bulk of the worlds reefs, except perhaps for a short pulse in the Cretaceous period. In fact, paleontologists had long thought the group had gone extinct in the Mesozoic,

Figure 14.8 Stromatoporoid sponge. Fossil is in cross-section. Image from C. Stock, University of Alabama.

although now it appears that relatively closely related sponges persist today; these relatives still play a minor role in building some modern reefs.

Paleozoic Cnidarian Reefs

During the Ordovician two other important reef-forming organisms evolved: these were the tabulate and rugosan cnidarians (the latter, those of the lengthening day saga already discussed). Tabulates were colonial (Figure 14-10) cnidarians, while the rugosans included both solitary and colonial forms. As the stromatoporoids generally waned, these became the dominant reef-forming organisms, particularly of the mid-Paleozoic. They too, however, went extinct, in this case, at the end of the Permian.

The End Paleozoic and Mesozoic: Reefs get Stranger and Stranger

By the end of the Paleozoic another group joined the lengthening litany of reef-building animals: the brachiopods (Figure 14-11). Permian brachiopod reefs occur prominently in western Texas. There the Capitan Reef complex in the Glass Mountains and Guadeloupe National Park preserves immense reefs 300 feet high. These are made up of a diverse set of brachiopod species and include some, by brachiopod standards, particularly bizarre morphologies. These reef-building brachiopods (along with many of their brachiopod kin) succumbed at the end-Permian extinction.

Figure 14.9 Modern sponges. The large pipe-like structures. Image from iStockPhoto.com © M. Stubblefield.

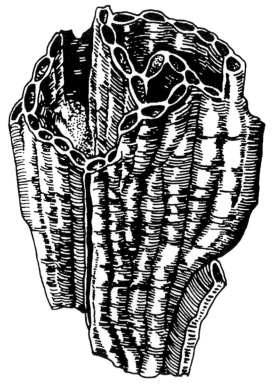

Figure 14.10 Tabulate coral. From the Ordovician and Silurian. Diagram from E. N. K. Clarkson (1998) *Invertebrate Palaeontology and Evolution*, 4th Edn, Blackwell Publishing; redrawn from the *Treatise on Invertebrate Paleontology*.

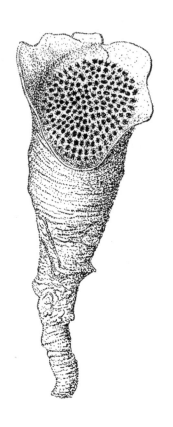

Figure 14.11 Reef-building brachiopod.
From the Permian of Texas. Diagram
from E. N. K. Clarkson (1998)
*Invertebrate Palaeontology and
Evolution*, 4th Edn, Blackwell
Publishing; redrawn from the *Treatise
on Invertebrate Paleontology*.

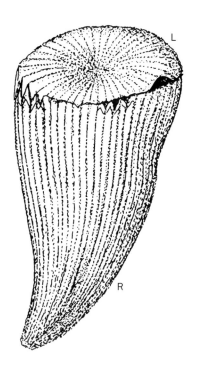

Figure 14.12 Reef-building clam.
A rudist from the Cretaceous.
Diagram from E. N. K. Clarkson
(1998) *Invertebrate Palaeontology
and Evolution*, 4th Edn, Blackwell
Publishing; based on Zittel.

Reefs are essentially absent from the fossil record for the next 20 million years
following that extinction. However, matching the familiar pattern, in the Mesozoic a
new type of reef-forming animal evolved and ascended to dominance; these new
organisms were clams, called rudists (Figure 14-12). Some particularly large

examples of Cretaceous rudist reefs are known from Mexico. It should be no surprise by now though that they too became extinct; it so happens close to the very end of the Cretaceous, and at about the same time the ammonites and dinosaurs went extinct. It was only then that the modern type of cnidarian reef-forming organisms truly ascended to dominance. The Great Barrier Reef complex that we already described is a spectacular example of a modern reef that has probably been growing for some 10 million years. In a very real sense, when we look back at a Silurian stromatoporoid reef, or a Devonian rugosan coral reef, or a Permian brachiopod reef, we are seeing an ancient analog of today's Great Barrier Reef. When it comes to reefs, it seems that over vast stretches of time the players have definitely changed, but perhaps the basic rules of the game have stayed the same. The organisms that formed the reef have come and gone, but typically after a hiatus a new organism evolves to fill that role.

Corals and the Biodiversity Crisis

It is in the context of the history of life that our understanding of, and current concerns about, coral reefs should be viewed.

Corals and Global Warming

One of the profound threats corals face today is global warming. As we described in Chapters 10 and 11, for a variety of reasons, both actual evidence for temperature rise, and increases in atmospheric CO_2 levels, global warming already seems to be here. Because of temperature increases and other stresses, corals have recently undergone several "bleaching" events. This happens when the corals release their photosynthetic algae. It is called bleaching because the symbiotic algae give corals a vaguely greenish cast; when the algae leave, the coral turns white. (Currently it is not entirely known why algae vacate their coral hosts as ocean temperatures climb.) Coral bleaching is not the instantaneous death knell of any given reef, for corals can still live without their algae, at least in the short term. However, they grow at a much slower rate, and are much less able to resist diseases. Therefore, they are more easily destroyed by storms and bacterial infection in the long term, and this rampant bleaching is one reason that corals are in decline today. Many individual reefs and the species that comprise them are threatened and in danger of being wiped out.

   In a way it might seem paradoxical that global warming is seen as a major threat to coral reefs. For instance, it might be thought that temperature increases *per se* would be favorable to reef communities, if temperatures did not climb too high, because the net result would be an increase in the area of the tropics. However, with global warming there appears to come greater variability in temperatures, and this would be very detrimental to corals. Further, if climate changes are very rapid, as anticipated, coral species will not be able to respond to the changes.

   This is actually important from the perspective of global warming itself because the photosynthetic algae inside corals are an important factor in removing CO_2 from the oceans and the atmosphere; they thus serve to keep the greenhouse effect in check. However, as reefs decline due to global warming their symbiotic algae will no longer be around removing CO_2; thus, CO_2 levels would rise, enhancing the greenhouse effect even further, causing a more precipitous drop in coral populations. (Another negative consequence that increasing atmospheric CO_2 levels has for modern reefs is it increases oceanic acidity, which makes it harder for cnidarians to build their skeletons.)

Consequences of Coral Destruction

Realize that killing off reef species would by itself lead to a mass extinction in the marine realm because so many other species depend on them for survival. Further, some of the fish species humans depend on for food shelter in and among reefs as juveniles, and thus they'll vanish too, along with one of humanity's important food sources. Given what we have described about corals in geological time, how do we view current threats to corals in the context of the history of life? It is true that throughout the history of life there have been several major extinction episodes, and at many of these times the dominant reef-forming organisms were displaced, yet ultimately new reef-forming organisms evolved and built new types of reefs; these in turn attracted a diverse host of other life forms. Despite their narrow ecological tolerances, reef communities are on the whole persistent throughout geological time. The cynic might say, "What's the big deal, even if we wipe out reefs, they'll come back, eventually?" But such would be a misinterpretation of the history of life: the key word is "eventually." Based on our understanding of the fossil record, it typically takes more than 10 million years for a new dominant reef-forming organism to come to the fore after an extinction event: little comfort for our children or grandchildren or even the next ten thousand generations of humans, if our species is lucky enough to survive that long.

The value of reefs to humans is tangible, not only because they provide shelter for the young of important fish species we eat, but also for providing us with joy and scenic beauty: the stuff that makes life worth living. People travel from all over the world and are willing to pay money to visit the reefs of Australia and the Caribbean Sea, but what if they were turned into a dead and barren pile of rubble? Who would pay to see that (and further, shouldn't we be willing to pay to make sure that doesn't happen)? In parts of the Caribbean Sea this profound level of destruction really is happening, as we describe more fully below. Nobody traveled for thousands of miles to visit Lake Erie, when it was in the depths of its pollution doldrums (though thankfully it appears to now be rebounding); who would want to see something so biologically dead, boring or depressing? How can we justify activities, running the gamut from climate change to pollution (discussed below), that risk turning our most spectacular underwater environments into environmental waste lands? The truth is, we cannot.

Lessons from Human Effects on Modern Reefs

Several interesting scientific studies have been conducted on modern reefs, particularly in the Caribbean, and these give us insight into the types of challenges that reefs face today, particularly where they are near large human populations. The greatest threat to modern coral reefs is habitat degradation; this mass extinction is being caused by a growing human population which needs space to live and food to feed itself, thereby degrading the environment for other life forms. There comes a point when increases in human population size come at the expense of other living things, especially plants and animals. This trade-off is particularly well exemplified in the coral reefs of Jamaica (Figure 14-13) which have been under intense scientific

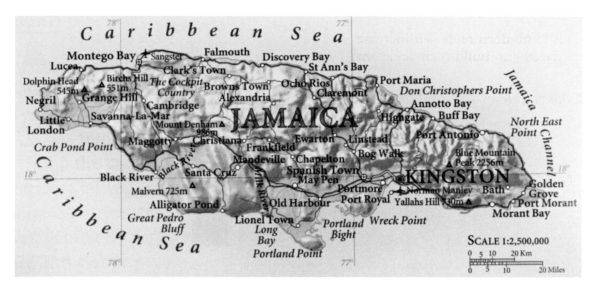

Figure 14.13 Map of Jamaica. Diagram from iStockPhoto.com.

study for the past 20 years. Jamaican reefs fringe most of the north coast of the island in a narrow belt 1–2 km wide that extends over about 350 kilometers from east to west.

The reefs are built up of several cnidarian species and many, many more species depend on the reef for shelter and food. Further, sea-grasses and mangrove trees occur along with the reefs, and the juveniles of many fish species live among the mangroves and sea-grass beds. The environment and economy of Jamaica is faced with significant problems due to population growth; population has increased 250 percent there in the last 70 years. This growing population has meant that almost all of the native Jamaican landscape has been cleared for agriculture and urban development.

Unintended Consequences of Over Fishing

The fisheries have also been heavily stressed in order to feed this growing population; recently, the total mass (biomass) of the fish caught off the waters of Jamaica has decreased by 80 percent. Basically, there is more demand for fish than available supply, and fish are being harvested at much higher than sustainable levels. The result is that most fish that are caught in this region are probably migrating in from elsewhere. In addition, over the past 30 years virtually all of the large predatory species of fish have been eliminated from these waters. Now most of the fish caught in the area are very small, with half the fish caught being pre-reproductive—the fish are killed before they can give birth to the young that would restock future generations.

The over fishing had many consequences. First, there was a large increase in the population size of a particular sea urchin (Figure 14-14), called *Diadema*. The animal is a relative of starfish, some sea urchins are even eaten in sushi bars, and its main predators are different species of fish. *Diadema* primarily feeds on algae (not the symbiotic algae of corals, but rather other types of algae that grow on or around the reef that were already described in the case of Ordovician reefs); many

Figure 14.14 Sea urchin. Image from iStockPhoto.com © Wendy Rood.

fish species also feed on algae, although some fish species feed directly on the reefs themselves. Viewed simplistically, one might think that increased rates of fishing should benefit the reef because fish were important predators of the reefs. But here the perils of human meddling (either purposeful or inadvertent) with biological complexity are well illustrated and the principal relevant organisms are the algae. As long as *Diadema* was around, algal population size remained low; typically fish would keep down algae population size too, but because few fish were around they were no longer an important buffer against population explosion by algae. It was by removing the many different buffering links of nature that we unintentionally set up an impending environmental catastrophe.

Disaster struck in 1983 when there was a massive die-off of *Diadema* due to a disease throughout its range; the disease may have been caused or hastened by human pollution, which also degraded the reefs themselves. The net result was a dramatic increase (more than 2000%) in the population of algae, with a so-called algal bloom that began in 1983 and continues through today (although *Diadema* may be making a bit of a come-back from the disease now). The algal bloom is very detrimental to corals, indeed far worse than fish predation on corals, because the algae grows over the corals, covering them, and smothering the reef. The result of this has been a precipitous decline in the Jamaican coral reefs and over the last 20 years as much as 90 percent of the more than 300 km long reef died as it was smothered by algae.

Concluding Remarks

The level of damage suffered by Jamaican reefs obviously is major, but thankfully it is not yet irreversible. This is because none of the individual species that make up the Jamaican reef have been completely driven to extinction; once that happens, of course, they can never come back, and the reef would permanently languish. So, as long as corals still live in the vicinity of Jamaica, and can recolonize Jamaican reefs, eventually the reef will recover. However, certain critical reforms also have to be taken for recovery to happen. These principally include reforming the Jamaican fishery such that over-fishing stops and fish stocks can rebound and once again happily munch on the coral smothering algae. In order for this to occur, the people of Jamaica need to be given other options besides over-fishing and other forms of rampant over-development. They also need to control the size of their population which has effectively gone beyond the size of what the island can support, and avoid critically endangering neighboring wild-life on land and in the seas.

We should state that of course we are not singling out Jamaica for particular criticism here. Many countries are guilty of similar phenomena; it is simply that Jamaican corals have been particularly well studied scientifically, probably because a lot of scientists want to get to do their research in Jamaica. The situation in Jamaica is in fact really a case where the developed industrialized nations like the United States can and should step in; relatively simply and inexpensively we could take steps that would address the problems in the Jamaican fishery and help mitigate the environmental damage that has occurred there and elsewhere on the island. It is also true that the floras and faunas, be they marine or terrestrial, of nearly all nations face similar problems because of excessive population growth. This is the fundamental problem that all nations need to face and we need to confront it soon before it is too late. Once the actual species themselves are eliminated, especially when they are species like corals that so many other species depend on, we and the biosphere cannot go back. Again, whatever mass extinction that we ultimately cause through these types of effects would not eliminate all the life on the planet, but it could eliminate a significant percentage, while degrading life for all. Another concern is that if enough things do go extinct then we could go down with the proverbial ship.

The current loss of biodiversity on this planet is akin to a phenomenon that we can perhaps all relate to: imagine rivets being removed from an airplane wing. Remove a few rivets from an airplane wing and perhaps nothing happens and it's still safe to fly, though we confess that we are not comfortable flying in an airplane lacking any wing rivets—that's why we avoid certain budget air carriers. Remove too many rivets from a plane wing though, and you've got a serious problem and by serious we mean disaster in the making; further, the transition between all situations green for flying and something disastrous might represent a pretty sharp cutoff. It turns out that our planet's biosphere is in a similar situation now. Imagine that the biosphere is the plane we're flying in. We depend on that biosphere for food, oxygen, water, and a host of other things. The biosphere is akin to the body of the plane and the wings are supported by the millions of species that populate our planet. We depend on our biosphere, yet it is under significant stress thanks to the great success of humankind. Solely because of us, many species have already gone extinct. Therefore recognize that already a few rivets have been removed from the biosphere. Still, it is not too late to check population growth and other activities like global warming and get things back on track by living more in harmony with nature. If we go too far, however, in terms of population growth and global warming, it may never be possible to go back, and it will take the Earth's biosphere as long as 10 million years to recover. Controlling the unchecked growth and spread of the human species, with the associated habitat degradation we are causing, will be the greatest challenge that humans must face in the coming decades. Still, thankfully, we already have the knowledge and technology to implement most of the necessary fixes including population control and minimizing climate change; that is why we believe the solution to this problem is well within reach and we are more sanguine about our future prospects than might be supposed.

Additional Reading

Aronson, R. B. (ed.). 2006. *Geological Approaches to Coral Reef Ecology*. Springer, New York; 442 pp.

Birkeland, C. (ed.). 1997. *Life and Death of Coral Reefs*. Springer, New York; 560 pp.

Ferrari, A. 2002. *Reef Life*. Firefly Books, London; 288 pp.

Hallam, A., and Wignall, P. B. 1992. *Mass Extinctions and Their Aftermath*. Oxford University Press, Oxford; 328 pp.

Hardin, G. 1995. *Living within Limits: Ecology, Economics, and Population Taboos*. Oxford University Press, Oxford; 352 pp.

Hughes, T. P. 1994. Catastrophes, phase shifts, and large-scale degradation of a Caribbean coral reef. *Science* **265**: 1547–1551.

Jones, G. P., McCormick, M. I., Srinivasan, M., and Eagle, J. V. 2004. Coral decline threatens fish biodiversity in marine reserves. *Proceedings of the National Academy of Sciences, U.S.A.* **101**: 8251–8253.

Knowlton, N. 2001. The future of coral reefs. *Proceedings of the National Academy of Sciences, U.S.A.* **98**: 5419–5425.

McMichael, A. J. 1993. *Planetary Overload: Global Environmental Change and the Health of the Human Species*. Cambidge University Press, New York; 372 pp.

Pandolfi, J. M., Bradbury, R. H., Sale, E., *et al.* 2003. Global trajectories of the long-term decline of coral reef ecosystems. *Science* **301**: 955–958.

Ponting, C. 2007. *A New Green History of the World: The Environment and the Collapse of Great Civilizations*. Penguin, New York; 464 pp.

Stanley, G. D., Jr. (ed.). 2001. *The History and Sedimentology of Ancient Reef Systems*. Springer, New York; 468 pp.

Stanley, G. D., Jr. 2003. The evolution of modern corals and their early history. *Earth-Science Reviews* **60**: 195–225.

Veron, J. E. N. 2008. *A Reef in Time: The Great Barrier Reef from Beginning to End*. Harvard University Press, Cambridge, MA; 304 pp.

Wood, R. 1999. *Reef Evolution*. Oxford University Press. Oxford; 432 pp.

Chapter 15

Key Evolutionary Transitions: The Origins of Multicellularity and the Evolution of the Vertebrate Brain

Outline

- Introduction
- Origins of Multicellularity
- The Evolution of the Vertebrate Brain
- Trends in Brain Size Within Primates and Hominids
- Concluding Remarks
- Additional Reading

Introduction

As we have mentioned repeatedly throughout this book, the scientific study of the history of life is about using the evolutionary patterns preserved in the fossil record to study evolutionary processes and mechanisms. Here we will focus on two key evolutionary patterns, the origins of multicellularity and the evolution of the vertebrate brain, and use these to consider how evolution works. Each of these evolutionary episodes can rightly be viewed as important in the broader context of the history of life, but they were also pivotal from the perspective of humanity. If multicellularity had not evolved, the Earth would be devoid of animals, and thus humans; further, without our large, complex brains, we would of course not be able to ponder such questions in the first place.

Prehistoric Life: Evolution and the Fossil Record. 1st edition. By Bruce S. Lieberman and Roger Kaesler. Published 2010 by Blackwell Publishing.

Origins of Multicellularity

We partly considered the question of animal origins in detail in Chapter 13 where we described the Ediacaran fauna (also considered in Chapter 8) and the Cambrian radiation. Recall that for most of this planet's history animals were absent. Most Ediacarans, although large, were thin and flat, and they appeared near the end of the Proterozoic. With a few exceptions, they went extinct right before the start of the Cambrian radiation. Paleontologists still debate whether they are directly related to the Cambrian animals that followed them. The Cambrian radiation describes the time when most of the animal phyla appeared in the fossil record over a geologically short period of time: 10–20 million years or so. Of course, that is a long stretch of time, but it is rapid compared to the relative evolutionary quiescence that persisted before. Although we discussed this in Chapter 13, we did not focus on the evolutionary mechanisms tied up with the origins of multicellularity. This is important since by definition all animals are multicellular. (There are other life forms not directly related to animals that are also multicellular, for instance, plants, fungi, and slime molds, such that it is clear that multicellularity evolved multiple times.)

Part and parcel with the issue of the evolution of multicellularity is the evolution of large size. Thus, Galileo's principle, which we discussed in Chapter 8, is an important concept. A single cell obtains nutrients and oxygen from its exterior. As a single cell grows, its volume increases much more rapidly than its surface area. Regions in a large cell would quickly become separated from the external milieu that it relies on to obtain nutrition and also to eliminate wastes. If a single cell grows too large it becomes very difficult to transport nutrients and wastes to and from different parts of the cell. There are very few single-celled organisms that grow large enough to view with the naked eye; more than 99.9999 percent of all single-celled organisms are extremely small and only visible under a microscope. Thus it turns out that, typically, for life to become large, it must evolve a multicellular habit, and we will treat multicellularity and large size as largely synonymous.

Evolutionary Advantages and Disadvantages of Large Size

It is worth pondering some of the reasons why large body size might be favored evolutionarily. First, large body size allows organisms to exploit and eat different and larger food sources. Also, evolving large body size makes it harder to get eaten. Finally, in certain types of substances it can be difficult to move when an organism is small because of forces like surface tension. A case in point: many of you may be familiar with how tiny insects (excepting water striders) struggle to escape from a body of water. Larger organisms are less negatively influenced by surface tension.

There are certain requirements that need to be met for an organism to become large. Even in multicellular organisms, the cells have to be arranged to facilitate exchange of nutrients and oxygen with the environment, just as in single-celled organisms. One way some organisms do (or did) this, like tapeworms (and Ediacarans), is to maximize their surface area relative to their volume by being thin and flat. However, the way adopted by most animals is to become complex and develop a circulatory and digestive system, as well as specialized organs. This allows them to take in nutrients and oxygen and distribute these essential things to all of their cells. To perform this type of activity efficiently, cells must be differentiated into various cell types that perform different functions. For instance, in the animal body there are two broad types of cells, somatic and germ cells. The germ cells are those that eventually become the gametes (sperm and eggs). The somatic cells are all the other cells in your body. Although certain aspects of being multicellular and large are evolutionarily favored, there are also evolutionary difficulties associated with the acquisition and maintenance of multicellularity and these were nicely elucidated by the evolutionary biologist Leo Buss. In particular, different cells within an organism, such as the germ cells and the somatic cells, might compete with one another; if this were to happen, it would be difficult for multicellularity to persist.

To consider how the interactions between somatic and germ-line cells create challenges to the evolution of multicellularity it is worth reconsidering Darwin's theory of natural selection, which was discussed in detail in Chapter 4. Recall that natural selection was a statistical law derived from the following facts: (i) individuals vary; (ii) that variation is heritable; (iii) sometimes different variants are selected. Also recall from Chapters 4 and 5 that a gene, an organism, or a species can be selected. A cell should also be added to that list.

Selection at the Cellular Level

Leo Buss recognized that the evolutionary challenge of multicellularity is that it requires cooperation among different cells, yet before multicellularity evolved individual cells were working independently for their own evolutionary purposes. This principle can be illustrated nicely by considering a microscopic colony that is made up of many simple single-celled eukaryotic organisms called *Proterospongia*. Such colonies exist in the real world, and each cell can actually function by itself as an individual organism, but can also function in a colonial arrangement where they all live together (Figure 15-1). In such a colony, cells are broadly arranged into two types. There are cells that live inside the colony in a gelatinous matrix that are called amoeboid cells. There are also cells on the outside of the colony, and these are endowed with small whip like structures called flagella; these flagellated cells can act in concert to allow the colony to move. Locomotion is important for the colony because it allows movement towards food and its preferred environment. Although the flagellated cells move the colony, and are thereby performing essential labor on behalf of the colony, they cannot reproduce. By contrast, the amoeboid cells do little of the actual work of the colony, yet they can reproduce. These two types of cells are directly analogous to the somatic and germ-line cells, respectively, we have just described. Consider that the flagellated cells and amoeboid cells in this colony are not genetically identical. The flagellated cells risk being excluded from contributing to the next generation since they do not reproduce; in an evolutionary sense they are being "taken advantage of"—providing the labor necessary to feed the amoeboid cells, yet it is the amoeboid cells that are passing on their genes to subsequent generations. Amoeboid cells, in an important way, act as cellular parasites on the larger colony. At the cellular level, they are being favored evolutionarily because they are leaving descendants behind while the flagellated cells are not. Because some cells are doing all of the work and other cells are doing all of the reproduction, and these cells differ genetically, the colony cannot persist for long and eventually breaks apart into its component cells.

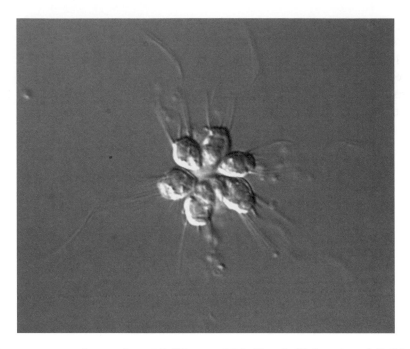

Figure 15.1 *Proterospongia*. Image from N. King and M. Dayel, University of California, Berkeley.

Cellular Selection and Multicellular Organisms

Let's leave this colony now and turn to animals like ourselves. Our bodies consist of billions of cells. In our body there are lots of somatic cells that are responsible for basic body maintenance. For example, the cells of the heart act to pump the blood, the cells of the gut act to digest food, etc. There are, by contrast, relatively few germ-line cells in the human body that explicitly govern reproduction; these are found in the gonads of both males and females. Obviously, from the perspective of the organism it would be catastrophic if something akin to what occurs in the *Proterospongia* colony happens, that is, if the cells were only looking out for their own reproductive interests and disbanded. The cells in an organism must work in concert or the organism will not persist for long. To persist, every multicellular organism must have some means of sublimating selection processes operating at the cellular level to those operating at the level of the entire organism.

This points out one of the evolutionary disadvantages of multicellularity. Selection at the level of the cell could favor cellular parasites within an organism: genetically variant cells that do not do any work, but only reproduce. As long as the germ-line cells and somatic cells are genetically identical, this will not be a problem. However, imagine if genetically variant cells arise within a body and start to proliferate; if these were somatic cells, they would outcompete normal somatic cells because they reproduce and quickly come to dominate. An organism containing such selfishly proliferating cells would quickly degenerate to the unicellular state, thereby killing the organism: selection at the level of the cell will favor variant cell parasites, but selection at the level of the organism will not.

This is actually not a hypothetical example. Consider the case of cancer in mammals. A single, genetically variant somatic cell starts making many copies of itself; in the short term, this would be selectively favored at the cell level, but this has very grave effects on the organism. There also may be analogies in human society. Imagine if everyone was totally selfish, and only looked out for their own interests: society would fall apart (no matter what Ayn Rand said).

How Organisms Control Selection at the Cellular Level

A potential problem in multicellular organisms is if the germ-line cells came to differ genetically from the somatic cells. Again, if this were the case, the somatic cells would be doing all the work of the organism, but the genetically different germ-line cells would be passing their genes on to subsequent generations. They would be genetic parasites, winning the evolutionary game, while avoiding any hard work.

As Leo Buss argued in his book "*The Evolution of Individuality*", every multicellular organism must square the demands of selection at the level of the cell with the demands of selection at the level of the organism. In many animals, especially those considered more complex, the chance of selfish variant cells arising in the germ-line is reduced by what is called "germ-line sequestration." In effect, germ-line cells are sequestered or set aside from other somatic cells and can undergo only a limited number of cell divisions. For example, in humans the germ-line cells that make the eggs in a female or the sperm in a male are set aside around day 50 in a developing embryo, and thus decades before use. Germ-line sequestration greatly lowers the probability of mutations occurring because most mutations arise during cell division, and these cells remain essentially quiescent until sexual maturation. This is important because such mutations would cause the germ-line cells to come to differ from their associated somatic cells. It is an interesting fact that germ-line sequestration does not happen in all animals: in cnidarians (jellyfish and corals) and sponges the germ-line cells are not set aside and instead can turn into somatic and germ-line cells throughout the life of the organism. This suggests that in terms of the way the different cells in the body interact, what we call a cnidarian organism is not exactly akin to a human organism. It also suggests that uncovering some of the mechanisms responsible for the conversion between somatic and germ-line cells in creatures distantly related to humans, like cnidarians and sponges, might someday help us find a cure for cancer, because it might be useful in stopping or arresting the proliferation of wayward cells.

The Role of Cooperation in Evolution

We have seen that in the short term, some of the cells in a multicellular organism can proliferate, but in the long term this will have deleterious effects on the organism, leading to its death; this will in turn cause the death of all the cells inside the organism. Thus, what in the short term is selectively advantageous for a proliferating cell will not be selectively advantageous in the long term. If the different cells in an animal body are to function successfully as a unit, there must be some sort of cooperation whereby the short-term demands of cellular selection are set aside for the long-term benefit of these cells and the organism. This finds resonance with our discussion in Chapter 12 of the origins of the eukaryotic cell;

recall that cooperation among different bacterial lineages played an important role in that key episode in the history of life. The same must be true of the origins of multicellularity. Different cells have to cooperate rather than compete with one another in order for there to be a viable, successful animal. This is one more bit of evidence indicating that the history of life is not simply about competition. Instead, many of the key events involved some form of cooperation too, and ultimately that in turn provided some sort of competitive advantage.

The Origins of Multicellularity: A Possible Scenario

We know that multicellular organisms evolved from single celled eukaryotic organisms, but because tiny multicellular organisms consisting of just a few cells are certain to be rare as fossils (though not unknown, see our discussion of the earliest fossil embryos in Chapter 13), paleontological evidence for the fateful origins of multicellularity is lacking. However, evolutionary biologists do have some idea of how this transition happened. This is because all animals, including humans, pass from a single cell stage to a multicellular stage as they develop from a fertilized egg to an adult. Scientists can potentially gain some insight into the evolutionary transition from unicellularity to multicellularity as they watch it happen in developing organisms. Every animal, early in its development, passes through a stage when it is a tiny, simple, hollow ball of cells called a blastula (Figure 15-2). This has led scientists to conclude that probably the earliest multicellular animals

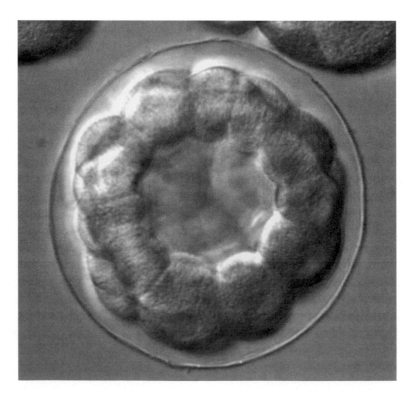

Figure 15.2 Blastula. Of a sea urchin. Image from P. Cartwright, University of Kansas.

were just simple, hollow balls of cells. Such a simple, hollow ball of cells could form in a few different ways. Perhaps several separate unicellular organisms agglomerated into a ball of cells and became a colony; then, eventually the separate cells in this colony became so tightly integrated that they became a single animal. In this case, the origins of multicellularity would be analogous to the origins of the eukaryotic cell. Another possibility is that perhaps a single-celled organism underwent cell division, which is how unicellular organisms reproduce, yet instead of splitting apart the dividing cells remained in contact. In either case, selection both at the cellular level and the newly formed organismal level played a crucial role in this key episode in the history of life. Ultimately, the origin and proliferation of animals in the fossil record, what we have referred to as the Cambrian radiation, follows on the heels of the origins of multicellularity.

The Evolution of the Vertebrate Brain

Vertebrates are one of the many groups that first appear in the fossil record during the Cambrian radiation. If we run the origins of multicellularity forward they're one of the groups that evolve, and they are a group with special significance to us humans, being that we too are vertebrates. Moreover, the brain is one of the group's key traits, again finding special meaning to us because of the consciousness it ultimately provided in our own lineage. Many other vertebrate species also display exceptional intelligence (possibly even consciousness) including whales and dolphins, elephants, dogs, cats, and various apes and primates. We believe it is perhaps self-centered, but also understandable, that evolutionary patterns and trends within this feature have intrigued humans for so long. People are especially interested in the brain's connotations for intelligence. Obviously intelligence is a nebulous concept. Its meaning has been debated for hundreds of years; it is also a slippery concept and difficult to measure and we are not able to resolve that debate in this chapter. Still, on the whole, humans do have some idea what they mean when they invoke the term. A tree shrew is intelligent enough to survive and it may be much smarter at being a tree shrew than a human would be. However, no tree shrews have rivaled the great human intellectuals when it comes to producing inspiring works of art, literature, musical pieces, or scientific discoveries. It is for this reason that humans are held to be more intelligent than tree shrews, and therefore we might ask what traits correlate with these perceived differences in intelligence, and how they might have evolved. Here we focus in greater detail on what we can say about the evolution of intelligence, discuss differences in brain size among living vertebrate groups, and also changes in brain size through time, and finally consider what may be special about our own brains.

Brain Size as a Proxy for Intelligence Within and Among Vertebrate Species

We can tell much about the evolution of the vertebrate brain just by considering the vertebrate groups alive today; these include a diverse variety of animals including species that are popularly referred to as sharks, fish, amphibians, reptiles, birds, and mammals (Figure 15-3). The fossil record is important in this regard too because it provides our one opportunity to trace changes in brains through time, while making it possible to consider groups that are now extinct. However, as you probably have guessed, brains are things that are rarely if ever preserved as fossils. What we have to

Figure 15.3 **Vertebrates**. (a) A shark, image from U.S. Fish and Wildlife Service. (b) A fish, from F. Abe and M. Davis, University of Kansas. (c) A frog, image from W. Duellman and L. Trueb, University of Kansas. (d) A snake, image from W. Duellman and L. Trueb, University of Kansas. (e) A bird, image from M. Robbins, University of Kansas. (f) A mammal, image from U.S. Fish and Wildlife Service.

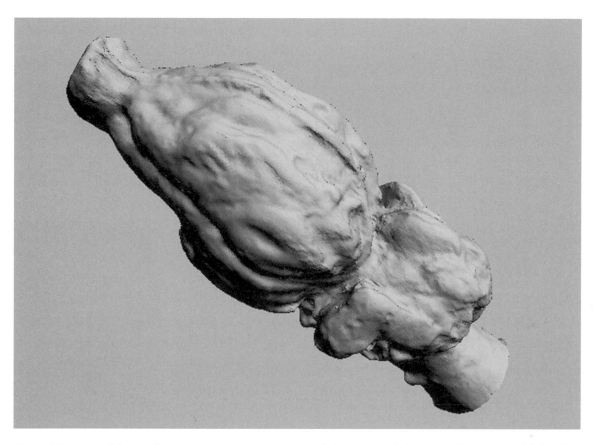

Figure 15.4 Fossil brain. A computer reconstruction of the brain of the 30 million year old fossil mammal *Anoplotherium*. Image from H. Jerison, University of California, Los Angeles.

be content with is a fossil skull, yet these are valuable because a skull can be used to make a cast of the brain (Figure 15-4). Such casts provide a bit of information about the shape of the brain, but are particularly important because of what they can tell us about the size of the brain. It so happens that the size of the brain can be used as a proxy for overall intelligence, but only with certain caveats. In particular, it does seem that among different species, those with larger brains tend to display more of what humans refer to as basic intelligence, but only when we correct for differences in brain size relative to differences in body size. This is because bigger animals tend to have bigger brains, because they need more neural matter to coordinate their activities and movements. Further, brain size does not increase as quickly as body size, such that big animals have relatively smaller brains, while small animals have relatively bigger brains, at least if you compare brain weight to body weight. In short, brain size may be correlated with intellect, but it's also powerfully correlated with body size.

Also, while differences in brain size, when corrected for body size, do appear to correlate with differences in intelligence among different species, they do not appear to correlate with differences in intellect within any given species, particularly humans. For instance, within our own species, typical brain size (disregarding obvious skeletal and genetic anomalies) ranges from 1,000 to 2,000 cubic

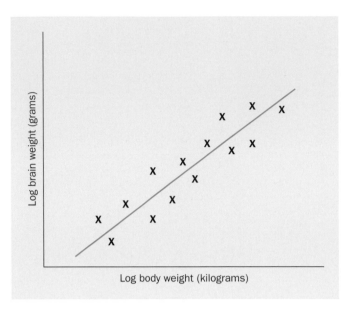

Figure 15.5 A graph relating the logarithm of brain weight versus body weight. The "X" symbols represent the values for different vertebrate species.

centimeters; Einstein, or other figures regarded as being at the apogee of human intellectual achievement, do not happen to have the largest brains on record. Indeed, Oliver Cromwell, who was one of the most notorious figures in British history, particularly for his subjugation of the Irish, apparently has one of the largest brains on record, yet no one would argue that he was the smartest person that ever lived. So, brain size is only a valid measure of intelligence across different species and only if it is corrected for body size.

Plotting Brain Size Versus Body Size

A pioneer in this area of research was Harry Jerison and he and Stephen Jay Gould have described the significance of this work in some detail. Jerison recognized that a good way to study the relationship between brain size and body size is to obtain a list of average adult brain and body weights in many different species and then plot them as points on a graph (Figure 15-5). The points happen to cluster around a line, and demonstrate that brain weight increases about two-thirds as quickly as body weight. Further, recognize that any species that plots above the line has a relatively larger brain than the average vertebrate; any species below the line has a relatively smaller brain. As an example, among living vertebrates, consider chimpanzees (Figure 15-6). They have an average brain weight of 400 grams; the average mammal at their body weight has a brain that weighs 150 grams. Thus, a chimp's brain is about 2.5 times as heavy as it should be, signifying that chimps are highly intelligent animals.

You can probably recognize that there are certain difficulties with applying this approach to long extinct fossil organisms. First, we need a measure of body weight, and it is not possible just to throw a fossil specimen on a scale and obtain an accurate body weight. However, thankfully there are several skeletal traits, including the size of the thigh bone and the molars, that in living mammals correlate quite closely with body weight. Therefore these can be used as a proxy for body weight in fossil animals.

Figure 15.6 Chimpanzees. A mother and infant from Uganda, Africa. Image from iStockPhoto.com © D. Parsons.

Differences in Brain Size Across Different Vertebrate Groups

When brain size versus body size is plotted for different fossil and living vertebrates, Jerison and Gould recognized that there are two major clusters of points. Cold-blooded fish, amphibians, turtles, snakes, and lizards cluster towards the low end of the spectrum while the warm-blooded vertebrates—the birds and mammals—cluster towards the high end of the spectrum, along with, perhaps surprisingly, the cold-blooded sharks. As we described in Chapter 5, sometimes evolution is inaccurately portrayed as following a ladder-like pattern of progressive increase and improvement through time. Again, when it comes to brain size, this presumed pattern is not actually recovered. Cultural bias might mistakenly suppose that there would be a ladder-like increase in brain size from what are thought to be "less advanced" organisms like fish up through amphibians, reptiles, and birds, culminating in mammals, but this is not the case. Instead, there are two different classes of brain size and these are largely associated with differing physiologies like whether species are warm or cold blooded, with sharks comprising the important exception.

Why are Shark Brains Relatively Large?

It is worth exploring some of the patterns in brain size in greater detail. For instance, why might sharks have relatively large brains? The answer appears to lie in the fact that most sharks are predators, and being a predator requires intelligence. In particular, it takes more smarts to hunt an animal than it does to hunt grass; therefore, carnivores typically have relatively larger brains, when corrected for body size, than herbivores. Interestingly, through time the relative brain size of both predators and their prey has increased, implying possibly that increased intelligence was somehow favored evolutionarily both in the hunters and the hunted. (Still, something more than just predatory habit must explain brain size in sharks considering that predatory snakes and lizards do not have particularly large brains, at least when compared to other vertebrates.)

Why do Mammals have Large Brains?

Another interesting issue is the reason why mammals might have relatively large brains. Recall our discussion of evolution and extinction in reference to mammals and dinosaurs in Chapter 6. It was once suggested that the reason large, non-flying dinosaurs are no longer with us is because they were driven to extinction by the smarter, competitively superior mammals. As we discussed, this reasoning is no longer accepted. Indeed, the argument as it relates to intelligence has been somewhat turned on its head. In particular, Stephen Jay Gould and others have argued cogently that mammals evolved large brains partly because they were being outcompeted by dinosaurs. This meant that during the Mesozoic mammals were relegated to marginal environments and nocturnal habits where they lived in the shadows of a world that dinosaurs ruled. Many early mammal fossils (Figure 15-7) of species that lived during the time of the dinosaurs bear large eye orbits and were probably nocturnal. We can reach this conclusion because nocturnal vertebrates tend to have large eyes, since they need heightened visual acuity. In turn, vertebrates with large eyes tend to have more nerves and larger brains which they use to process the information coming in from their low light surroundings. Isn't it ironic that the large mammalian brain was not a structure that allowed us to outcompete dinosaurs, but instead may represent tangible evidence that mammals were being outcompeted by dinosaurs?

Were Dinosaurs Unusually Dumb?

Given that the traditional views on mammals and brain size seem to be wrong, it is worth exploring whether the traditional view of dinosaurs, which are portrayed as having unusually small brains, is also wrong. In fact, Jerison and Gould have shown that this traditional view is indeed wrong. Basically, even a large dinosaur like *Apatosaurus* (the dinosaur formerly known as *Brontosaurus*), which did have a rather

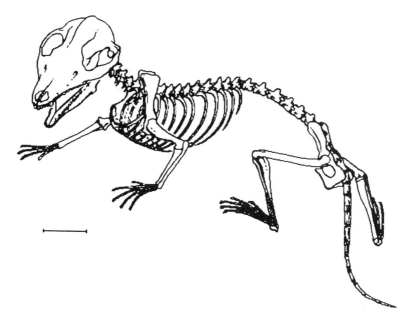

Figure 15.7 **Skeleton of a Mesozoic mammal**. A morganucodont, scale bar equals 2 cm. Image from R. Cowen (2005) *History of Life*, 4th Edn, Blackwell Publishing; after Jenkins and Parrington, and Jenkins.

Figure 15.8 *Apatosaurus* (formerly *Brontosaurus*). A reconstruction of this large Jurassic dinosaur. Image from iStockPhoto.com © A. Tooley.

tiny head and certainly a tiny brain (Figure 15-8) relative to its body size, had a brain that was just the right size. Reptiles have small brains anyways and as animals increase in size their brain size increases more slowly than their body size. *Apatosaurus* shouldn't have had a large brain because large animals have relatively small brains and reptiles at any body weight have smaller brains than mammals. It is worth adding that carnivorous dinosaurs like *T. rex* (Figure 7-14) also had a relatively larger brain than the herbivores they fed on and this makes sense given the pattern that we already described with sharks. Of course, the skull of *T. rex* is huge, but much of that bulk supported the impressive set of jaw musculature that this fearsome carnivore employed to rend and rip its prey. There are other carnivorous dinosaurs that were apparently as big as *T. rex*, or maybe even a bit bigger, like

Figure 15.9 *Carcharodontosaurus*. The skull of this large, carnivorous dinosaur, with a human skull for scale. Image from Project Exploration, Chicago, with assistance of P. Sereno and M. Her.

Carcharodontosaurus and *Giganotosaurus* (Figure 15-9). However, current analyses indicate that these beasts had smaller brains than *T. rex*. Maybe *T. rex*, whose name translates as "Tyrant lizard, king" dominated not simply because of its ferocious aspect but also because it was smarter than the other carnivorous dinosaurs that were of equivalent size. Among dinosaurs though, the actual king when it came to relative brain size was the tiny *Troodon* (Figure 15-10). Of course, the dinosaurs gave rise to birds, which are really just flying dinosaurs, and birds have relatively large brains. Further, some bird species display exceptional intelligence. For instance, crows can use tools, parrots can imitate human speech, and bower birds can create dramatic bowers to attract mates which are marvels of exterior and even interior design (the birds decorate the exterior and sometimes the interior of their bowers).

Figure 15.10 *Troodon*. A reconstruction of this presumed highly intelligent dinosaur. Diagram from R. Cowen (2005) *History of Life*, 4th Edn, Blackwell Publishing; derived from B. Giuliani © Dover Publications, Inc., New York.

Trends in Brain Size Within Primates and Hominids

A final pattern worth considering is what happens to brain size in the primates (Figure 15-11), a group that includes lemurs, monkeys, baboons, gorillas, chimps, humans, and many fossil species. It turns out that large brain size is a characteristic feature of our lineage and the earliest definitive primate fossils, which date from 55 million years ago, already have very large brains. For their body size, primates have the largest brains of any mammal; their brains, when corrected for body size, are three times larger than an average mammal (Figure 15-12). Also, recall as we discussed already, mammals have large brains relative to other vertebrates in general.

Trends in Brain Size: Humans and Our Closest Relatives

Furthermore, paleontologists and evolutionary biologists Niles Eldredge, Ian Tattersall and Stephen Jay Gould have demonstrated that even within primates there appears to have been an active trend towards increasing brain size. This trend culminates in our immediate relatives and our own species, *Homo sapiens*. This group is referred to as the hominids; its closest living relative is the chimpanzee, and the group includes upwards of 20 extinct fossil species more or less closely related to humans (Figure 3-11). Of these hominids, *Australopithecus africanus* (Figure 15-13) was one of the early species that diverged. Its average brain size is indicated in Figure 15-12; note that the average brain size of *Australopithecus africanus* was slightly larger (when corrected for body size) than the average brain of a chimp. Chimps in turn have brains that were slightly larger than those of gorillas (when corrected for body size). Sitting above all of these in terms of its relative brain size (Figure 15-12) is our even closer relative, *Homo erectus* (Figure 3-14).

Figure 15.11 **Primates**. (a) A lemur, image from I. Tattersall, American Museum of Natural History. (b) A monkey, in this case, a gibbon, image from iStockPhoto.com © A. Yu. (c) A baby baboon, image from iStockPhoto.com © H. Bentlage. (d) A gorilla, image from iStockPhoto.com © E. Isselée. (e) Human children, image from P. Cartwright.

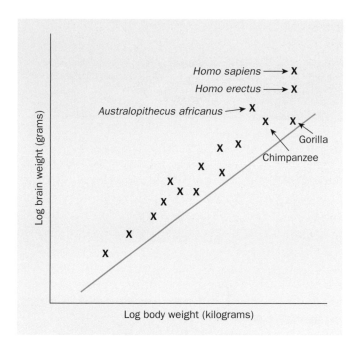

Figure 15.12 A plot of the logarithm of brain weight versus body weight. The "X" symbols represent species of primates (some species are labeled), and the line represents the average values for mammals. Based on the work of S. Szalay.

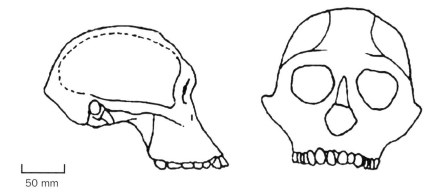

Figure 15.13 The skull of *Australopithecus africanus*. Diagram from M. J. Benton and D. A. T. Harper (2009) *Introduction to Paleobiology and the Fossil Record*, Blackwell Publishing; based on Lewin.

This species evolved after *Australopithecus africanus*, shares a more recent common ancestry with *Homo sapiens*, and went extinct only 100,000 years ago. Finally, our own species, *Homo sapiens*, possesses by far the largest brain (when corrected for body size) of all the living primates and thus of all mammals and vertebrates. It may sound immodest, but we are clearly the smartest species that has ever evolved. (We say this with one caveat: Neanderthals [Figure 3-15] arose at the same time as archaic *Homo sapiens*: both are descended from a single common ancestor. Archaic *Homo sapiens* ultimately gave rise to modern humans. Average Neanderthal brain size was basically equal to that of *Homo sapiens* and interestingly Neanderthals were probably much stronger than typical humans. They also buried their dead and created tools, yet they vanished some 30,000 years ago while we survived. The reasons why they vanished while we persisted may never be known, but are nicely explored in a book by William Golding, he of *Lord of the Flies* fame, called

The Inheritors.) Thus, there is a clear trend in our lineage whereby through time the relative, and absolute, size of the brain increases as new hominid species evolved in Africa. This suggests that aspects of intelligence were highly favored and were related in an important way to the evolutionary changes that occurred within the hominid lineage.

Concluding Remarks

Throughout this book we have debunked some of society's prevailing views about evolution. For example, although it has often been assumed that those lineages that are more complex or progressive are favored evolutionarily, or that the history of life is a competitive struggle, writ large, little evidence supports these conjectures. An important case was explored here related to how the origins of multicellularity may have occurred. However, it appears that one of the frequently cited trends does indeed seem to be true. In particular, it has often been assumed that one of the key patterns within our own lineage was an ever expanding brain. In fact, there is a profound increase in relative brain size, first in primates, later in hominids, finally culminating in our own species, *Homo sapiens*. We suggest that this means that our intelligence, the key to our evolutionary origins, will also be the key to our long-term survival. If we use our intelligence, we are likely to be able to surmount many of the problems and challenges that we, and the biota of this planet, face.

Additional Reading

Beard, C. 2004. *The Hunt for the Dawn Monkey: Unearthing the Origins of Monkeys, Apes, and Humans.* University of California Press, Berkeley, CA; 363 pp.

Bonner, J. T. 1988. *The Evolution of Complexity.* Princeton University Press. Princeton, NJ; 272 pp.

Bonner, J. T. 2006. *First Signals: The Evolution of Multicellular Development.* Princeton University Press, Princeton, NJ; 156 pp.

Buss, L. W. 1987. *Evolution of Individuality.* Princeton University Press, Princeton, NJ; 224 pp.

DeSalle, R., and Tattersall, I. 2008. *Human Origins: What Bones and Genomes Tell us about Ourselves.* Texas A & M Press, College Station, TX; 240 pp.

Eldredge, N., and Tattersall, I. 1982. *The Myths of Human Evolution.* Columbia University Press, New York; 197 pp.

Fleagle, J. G. 1998. *Primate Adaptation and Evolution*, 2nd edition. Academic Press, New York; 608 pp.

Golding, W. 1963. *The Inheritors.* Harvest Books, New York; 240 pp.

Gould, S. J. 1992. *Ever Since Darwin.* W. W. Norton, New York; 288 pp.

Gould, S. J. 1996. *The Mismeasure of Man.* W. W. Norton, New York; 444 pp.

Jerison, H. J. 1973. *Evolution of the Brain and Intelligence.* Academic Press, New York; 482 pp.

Jerison, H. J. 1991. *Brain Size and the Evolution of Mind.* American Museum of Natural History, New York; 99 pp.

Tattersall, I. 2001. *The Human Odyssey: Four Million Years of Human Evolution.* iUniverse, New York; 195 pp.

Chapter 16

Key Events in Vertebrate Evolution

Outline

- Introduction
- Cambrian Origins
- Major Groups of Chordates
- The Vertebrates
- Lobe-Finned Aquatic Vertebrates
- The Origins and Evolution of the Tetrapods
- Concluding Remarks
- Additional Reading

Introduction

The evolutionary history of our own group, the vertebrates, has fascinated paleontologists and the general public alike for more than a century, and with good reason. We are, after all, vertebrates, and thus are going to be keenly interested in our own origins: where we came from in a sense. It is true that when comparing their extant diversity with other groups like arthropods, vertebrates pale in comparison. Recall that more than a million species of arthropods have been described, while there are only about 50,000 living vertebrate species, and most of these are fish. Still, vertebrates are not just a marginal group, simply fascinating from our own myopic, anthropocentric perspective. They have evolved a tremendous range of body forms, and included amongst these are the largest animal, the blue whale (Figure 16-1), and the largest land animals, the dinosaurs.

Vertebrates also occupy an extensive number and diverse variety of habitats. For instance, not only are they still successful in the oceans that they originally sprang

Figure 16.1 **Blue whale**. The largest animal. Image from NOAA Fisheries, National Marine Mammal Laboratory, Alaska Fisheries Science Center, with assistance of S. Calderon.

from, both in the shallowest and deepest parts, but they also have managed to successfully colonize both land and the air during their history. Powered flight evolved independently in the group at least three separate times: in pterosaurs, birds, and bats (Figures 4-9, 6-2 and 16-2). (Note, this is not counting a gliding habit, which also has evolved independently several times, e.g., lizards, squirrels, and fish). Terrestrial vertebrates have even repeatedly reinvaded the sea. The whales are one such example, as are the spectacular ichthyosaurs and plesiosaurs, along with the mosasaurs. The vertebrates as a group also contain numerous species that are highly intelligent, including the only organisms on this planet that have definitely evolved conscious thought (with the octopus perhaps representing a notable exception); we speak of course of *Homo sapiens* but also several other species including elephants, chimps, porpoises and whales, and perhaps even dogs and cats. Finally, there is an exceptional record of vertebrate evolution preserved in the fossil record. Indeed, it is the stunning examples of that record, like the preserved remains of the dinosaur *Tyrannosaurus rex*, that inspired one of us (BSL) to become a paleontologist.

In this chapter we consider some of the key events in vertebrate evolution, concentrating primarily on the paleontological record of that evolution from the Cambrian up to the Cenozoic. We also discuss the broader evolutionary origins of the vertebrates in the context of the group they belong to, the phylum Chordata.

Figure 16.2 Bat. Image from B. Timm, University of Kansas.

Cambrian Origins

The Cambrian radiation was the subject of Chapter 13. Recall that it marked the time when most of the major animal groups first appeared and started diversifying in the fossil record. The vertebrates are no exception to that general pattern. They and the larger group they belong to, the chordates, also appeared during this key episode in the history of life. In addition, recall from the discussions in Chapter 13 and Chapter 6 that one of the key aspects of the Cambrian radiation, from an evolutionary perspective, was that it illustrated the significance of the phenomenon of contingency. This was the idea put forward most forcefully by paleontologist and evolutionary biologist Stephen Jay Gould that the history of life offers few prospects for predictability. When it comes to considering a group at any one point in time in the fossil record, we usually have no scientific basis for being able to extrapolate forward about its long-term evolutionary prospects. Certainly a consideration of the Cambrian record of chordates (and vertebrates) provides us with no basis for predicting that the then humble representatives of our own phylum would go on to enjoy great success or evolve conscious thought.

Chordates started out rare and small in the Cambrian. The famous Burgess Shale deposit from the Middle Cambrian, of British Columbia, Canada, has produced only very limited remains of the early chordate *Pikaia* (Figure 13-21). Chordates are more abundant in the now perhaps equally famous Chengjiang fossil deposit

Figure 16.3 Hemichordate. Image from C. Cameron, Université de Montreal.

from the Early Cambrian of China, though they would still not be considered common and are rare in comparison with other groups such as arthropods.

Features that allow us to recognize these Cambrian organisms as fossil chordates include the fact that they, and all modern chordates, possess a major dorsal (that is, on the back) hollow nerve cord that is surrounded by cartilage, called the notochord. They also possess chevron shaped bands of muscles that have a segmental character. Alas, these are not always visible in most humans, at least externally, with the notable exception of those that are extremely fit, including body builders. The closest living relatives of the chordates are the hemichordates, sometimes called the acorn worms, though they are very distantly related from true worms like earthworms (Figure 16-3).

Major Groups of Chordates

Within the chordates, scientists have recognized three main groups. One of these are the familiar vertebrates, the group we belong to, and these have teeth and often bones, including a skeletonized backbone or vertebral column (hence the name), made out of the hard mineral calcium phosphate. These will be the principal focus of this chapter. However, the two other groups are interesting and merit some discussion. One group, the urochordates or tunicates, is rather diverse today, and their fossil record probably extends back to the early Cambrian (Figure 16-4). (We say probably because some paleontologists have disputed the status of some

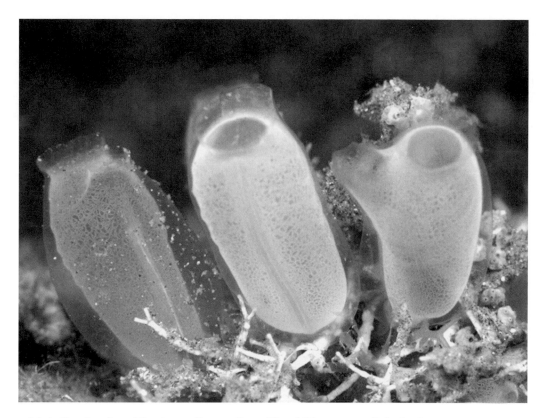

Figure 16.4 **Urochordate**. Tunicate. Image from iStockPhoto.com © J. Anderson.

Early Cambrian fossils that have been called by others tunicates, although their occurrence later in the Cambrian seems much more straightforward.)

The condition of the notochord in some urochordates is rather fascinating. Recall that all chordates possess a notochord and urochordates are no exception. However, urochordates only have a notochord early in their life span during the larval phase, when they are active swimmers (Figure 16-5). The notochord is lost as they age and develop into adults. The result is that some urochordates start out looking much like *Pikaia*, but as they age they turn into a sac-like blob, a fate that hopefully awaits none of us. In important respects vertebrates, the group we belong to, are just larval urochordates. This indicates that perhaps the evolutionary transition between the last common ancestor of urochordates and vertebrates involved a relatively subtle change in the timing and rate of development of the organism. As we shall see in Chapter 18, many significant evolutionary changes may be related to changes in developmental timing.

The final group of chordates are the cephalochordates (Figure 16-6). These are relatively small and rare today, yet they bear substantial resemblance to the Cambrian *Pikaia*.

The Vertebrates

The evolutionary history of the vertebrates is characterized by the acquisition of a series of key characters in a sequence, e.g., bony skeleton, jaws, etc. Moreover, the

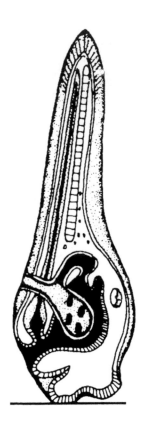

Figure 16.5 Larva of a urochordate. Still possessing a notochord. Diagram from R. Cowen (2005) *History of Life*, 4th Edn, Blackwell Publishing; derived from Barnes et al., *The Invertebrates: A Synthesis*, 3rd edn. © Wiley-Blackwell.

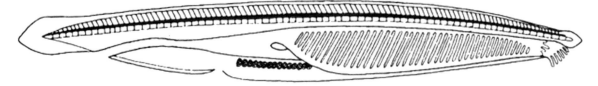

Figure 16.6 Cephalochordate. Diagram from R. Cowen (2005) *History of Life*, 4th Edn, Blackwell Publishing; derived from Barnes et al., *The Invertebrates: A Synthesis*, 3rd edn. © Wiley-Blackwell.

appearance of the forms in the fossil record closely matches the predicted order in which the evolutionary features were acquired. For example, at a general level, *Pikaia*, a chordate, not a vertebrate, appears before the earliest bony armored "fish" (discussed more fully below), as we might predict given that evolution has happened, and further, this pattern actually holds up in much greater detail. Here we will focus on some of these key characters, and consider some of the major living and extinct vertebrate groups.

Early Vertebrates: Animals Without a Lower Jaw

The vertebrates are also known back to the early Cambrian, and they become relatively more diverse and common in the later part of the Cambrian, roughly 495 million years ago, although even then they are still rather rare; vertebrates did not become common until the Silurian Period. Furthermore, the earliest vertebrates

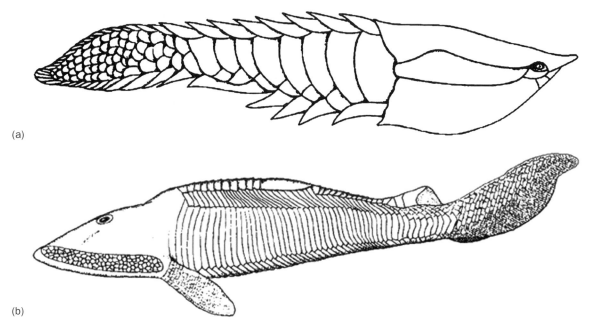

(a)

(b)

Figure 16.7 Jawless fish. Two different types of these armored early vertebrates. Diagrams from R. Cowen (2005) *History of Life*, 4th Edn, Blackwell Publishing; with (a) after Kiaer, and (b) after White.

are found in marine rocks and were ocean dwelling and swimming forms, though by the Silurian they lived in freshwater lakes as well. Although these early vertebrates swam around in watery environments, they differed in several crucial respects from modern fish. For instance, they were covered with an armor of bony plates, while much of their internal skeleton was made of cartilage, like sharks. They also lacked a rigid, tooth-bearing lower jaw. For this reason, and because they dwelled in aquatic environments, these animals are often referred to as jawless fish (Figure 16-7). The lack of a lower jaw suggests they would not have been effective predators, at least of moderately sized organisms, though they could perhaps have eaten smaller, soft organisms like jellyfish and worms. Instead, they were probably scavengers. These armored, jawless vertebrates went extinct at the end of the Devonian, 360 million years ago; two closely related and somewhat comparable groups of vertebrates are still alive today. One of these is the hagfish, an important ocean dwelling scavenger; another is the lamprey (Figure 16-8), a parasite of modern fish. Lampreys are extremely abundant in the Great Lakes of the United States today. Indeed, as parasites they limit the population of some of the fish (with jaws) that live in the Great Lakes. Thus, although we might tend to think of them as more primitive vertebrates, at least relative to their jawed kin, their possession of this primitive trait has not necessarily stood in the way of their success.

The Evolution of the Lower Jaw

The lower jaw first evolved in the late Silurian and is rightly seen as a major evolutionary breakthrough. A lower jaw with sharp teeth is important because it greatly increases the variety and size of prey that can be eaten. The evolution of the lower jaw is also instructive because it turns out that it evolved from a structure that

Figure 16.8 **Lamprey**. Image from iStockPhoto.com © A. Nekrassov.

was already present. This is actually how most features typically evolve. In particular, most structures do not evolve all at once or *de novo*, but rather by the modification of a pre-existing structure. Further, that pre-existing structure usually played a different role or function than the modified structure. In the case of the lower jaw, it was derived from and represents transformed gill arches, a structure already present in the ancestors of Silurian marine vertebrates. Gill arches are structures made out of cartilage or bone that support the gills, and back in the Silurian, and today for instance in bony fishes and sharks, these were/are used to extract oxygen from water. In particular, water enters the fish, passes over the gill, and this is how they breathe. In the earliest vertebrates with lower jaws, the jaw is in the same position as the first (furthest forward) gill arch, it looks almost identical to the other gill arches, and it is not even attached to the upper jaw (Figure 16-9). The only crucial difference in fact between this early "lower jaw" and a gill arch is that it bears teeth.

Fairly soon after the lower jaw evolved there was a dramatic increase in the size of vertebrate species. Prior to the Devonian, the largest vertebrates were 2–3 feet in length, yet during this period some vertebrates began to reach impressive sizes. These animals, called placoderms, were ferocious predators, and some grew over

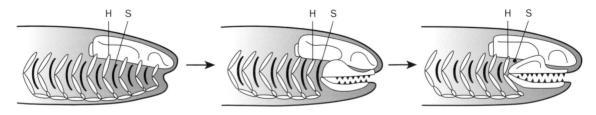

Figure 16.9 **Transformation of gill arches into jaws**. A schematic illustration showing how this may have happened. Diagram from M. J. Benton (2004) *Vertebrate Palaeontology*, Blackwell Publishing; after Romer.

Figure 16.10 Giant armored fish. The head of *Dunkleosteus* from the Late Devonian. Image from R. Cowen (2005) *History of Life*, 4th Edn, Blackwell Publishing.

20 feet in length; they were still wrapped in a curtain of armor, and developed an impressive, battery of slicing teeth (Figure 16-10).

Shark Evolution

Placoderms were not the only group of jawed vertebrates to undergo an evolutionary expansion during the Devonian. Sharks started to diversify in the Devonian and are still important marine predators to this day. In fact, there is a spectacular array of fossil sharks. Included among these is a shark with a bizarre fin on its back shaped like an ironing board (Figure 16-11).

Figure 16.11 Ironing board shark. A bizarre Carboniferous shark. Diagram from M. Coates, University of Chicago.

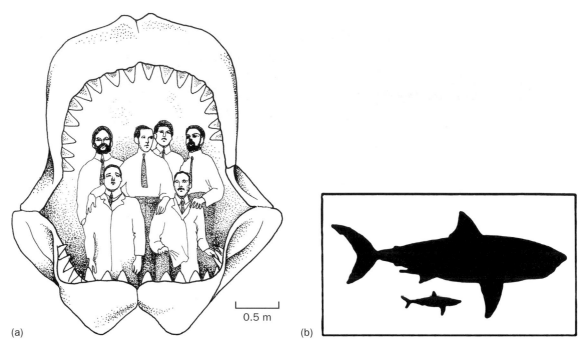

0.5 m

(a)   (b)

Figure 16.12 *Carcharodon megalodon*. (a) A reconstructed jaw of "Megalodon," diagram from M. J. Benton (2004) *Vertebrate Palaeontology*, Blackwell Publishing; based on Pough et al., 2002. (b) Silhouette of "Megalodon" (top) compared with an average modern great white shark, diagram from M. Gottfried, Michigan State University.

Perhaps no fossil shark was as spectacular as the famous "Megalodon" (scientific name *Carcharodon megalodon*). Megalodon was the largest shark, and likely the largest carnivorous fish, that ever lived. This shark roamed the oceans until as recently as 1.8 million years ago. The teeth are the only fossilized remains of this shark, and the teeth indicate that the animal may have been a close relative of the modern great white shark, yet it would have dwarfed it in scale (Figure 16-12). We can use these teeth, common fossils on the eastern seaboard of the United States, to extrapolate that the animal grew to more than 60 feet in length. Fossil whales bones with Megalodon tooth marks are also known, suggesting that Megalodon hunted these large marine mammals. (It is conceivable that they could have scavenged whale carcasses as well; when you're 60 feet long and have teeth like Megalodon's you can pretty much do whatever you want to.)

The Ray Finned Fishes

Modern fish also first appeared in the Devonian. These lacked the extensive armor of placoderms and thus were rather light by comparison. Their fins are basically made of skin supported by internal, thin bones. Because of this they are called "ray-finned" or "rayfin" fishes (Figure 16-13), and they are arguably the most successful of all living vertebrate groups, at least when considered in terms of their total species diversity. For instance, currently there are seven times as many fish species as mammal species (the vertebrate group we belong to). One of the most well known fossil ray-finned fishes is *Xiphactinus*, a relatively common fossil from the Cretaceous of the central United States. This rather ugly, to human eyes, carnivore

Figure 16.13 Skeleton of modern ray-finned fish. Image from F. Abe and M. Davis, University of Kansas.

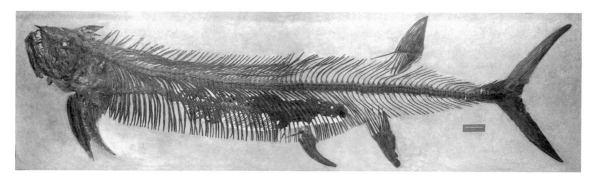

Figure 16.14 Fish within a fish. A Cretaceous *Xiphactinus* that was a victim of indigestion, with its last meal preserved, from western Kansas. Image from M. Everhart, Sternberg Museum of Natural History.

grew upwards of 15 feet in length. The most famous and noteworthy example of *Xiphactinus* was collected by the renowned fossil hunting family the Sternberg's, and is housed today in the Sternberg Museum of Natural History in Hays, Kansas. This spectacular fossil is paleontology's equivalent of a hole-in-one, or better yet, a midnight snack gone terribly awry. Inside the *Xiphactinus* is a smaller, yet still relatively large (4+ feet) and complete ray-finned fish (Figure 16-14). Clearly, this *Xiphactinus* bit off more than it could chew. Rarely can we definitively determine a cause of death for animals extinct that died many tens of millions of years ago, but this *Xiphactinus* represents just such a case: the animal effectively died from indigestion. This particular *Xiphactinus* would have benefited from being around parents like ours who frequently intoned "chew your food!"

Lobe-Finned Aquatic Vertebrates

Given the background on the various vertebrate groups provided thus far it is worth taking a step back to consider broader scale evolutionary patterns within the group (Figure 16-15). Although the different vertebrates described thus far are often called "fish," each represents a separate evolutionary line. If we were to dignify any one of these groups with the moniker "fish," it should probably be the ray finned fish, simply because they are so abundant today and human's are familiar

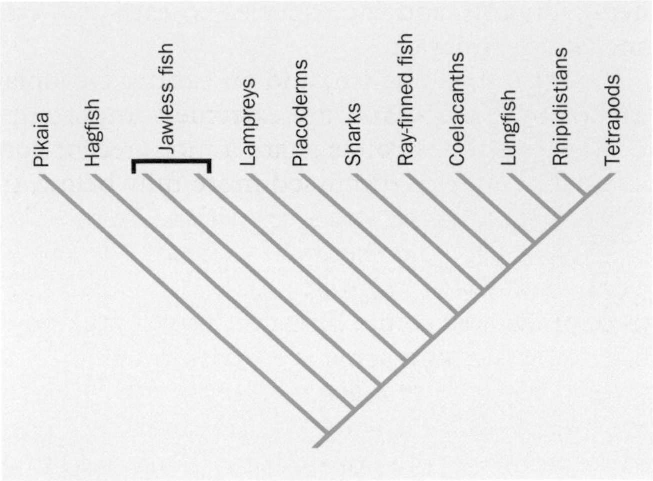

Figure 16.15 Cladogram relating major vertebrate groups. Diagram from F. Abe, University of Kansas.

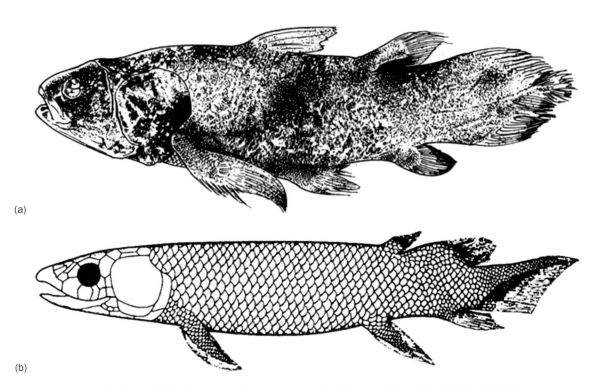

Figure 16.16 Lobe-finned fish. (a) A modern coelacanth, diagram from B. Giuliani © Dover Publications Inc., New York. (b) A Devonian lungfish, diagram from R. Cowen (2005) *History of Life*, 4th Edn, Blackwell Publishing; after Traquair.

with them as protein sources, pets, etc. However, there are two other closely related but distinct groups of aquatic vertebrates that still persist today, the coelacanths and the lungfish (Figure 16-16). They seem to be only just hanging on, yet they were diverse and abundant in Mesozoic and Paleozoic rocks. For this reason they are often called "living fossils". In fact, coelacanths were actually thought to be extinct until fishermen recovered them off the coast of Madagascar, Africa, in the 1930s. Ironically, today they are in danger of truly going extinct because their "rediscovery" has spurred a lucrative trade in coelacanth specimens. There are three species of living lungfish, and each species is restricted, respectively, to Africa, Australia, and South America.

Coelacanths, lungfish, and an extinct Devonian group closely related to them, called the rhipidistians, are extremely important from an evolutionary perspective. These vertebrate groups share a more recent common ancestry with the tetrapods, a vertebrate group discussed more fully below, that made the switch to life on land. Their fins are much stronger, with more prominent bony elements, than the fins of ray-finned fish, and therefore they are referred to as "lobe-finned" or "lobefin" fish. Their evolutionary significance is derived from the fact that these more rigid fins were precursors to the limbs we use to walk on land and grasp things: the limbs we refer to as legs and arms are transformed and modified fins. Of course when this type of lobed fin first evolved it was not because evolution could in any sense see ahead and realize it would be advantageous many tens of millions of years later to use them to waddle out on land. Thus, the limbs did not evolve specifically to allow walking on dry land or swinging through the trees, at least originally, but they were later co-opted to do just that. In a sense they are much like the spandrels of St Mark's Cathedral in Venice, Italy we described in Chapter 7. This is further evidence that evolution can generally be described as a process that involves the modification of structures that are already present (think again of the example of the lower jaw that we just described).

The lobe-finned vertebrates are in fact the closest relatives of the vertebrate group that made the shift to land, the tetrapods, and these we will discuss more extensively below. There are many commonalities between the elements of the fin bones of coelacanths and lungfish, and the limbs of tetrapods (Figure 16-17). Ultimately these are the limbs that allowed vertebrates to make the shift to land and perhaps rhipidistians were among the first vertebrates to have the ability to move around on land and breathe air. The structure of their nasal region suggests

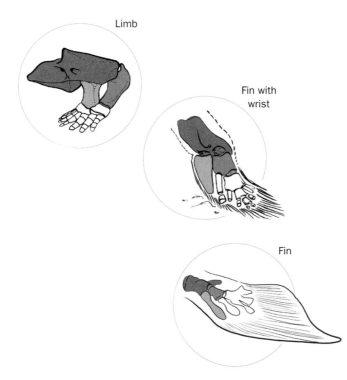

Figure 16.17 From a fin to a limb. A lobe-finned fish fin (bottom) and a tetrapod limb (top) with a transitional form sandwiched in between; shading indicates equivalent bones. Diagram from K. Monoyios and N. Shubin, University of Chicago.

they could breathe air, though perhaps not efficiently, and their limbs were slightly more prominently developed than those of coelacanths and lungfish, yet they were still rather "fish-like." Thus, their movements on land would have probably been restricted to very short sorties, perhaps between one small pond and another.

We might consider why it would be evolutionarily advantageous for a vertebrate to breathe oxygen from air rather than water such that it would give them an advantage either in the ability to survive or reproduce relative to those fish that could not breath oxygen from air. One reason breathing oxygen from air is advantageous is that oxygen concentrations are much higher in air than in water. In fact, that's what drowning is all about: oxygen exists in higher concentrations and is more easily extracted from air than from water. Especially when water is warm and shallow, oxygen concentrations can become dangerously low, even for fish. Any fish that could breathe oxygen from air, especially in such circumstances, would have a distinct advantage. This is a reasonable scenario for why such a trait would be perpetuated, after it evolved.

The Origins and Evolution of the Tetrapods

Relatively shortly on the heels of the rhipidistians, at the end of the Devonian, the first vertebrates bearing four differentiated limbs that could be used to truly walk on land evolved. This group is called the tetrapods and they are the focus of the remainder of the chapter. All mammals, reptiles, birds, and amphibians are tetrapods. (Again, we call our four limbs arms and legs.) Even snakes, which lack limbs as adults, are tetrapods. The name is appropriately applied to snakes because not only were their ancestors tetrapods, but also there are fossil snakes with limbs. Furthermore, certain living snakes possess rudimentary limbs during their development inside an egg that are lost as they grow.

The earliest tetrapods were amphibians, and amphibians are still with us today: consider the frogs and salamanders (each now threatened by human activities).

Early Amphibians

In the evolutionary skein connecting lobe finned aquatic vertebrates like rhipidistians and the tetrapod amphibians, paleontologists have recently identified an exceptional transitional fossil: *Tiktaalik* (Figure 16-18). This animal with the funny name was found in Late Devonian rocks from Greenland and preserves a beautiful combination of "fish-like" and amphibian-like characteristics.

Another important fossil from the Late Devonian is *Ichthyostega*, one of the oldest tetrapods and oldest amphibians (Figure 16-19). This animal probably spent much of its time in the water, and like modern amphibians it would have been tied to freshwater for the purposes of reproduction and development of its eggs and presumed tadpole-like larvae. It, however, could also walk around on land. As described in Chapter 11, by the late Devonian terrestrial habitats were no longer barren; instead, there were extensive forests that harbored abundant plant matter ripe for the proverbial picking. There also were plentiful arthropods (Figure 16-20) on land, including insects and predators on these, such as spiders. In short, food for vertebrates would have been plentiful, and it is certainly no coincidence that the vertebrate transition to land occurred at this time and not, for instance, significantly earlier.

Figure 16.18 *Tiktaalik*. A reconstruction, from T. Keillor and N. Shubin, University of Chicago.

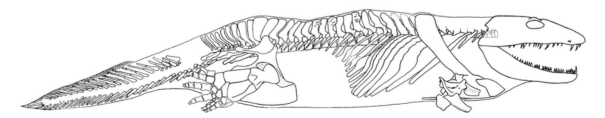

Figure 16.19 *Ichthyostega*. Diagram from J. Clack, University of Cambridge.

The Significance of the Amniotic Egg

The other living group of tetrapods is the amniotes. The amniotes are tetrapods that are no longer tied to the water for purposes of reproduction, whereas by contrast amphibians and the other aquatic vertebrates we have described need water to lay their eggs; further their young (minimally, in the case of amphibians) need to live in

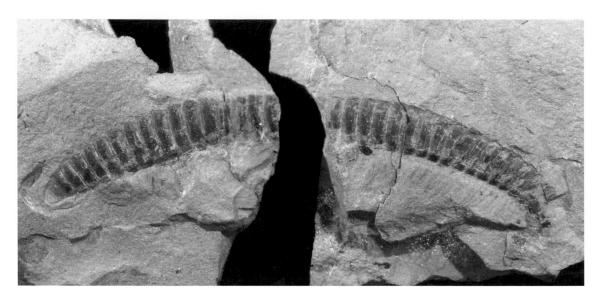

Figure 16.20 Fossil millipede. Image from P. Selden, University of Kansas.

water. A chicken egg is a classic example of an amniotic egg. What is special about an amniotic egg, in comparison with an amphibian egg, is that it does not dry out (at least easily), yet water can pass through the egg shell thereby allowing the developing embryo to breathe. The extant amniotes comprise the reptiles and the mammals and these are groups that first appear in the fossil record in the Carboniferous period.

Synapsids and Mammals

Mammals are defined by several features not easily identified in fossils: for example, they are warm blooded, possess hair and they can lactate and produce milk to nourish their young. In spite of the fact that these characters are rarely if ever preserved in the fossil record, it is possible to identify those fossil lineages that were on the line leading to mammals. The mammal line is part of a larger group called the synapsids. This term refers to the fact that when you look at a synapsid skull in side profile, in addition to the holes for the eye sockets and the nose, there is one (the Greek for one is "syn") other prominent hole (Figure 16-21). Among the most famous early fossil synapsids is the Permian carnivore *Dimetrodon* (Figure 16-21), although there were also many herbivorous synapsids (Figure 16-22) too. We do not know if *Dimetrodon* could suckle its young, nor do we know if it was warm blooded and possessed hair. Most paleontologists guess no on all counts. However, by the Mesozoic mammals had evolved. Throughout most of the Mesozoic mammals were very small, mostly rodent sized, though a few may have reached the size of a modest dog, and occupied the shadows of the dinosaurian world (see Figure 15-7 and also discussion in Chapter 15). Upon the demise of the dinosaurs (these are discussed more fully immediately below), the mammals underwent a spectacular evolutionary radiation, evolving into forms as distinct as bats and whales within perhaps as little as 10 million years.

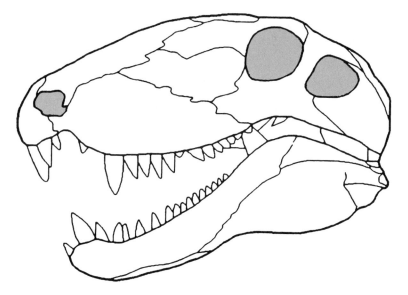

Figure 16.21 **Synapsid**. Skull of *Dimetrodon* from the Permian, diagram from R. Cowen (2005) *History of Life*, 4th Edn, Blackwell Publishing.

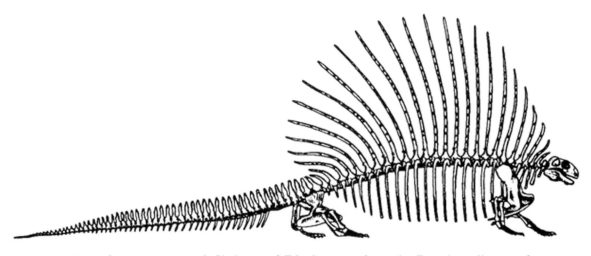

Figure 16.22 **Herbivorous synapsid**. Skeleton of *Edaphosaurus* from the Permian, diagram from R. Cowen (2005) *History of Life*, 4th Edn, Blackwell Publishing; after Romer.

Reptiles, Anapsids, and Diapsids

The earliest reptiles had skulls rather different from synapsids in several important respects. One of these relates to the fact that the earliest reptiles were anapsids (Figure 16-23). In these, there are no prominent holes in the side of the head, other than the eyes and nasals. The earliest reptiles had skulls that looked much like the skulls of living turtles, although their bodies were very different.

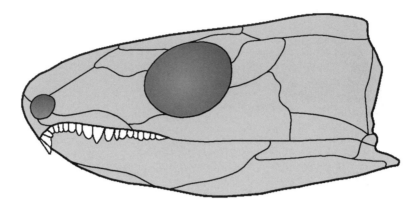

Figure 16.23 Anapsid skull. *Captorhinus* from the Permian. Diagram from R. Cowen (2005) *History of Life*, 4th Edn, Blackwell Publishing.

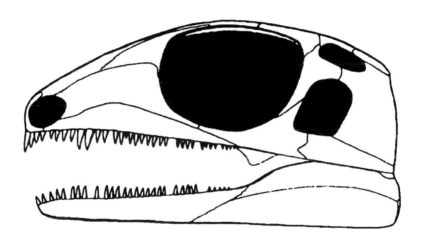

Figure 16.24 Diapsid skull. *Petrolacosaurus* from the late Carboniferous. Diagram from R. Cowen (2005) *History of Life*, 4th Edn, Blackwell Publishing.

The significance of the hole in the side of the head is that this is a site where muscles to the jaw attach. On average, the more holes in the side of an animals skull, the greater the area for muscle attachment and thus the stronger the potential bite. No one wants to be bitten by a snapping turtle, but hands down a snapping turtle bite is a bargain any day when compared with the bite of an equivalently sized carnivorous synapsid.

This makes one other group of reptiles that evolved in the late Carboniferous seem particularly impressive: the diapsids. These reptiles have two holes (in addition to those representing the eyes and nose) in the side of their skulls. Although the earliest diapsids were not formidable predators (Figure 16-24), and perhaps fed on insects, diapsids ultimately became the largest and most fearsome carnivores ever to have walked the Earth: the group traditionally called the dinosaurs. *Tyrannosaurus rex* and other giant dinosaurs would have been endowed with excessive batteries of jaw muscles, facilitated not only by the dramatic size of their skulls, but also by their

Figure 16.25 Skull of *T. rex*. Image courtesy of Library Special Collections, American Museum of Natural History.

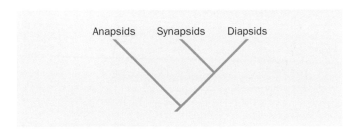

Figure 16.26 Cladogram of major tetrapod groups. Excluding amphibians. Diagram from F. Abe, University of Kansas.

large areas for muscle attachment in the two holes on the side of their skull, associated with their diapsid lifestyle (Figure 16-25). These dinosaurs would have had a tremendously powerful bite, making them particularly ferocious and effective predators.

Many modern reptiles have scales and are cold blooded but the reptiles also eventually gave rise to warm blooded groups (Figure 16-26). Just as the synapsids ultimately evolved warm bloodedness in the mammal lineage, some reptiles ultimately become warm blooded: namely the birds. Indeed, the fact that birds are still around is the reason why we should not say dinosaurs are extinct. It is true that there are no large, nonflying dinosaurs alive today (this is viewing ostriches as on the "small side"), but the birds, the flying dinosaurs, are a remarkably successful and diverse group that survived the end Cretaceous mass extinction and are currently flourishing today.

One key question is when in an evolutionary sense did the transition to warm bloodedness occur. All living birds are warm blooded, and even the early bird *Archaeopteryx* was warm blooded. Therefore, the transition to warm bloodedness

Figure 16.27 Older reconstruction of *T. rex*. Image from iStockPhoto.com © C. Ermel.

likely happened among the nonflying dinosaurs. Thirty years ago, if one had asked most paleontologists, they would have told you that dinosaurs were cold blooded and covered with scales (Figure 16-27). With long extinct organisms it is often not easy to discern whether they were warm or cold blooded; however, there are certain hallmark features associated with warm bloodedness in reptiles. In particular, certain features of how the skeleton grows and the distribution of blood vessels in bones, and also the possession of feathers can provide indications of whether a species was warm blooded or not. Indeed, there are several examples of fossilized dinosaur skin that clearly show the possession of scales, without feathers, and there are also several dinosaurs that do appear to have patterns of skeletal growth most compatible with a cold-blooded habit. Thus, the 30 year old view of the large nonflying dinosaurs is not entirely wrong. However, there are also several recently discovered, spectacular dinosaur fossils from China that show that some dinosaurs did indeed possess feathers. These fossils belong to dinosaurs quite closely related to birds, including Tyrannosaurs. *Tyrannosaurus rex* probably did not look like a giant chicken. But it was probably more bird-like than recently thought, and its representation in the famous and classic movie *Jurassic Park* is rather inaccurate.

Figure 16.28 Moa. The giant bird from New Zealand that went extinct less than 1,000 years ago. Diagram from R. Cowen (2005) *History of Life*, 4th Edn, Blackwell Publishing; after Frohawk.

Tyrannosaurus rex also likely cared for its young, just like birds. Further, thanks to recent fossil finds, like spectacular fossilized blood vessels from *T. rex*, we can surmise that this large non-flying dinosaur did have a physiology much like that of modern warm-blooded birds, and rather different from that of living cold-blooded reptiles.

Concluding Remarks

Vertebrates are quite a diverse living group with an exceptional fossil record that has been studied by paleontologists for centuries. Over the course of many hundreds of years, remarkable discoveries have been made. These have shed light on numerous key aspects of vertebrate evolution, for instance, the evolution of dinosaurs and mammals, and have also brought spectacular extinct animals to light, while sparking the curiosity and interest of countless "children" old and young; new discoveries continue to be made by vertebrate paleontologists that enhance our understanding of evolution, and sometimes even change what were established views. For example, the old paleontological view was that the faster, smarter, and better mammals outcompeted the sluggish and stupid dinosaurs, but we already debunked this view in earlier chapters. Given that dinosaurs live on in the form of birds, such a viewpoint is even more wildly inaccurate. In fact, consider that when one adds up the total number of species of mammals relative to dinosaurs, both today and in the distant past, and bird species are counted as dinosaurs, as is proper given their evolutionary origins, dinosaur diversity has always exceeded mammalian diversity. It may be hard for some to visualize bird superiority relative to mammals, given their typically small stature and seemingly mild demeanor; yet consider the fact that until recently relatively monstrous birds walked the planet. Some of these, like the 10 foot tall Moa (Figure 16-28) from New Zealand, were herbivores, yet surely they could have delivered a deadly kick to any human within range (the Moa was driven to extinction by humans in New Zealand less than 1,000 years ago; apparently there was too much demand for omelet's and human hunting tools won out against deadly Moa kicks). However, there were also highly successful lineages of large, fleet, carnivorous birds (Figure 10-9). Some of these assuredly hunted mammals in South America as recently as two million years ago. In a way, it is fascinating to recognize that although the largest terrestrial herbivores and carnivores ever to have lived, in the form of dinosaurs, have vanished from the face of the Earth, their direct evolutionary descendants, the birds, live on.

Additional Reading

Benton, M. J. 2004. *Vertebrate Palaeontology*. Wiley-Blackwell, New York; 472 pp.

Chiappe, L. M. 2007. *Glorified Dinosaurs: The Origin and Early Evolution of Birds*. Wiley-Liss, New York; 192 pp.

Cowen, R. 2005. *The History of Life*, 4th edition. Wiley-Blackwell, New York; 336 pp.

Fastovsky, D. E., and Weishampel, D. B. 2005. *The Evolution and Extinction of the Dinosaurs*, 2nd edition. Cambridge University Press, New York; 500 pp.

Martin, A. J. 2006. *Introduction to the Study of Dinosaurs*, 2nd edition. Wiley-Blackwell, Oxford; 560 pp.

Prothero, D. R. 2007. *Evolution: What the Fossils Say and Why it Matters*. Columbia University Press, New York; 381 pp.

Rose, K. D. 2006. *The Beginning of the Age of Mammals*. Johns Hopkins University Press, Baltimore, MD; 448 pp.

Shubin, N. 2000. *Your Inner Fish: A Journey into the 3.5 Billion Year History of the Human Body*. Pantheon, New York; 240 pp.

Chapter 17

Are We Alone in the Universe?

Outline

- Introduction
- What is the Potential that Humans will Encounter Extraterrestrial Civilizations?
- Radio Waves and the Search for Extra-Terrestrial Intelligence
- Possible Evidence for Life in a Martian Meteorite?
- Concluding Remarks
- Additional Reading

Introduction

One of the most important questions ever posed by humankind, after "Why are we here?", is "Are we the only ones here?" In recent decades it has been recognized that all of the planets in this solar system are devoid of complex life, though simpler forms of life may exist, or may once have existed, in our solar system. Still, in a universe of infinite size it is not unreasonable to suppose that life may exist elsewhere, even if life and, further, intelligent life is very rare. In fact, as we already described in Chapter 12, life on this planet evolved relatively quickly. This suggests that the origin of life itself may not be such a difficult evolutionary transition; perhaps life might not be a rare phenomenon in the universe after all? Recall however from Chapters 12 and 13 that it took a very long time for complex multicellular life to evolve, suggesting by contrast that this evolutionary transition may be more difficult and this type of life rarer. Therefore, it almost goes without

Prehistoric Life: Evolution and the Fossil Record. 1st edition. By Bruce S. Lieberman and Roger Kaesler. Published 2010 by Blackwell Publishing.

Figure 17.1 Space alien?. Image from iStockPhoto.com © J. Hamm.

saying that there is no need to suppose that even if life exists elsewhere in the
universe it is a complex form of life capable of space travel. In spite of this,
millions of people in the United States and other countries believe in the
existence of aliens (Figure 17-1). In this chapter we will discuss the likelihood
that life, be it complex or simple, exists elsewhere in the universe. Further,
we will consider what is the likelihood that we would actually encounter such life,
while we describe some of the various scientific attempts to uncover evidence of
extraterrestrial life.

What is the Potential that Humans will Encounter Extraterrestrial Civilizations?

Although as we mentioned many humans are convinced of the existence of aliens
in our midst, their actual presence on Earth seems unlikely given that a space craft
traveling even at a pretty good clip—1 million miles an hour—which is 25 times
faster than the highest speed attained by an Earth-launched space craft, would take

over 3,000 years to reach Earth from the nearest star system; even more problematic, suitably habitable planets have not yet been found in that star system. Other star systems in our galaxy and even those in other galaxies are interminably further away. Still, anything is of course possible, and over the past few decades the search has been on to identify possible signs of extraterrestrial life.

Scientific estimates of the number of technological civilizations existing in the universe vary. The famous astronomer Carl Sagan argued that the number could exceed 1,000,000, while other astronomers have given lower estimates. It is noteworthy that even some of the ancient Greeks argued that the universe was infinite and thus likely had countless populated worlds; this reiterates their prescience, already shown in the case of their early thoughts on evolution (see Chapter 4).

The Drake Equation

The astronomer Frank Drake developed an equation, today called the Drake equation, that could be used to calculate the number of planets with intelligent life; some have used this as evidence that astronomers can and will develop equations for just about anything. Equations sometimes give the illusion of scientific rigor when none is present, and the Drake equation (shown below) may be just such an instance. Still, by the same token, expressing the terms involved in this equation, along with their values, may help give people a feel for the odds that intelligent extraterrestrial life does exist:

$$N = R^\star f_P n_e f_L f_i f_c L$$

In this equation, N is what we are interested in calculating: the number of intelligent civilizations that humankind will come into contact with or encounter. The other terms are: $R^\star$, the number of stars in the universe; f_P, the fraction of stars with planets; n_e, the number of planets with habitable environments; f_L, the fraction on which life has originated; f_i, the fraction on which intelligent life has evolved; f_c, the fraction of civilizations with technology to send signals; and finally, L, the longevity of civilizations.

When we try to assign concrete values to these terms the results are rather interesting. $R^\star$ is a prodigious number, considering there are more than 100 billion stars in our galaxy and there are more than a billion galaxies in the universe: it's roughly 10^{20}, a one followed by 20 zeroes. f_P is harder to estimate because at present we can only find very large planets based on our current methods of planetary detection equipment. Further, this work is relatively in the early stages, although already more than a hundred stars with planetary systems have been found; doubtless many more exist and that number will continue to climb.

n_e, the number of planets with habitable environments, is difficult to gauge. Many of the planets discovered thus far are rather close to stars; this is in fact because the way most planets are currently detected is by the gravitational wobble they induce in the stars they orbit about, favoring the discovery of large planets close to their suns. These planets thus would appear to have uninhabitable environments, by Earthly standards. Within our solar system, planets relatively close to the Sun have extremely hostile environments, for instance, Mercury and Venus. However, by the same token, we know that some life forms on Earth can tolerate hostile environments: bacteria live in the hot springs of Yellowstone National Park and several miles down in the crust; unusual "worm-like organisms" live thousands of feet below the surface of the ocean near superheated ocean vents, flourishing at tremendous depths, temperatures, and pressures. Thus, we must try to avoid defining the scope of habitable environments too narrowly. An interesting wrinkle is that just because a planet currently has an uninhabitable environment does not mean it always did. Consider the planet Mars: today its surface appears to have a rather poor quality environment with a limited atmosphere and very cold temperatures; however, billions of years ago its planetary surface was much more hospitable. There is good evidence for ancient running water on the planet's surface in the distant past, including topological features resembling river channels (Figure 17-2); the planet also once had a much richer atmosphere, again, a sign of a more tolerable environment. These more suitable environmental conditions, at least for Earth-based life, actually would have been in place on Mars right around the time life was originating on Earth. The truth is, at this time we cannot assign

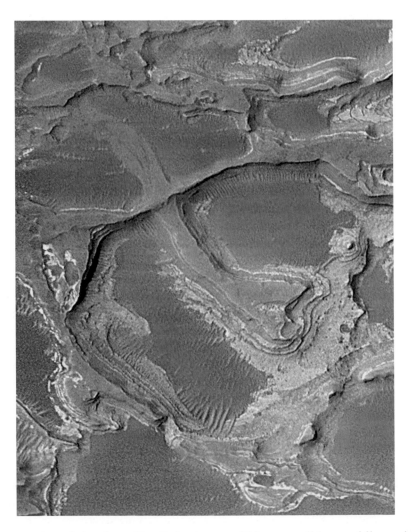

Figure 17.2 **Ancient river channels on Mars?**. A picture of Mars possibly providing evidence that there was once flowing water at the surface. Image from NASA.

a realistic value for n_e. It certainly is a number greater than zero, but how much greater is an open question that we don't know the answer to.

f_L is also difficult to gauge, but we do have some ideas as to its magnitude. As we have already described, the fossil record suggests that life evolved early on this planet; this indicates that the origin of life might not be such a difficult event and further life as a phenomenon might not be that rare. Other things that scientists have considered is what are the things that Earth-based life requires; the reasoning being, if these are plentiful it would again bode well for the notion that life exists elsewhere. We should say that it is of course not known if these are general requirement for all living things. We do know that Earth-based life needs water. In fact, the search for water is what has guided much of NASA's search for life within and outside of our solar system. Presently we do not have a great handle on how much water exists elsewhere in the universe, though the element hydrogen is extremely abundant; however, just to consider our own solar system, water is plentiful in the polar ice caps of Mars (Figure 8-16) and also appears to be concentrated in the soil at the Moon's poles: this may bode well for the existence of lots of water elsewhere in our universe. Carbon is also a requirement of Earth-based life and it too is plentiful in the universe. Each of these factors suggests that life in the universe might not be all that rare, and f_L could be close to one.

Returning to the Drake equation, f_i is meant to quantify how often intelligence evolves. This variable cannot be that well constrained because indeed "what is intelligence?" is a fair question that is often debated. Still, let us accept for the moment that what is often called human intelligence is a reasonable proxy for intelligence. Then, we can try to determine if there are other species on our own planet that show evidence of human-like intelligence. We can use this to broadly constrain f_i by considering how many times intelligence has evolved. Just as there is debate about the meaning of "intelligence" there is no precise consensus about the number of intelligent species on our own planet, but there are some general estimates. Further, considering that there are millions of species alive on our planet, and relatively few displaying human-like intelligence, it appears that intelligence is relatively rare, has evolved in only a small handful of lineages, and f_i is likely a very small number; we will describe our logic behind this conclusion. Consider that humans of course comprise one intelligent lineage but most agree we are not alone in this regard. Chimpanzees, gorillas, dogs, and cats (theirs clearly being a special kind of stubborn intelligence) can be added to that list, although intelligence evolved within a single human, chimp, gorilla common ancestor, so that should be treated as one evolutionary origin of intelligence. Whales and porpoises also represent a separate, independent origin of intelligence, as do crows. Some have argued that the phylum Mollusca contains another lineage where intelligence has evolved: in this case the octopus (Figure 17-3) which is able to perform several tasks evincing a highly coordinated and functioning intellect; these include navigating complex mazes. It has even been debated whether the social insects like bees, termites, and ants comprise other instances of intelligence. Although individual bees, termites, and ants do not display what we call intelligence, the coordinated activity of the hive or nest does perhaps signify some manifestation of intelligence. If we add such insect examples to the list, the number of intelligent species would grow somewhat, though not tremendously. Given that there is only a small fraction of life forms on this planet that display what most call intelligence, and if we make the assumption that life on Earth is in some way representative of other kinds of life (perhaps an unacceptable assumption), then f_i is likely a very small number.

Figure 17.3 An octopus. Image from iStockPhoto.com.

f_c is the fraction of civilizations with technology to send signals. Given that this has only evolved one time on our planet, it is perhaps not prudent to even guess at a value for f_c. However, consider the one episode comes out of a relatively limited number of intelligent lineages (based on the criteria for intelligence we developed above in our discussion of f_i). Therefore, maybe f_c is around 0.1.

Finally, the variable L is the longevity of civilizations. Here is another variable from the Drake equation that is difficult to constrain. The human species has been around roughly 100,000 years (based on our knowledge of the fossil record); agriculturally based humanity has been around roughly 10,000 years. By contrast, radio technology was developed in the 1920s and space travel in the 1950s. From the advanced technological perspective, thus far our civilization is very short-lived. We may wonder how long we will continue to persist and indeed how much our civilization and complex intelligence aids our long-term survival prospects. It is true that technology has brought humanity many benefits, and allowed human population to grow tremendously in the last 100 years. By the same token, our own species, and our planet's ecosystems, are under grave threat because of this population growth and attendant pollution. Technology has also provided us with nuclear, chemical, and biological weapons that imperil our long-term survival (Figure 17-4). In short, our brain does not necessarily aid our survival in the long term; might this be true of civilization as a general phenomenon? Of course

Figure 17.4 Mushroom cloud. Produced after dropping the atom bomb on Nagasaki, Japan in 1945. Image from U.S. Department of Defense.

there's no way to tell, and while we may hope from our own perspectives that L is a large number, we just don't know. Certainly, the larger L is, the longer we're around, and the more likely it will be that we contact other extraterrestrial civilizations.

When we combine all of these variables from the Drake equation together we're left with a lot of uncertainty, although some bits of promise. The uncertainty comes from the fact that many of the variables are so poorly constrained, particularly because we do not know how representative the Earth is. The promise is partly engendered by the $R\star$ term. Recall that that term was enormous: a one followed by 20 zeroes. Even if some of the other variables are rather small (and recall that not all of them may be, for instance, f_L could be close to one), then N could be a fairly large number. Even though there have been no confirmed sightings of aliens, their possible existence piques our curiosity and intrigues us. From the perspective of the Drake equation, contact with another alien civilization might not be so far fetched. Perhaps Carl Sagan was not so wrong when he espoused the view that much of the American public holds: aliens really are out there.

Radio Waves and the Search for Extra-Terrestrial Intelligence

The United States Air Force, NASA, and other organizations, in particular, SETI (Search for Extraterrestrial Intelligence, www.seti.org), have spent quite a bit of time searching for extraterrestrial civilizations, and one area that their activities have focused on is monitoring radio signals. Stars and other planets throughout the galaxy including the Sun and the planet Jupiter emit radio waves continuously and naturally. These signals can be translated as nonsensical groupings of numbers: for example, 3111, 2411, 43441, etc. These numbers are much different from the types of signals produced by radio broadcasts that initiate from humans on Earth. Thus far, the only reported extraplanetary radio signals captured appear to mirror these natural stellar and planetary sources, with one exception. In 1977 a NASA scientist identified a highly unusual and distinctive signal sequence: 6EQUJ5. This sequence looks artificial and akin to something made by humans, but it was not generated on this planet. Unfortunately, this signal was never repeated and thus it could not be verified if it was some systemic error, a chance phenomenon, or perhaps an actual signal from some other extraterrestrial civilization. For the present, scientists continue to tune in and monitor the skies for extraplanetary radio waves.

Possible Evidence for Life in a Martian Meteorite?

Towards the tail end of the 20th century, NASA made one of the most heralded scientific announcements ever. The date was August 7, 1996. There, at a press conference, NASA scientists introduced a set of findings that they argued suggested fossilized evidence for life in a Martian meteorite. No, not the little green people that some had expected but rather tiny gray worms. Understandably, the announcement generated significant excitement; eventually, however, it also generated quite a bit of controversy, as we shall describe.

The meteorite (Figure 17-5) containing the possible evidence for life has an interesting saga itself. It was discovered in Antarctica in 1984. Antarctica is actually one of the best places to find meteorites because there is little ground cover (other than snow) and meteorites tend to stand out prominently from the background landscape and thus are more easily detected. Although the meteorite was found in Antarctica, there is good evidence that it came from Mars. In particular, it contains minerals common to Mars and also other Martian meteorites. Radiometric dating confirmed an age typical of most meteorites: it is roughly 4.5 billion years old. However, there are fractures in the rock and these formed somewhat later, around 3.6 billion years ago. (Note, from 4.5 to much after 3.6 billion years ago it appears that the meteorite to be was "just a rock" and still part of the landscape on the planet Mars.) It is actually within the fractures that the NASA scientists found the possible evidence for life. Note that the age of the fractures corresponded to a time shortly after life evolved on our own planet. Further, this was a time when the Martian surface might have been quite hospitable, with an atmosphere and temperature range at the surface much like that of the Earth.

NASA scientists were also able to confirm that it was not until much later that this rock in question left the surface of Mars and became a meteorite. This is because there are certain radioactive isotopes only present in outer space that scientists could detect in the meteorite. Further, their presence is not only diagnostic of space travel, but also makes it possible to calculate the amount of time the meteorite spent in outer space: it turns out this was somewhere on the order of 16 million years. The transformation of this humble Martian rock into what was

Figure 17.5 Martian meteorite. Containing possible evidence for extraterrestrial life. Image from NASA.

eventually an internationally acclaimed meteorite began when another meteorite collided with Mars and sent this small chunk of rock into space. Eventually, the meteorite entered Earth's atmosphere and it crashed into the Antarctic ice cap 13,000 years ago, where it sat until scientists discovered it and brought it back.

As mentioned, it was inside the fractures that the possible evidence for life was recovered. In these, there are certain types of hydrocarbons, akin ultimately to the hydrocarbons found on Earth that form oil and gas. There are a couple of ways that such hydrocarbons can be formed. One way they form is due to standard, inorganic chemical processes (think nonlife based) that can occur in outer space. However, hydrocarbons can also be produced by the breakdown and decay of organisms. Indeed, that is how most hydrocarbons (and oil and gas) form on Earth. How could there be hydrocarbons in this meteorite? They could have formed naturally in space, though they also could have been generated by organisms that once lived inside the Martian soil that eventually became a rock. Finally, it is also conceivable that these compounds could be contaminants from Earth.

Amino acids were also found in the meteorite. These are the building blocks of proteins and are more diagnostic of life than hydrocarbons. Many of the amino

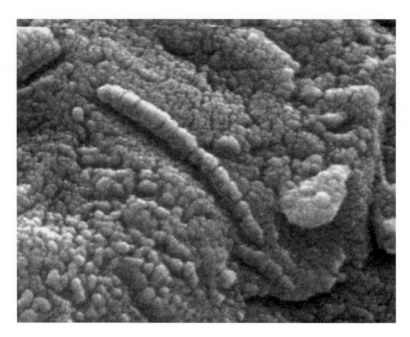

Figure 17.6 Fossilized extraterrestrial life? Tiny shapes inside of fractures in the Martian meteorite. Image from NASA.

acids found in the meteorite were confirmed to be recent contaminants acquired after the meteorite landed on Earth. However, there was a small component of the amino acids that did not appear to be contaminants from Earth, and these were offered up as possible evidence for past life preserved in the meteorite.

NASA scientists also found tiny minerals and globules in the meteorite. On Earth similar tiny globules can be formed by bacteria, although they do not have to be formed in this manner and can just be formed by simple inorganic processes. Finally, NASA scientists announced that they had also found tiny oval shapes in the fractures of the meteorite (Figure 17-6). These they interpreted as fossils of microscopic organisms, perhaps akin to fossils of small organisms found on Earth, like bacteria. NASA scientists argued that although some of the evidence might not be all that strong, in combination these results supported the notion that life had been found in a Martian meteorite and thus life once existed on Mars.

This was of course a revolutionary and fascinating announcement with far reaching implications. However, as often happens with revolutionary scientific announcements, they get scrutinized by other scientists. Indeed, this is the manner that science operates: there is a continual process of self-correction. If one scientist finds something exciting, there are always other scientists out there trying to prove them wrong. One scientist may make a mistake: hey it happens. But in the long term, large numbers of scientists won't make mistakes and the truth will be ferreted out. Unfortunately for the NASA scientists who initially made the announcement, as the truth got ferreted out their results did not always stand up, as we shall describe, although the jury is still out, and these are scientific results in progress.

Challenges to NASA's Claims

Since 1996, other scientists have very much challenged several of the initial conclusions reached by NASA scientists. Partly they focused on the fact that the hydrocarbon evidence was not very compelling, while also they raised the issue that most of the amino acids were contaminants. Criticisms especially focused on the contention that the small squiggles shown in Figure 17-6 were fossilized bacteria, or other such tiny organisms. The conclusions of NASA scientists on these were criticized for two reasons. First, squiggles of similar size and shape can be found in rocks on Earth, but they are often not fossilized bacteria and instead simply represent inorganic minerals. Further, the putative "fossil squiggles" are about a factor of ten smaller than modern bacteria. Based on our understanding of life on Earth they might be just too small to contain sufficient DNA, and thus they could not function as living organisms. Of course this does not guarantee they could not function at this size, but it certainly doesn't help NASA's argument. Thus, in spite of the initial fanfare, the current consensus is that the existence of evidence for life in the Martian meteorite is pretty shaky and we're back where we started from: no known life beyond what is found on Earth.

Beyond Mars to Europa

Although the challenges to the original NASA conclusions about the Martian meteorite were disappointing in one respect, in another respect they were beneficial, because they helped stimulate much additional research on the meteorite and the topic of the search for extraterrestrial life in general. They also forced scientists to come up with some detailed criteria that can be used to identify life on other planets. It also pointed out the type of equipment we need to outfit robotic explorers with if we hope to uncover living or fossilized traces of ancient life on other planets. One of the missions currently in the works that may hold great promise in the quest to find life in our solar system is a space ship targeting Jupiter's moon Europa; Europa's surface is encased in water ice (Figure 17-7). Beneath

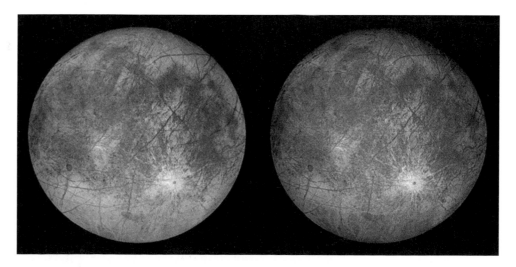

Figure 17.7 Europa. One of Jupiter's moons. Image from NASA.

the ice, there could well be a watery ocean, and such an ocean might be a very promising biological cradle. Several paleontologists contributed conceptually to the development and design of this lander, which aims to drill into the ice and down into the ocean in the quest for life.

Concluding Remarks

We return to the Drake equation described earlier. One term in that equation signifies there is great promise that life does exist in our universe: that is $R\star$, the term for the number of stars in the universe. There are 100,000,000,000,000,000,000 (10^{20}) stars, a prodigious number, and many of these have the potential to possess planetary systems. True, even if they have living inhabitants, including highly intelligent ones, we are unlikely to ever encounter them, or a signal from them, because they are so far away. Still, it is perhaps comforting to know that at least in some statistical sense it is indeed unlikely that we are truly alone in the universe. Be they tiny green people, little grey worms, or something that we cannot even possibly imagine, it's a good bet life is out there.

Additional Reading

Achenbach, J. 2000. Life beyond Earth. *National Geographic* (January Issue): 24–51.

Darling, D. J. 2002. *Life Everywhere: The Maverick Science of Astrobiology*. Perseus Books, London; 224 pp.

Dick, S. J. 2005. *The Living Universe: NASA and the Development of Astrobiology*. Rutgers University Press, New Brunswick, NJ; 308 pp.

Michaud, M. A. G. 2006. *Contact with Alien Civilizations*. Springer, Berlin; 466 pp.

Sagan, C. 1986. *The Dragons of Eden*. Ballantine Books, New York; 288 pp.

Sagan, C. 2000. *Contact*. Pocket Books, New York; 448 pp.

Sagan, C. 2002. *Cosmos*. Random House, New York; 384 pp.

Ward, P. D. 2000. *Rare Earth*. Springer. Berlin, Germany. 368 pp.

Webb, S. 2002. *If the Universe is Teeming with Aliens—Where is Everybody?* Springer. Berlin; 288 pp.

Chapter 18

Humanity: Origins and Prospects

Outline

- Introduction
- How do New Species Evolve—
  The Shift from Chimps to Hominids
- Humans in a Changing Climate
- The Current Biodiversity Crisis
- Mapping a Course for Future
  Changes—Climate and Life
- Concluding Remarks
- Additional Reading

Introduction

Our discussion of prehistoric life began with a consideration of Time's Arrow and Time's Cycle, so this seems like a fitting way to conclude it. Basically, the history of life follows a pathway from start to finish and beginning to end, yet similar types of events have transpired at different times. With these repeating events we can understand more than just patterns in the history of life: we can come to understand the actual processes or mechanisms that governed and produced that history. In essence, the fossil record preserves a set of past experiments; these are events played out long ago. Some times we can use them to gauge whether certain events may come to fruition yet again. Here we synthesize what we've described about prehistoric life, considered in the context of our own lineage's origins and future prospects.

Prehistoric Life: Evolution and the Fossil Record. 1st edition. By Bruce S. Lieberman and Roger Kaesler. Published 2010 by Blackwell Publishing.

How do New Species Evolve—The Shift from Chimps to Hominids

Throughout most of their history, species are stable and do not change much; times of transition, when one species gives rise to another, are compressed into relatively short geological intervals such that evolution is not a process of continual adaptive change. This phenomenon, punctuated equilibria (described in Chapter 5), and the existence of mass extinctions, which we have discussed throughout the book (especially in Chapter 6, and will consider more fully below) makes it hard to predict in detail the pathway that life follows: the Time's Arrow of evolution.

This is not to say though that we have no idea how evolutionary transitions between closely related species happen, or what causes them. The differences between closely related species are typically small (because they're close evolutionary relatives) such that sometimes it is not hard to posit their causes. Differences in the appearance of closely related species can sometimes be explained by invoking a change in the rate that organisms in one of the species mature or develop from a fertilized egg to an adult, relative to the other species. (In fact, changes in these rates can even explain differences between major groups of organisms. Consider the difference between vertebrates and urochordates, described in Chapter 16. In an important respect vertebrates have slowed down their rate of development relative to urochordates, such that they retain a notochord throughout their lifespan, whereas urochordates lose a notochord after they pass through an early larval phase.) An excellent example of how changes in developmental rates cause evolutionary differences was described by Stephen Jay Gould in his book *Ontogeny and Phylogeny*; it comes from our own group, the primates (although the precise aspects of these changes in developmental timing are still being debated). Our closest living relative is the chimpanzee, but as we described in Chapter 3 there are actually many (more than 20) extinct species of hominids that share a more recent common ancestry with humans than they share with chimps. Thus, there are many smaller steps that connect humans to our last common ancestor with chimpanzees. Still, the similarities between chimpanzees and humans are impressive, including the shared possession of more than 98 percent DNA sequence identity. In spite of these profound genetic similarities, adult chimps and humans are of course not identical and instead differ in some aspects of their morphology. Just consider their respective skulls. Adult chimp brains, although big by mammalian standards, are relatively much smaller than our own; the lower jaw of an adult chimp is relatively larger than the lower jaw of an adult human (adult chimps also have a greater proliferation of facial [and body] hair). How can there be many anatomical differences in spite of such small genetic differences? The answer appears to lie in a simple change in developmental timing, which probably began after hominids and chimps split. Such a change probably requires only a very tiny genetic change.

Although adult human and chimp skulls differ, there is a much greater resemblance between adult human skulls and baby chimp skulls (Figures 18.1 and 18.2). Relative to humans, chimps accelerate their development as they grow from childhood to adulthood; this means that humans are in a juvenile state for a much longer relative period of time than chimps, and this has a whole host of consequences. One of the important ones relates to the fact that in both chimps and humans (and other mammals) there are several changes that happen during the

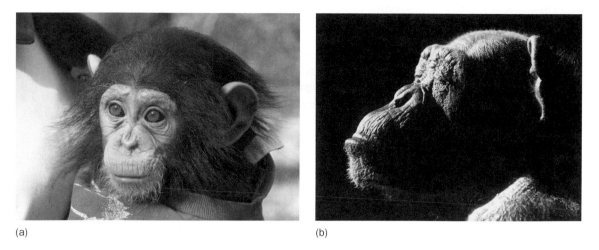

(a)   (b)

Figure 18.1 Baby and adult chimps. Emphasizing differences in the relative size of the lower jaw and the braincase. Images from iStockPhoto.com; (b) © M. Kolbe.

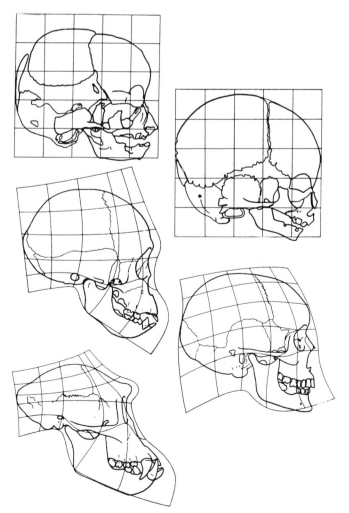

Figure 18.2 Developmental changes in chimp and human skulls. Juveniles are on the top, adults are on the bottom, and the chimp is on the left, the human is on the right. Diagram from http://www.schweizerbart.de, appeared in D. Starck and B. Kummer (1962) Zur ontogenese des Schimpansenschadels. *Anthropologischer Anzeiger* 25: 204–215.

transition from childhood to adulthood: one is that the brain comes to grow more slowly and then stops growing entirely; another is that the lower jaw begins to grow more quickly. The longer a mammal remains in the juvenile state, the longer its brain can continue to grow, and the more slowly its lower jaw expands. Because we humans stay juveniles much longer than chimps do, our brains are much larger. Human brains continue to grow for many years after a chimp's brain stops growing, and a growing brain bespeaks of an expanding and learning mind (think of how much easier children find it to learn a new language or play a musical instrument than adults). Just a simple switch in the rate of development can lead to a host of consequences: not only a relatively smaller lower jaw but a dramatically bigger brain with much greater potential for intelligence. Humans thus, in an important respect, are just juvenilized chimps.

Humans in a Changing Climate

One of the themes of this chapter involves experiments that have been done in the geological past, and global warming constitutes an excellent example of an experiment that has already been performed. Indeed, the results of these past experiments on global warming are preserved in ancient rocks and the fossil record. Many other times this planet has experienced average temperatures much warmer than those today. At every episode of past global warming, there was a concomitant and correlated shift towards rising CO_2 concentrations, just like the climb in CO_2 concentrations and temperature we see happening today. During times when it was warmer, polar ice caps were largely absent. With less water trapped as ice, global sea-level climbed much higher; indeed, large parts of the continents that today are above the waves were once submerged beneath water. In fact, we can say that if in the near future temperatures rise sufficiently, sea-level will rise again to these levels and all of the Earth's coastal cities will be flooded (Figure 18-3). Currently these cities harbor more than 50 percent of the Earth's population.

We can recognize what will happen if these events come to pass. First, there will be much less room for people to live (given that we cannot live under water), and also much less room available for us to grow our crops and other agricultural products. The consequences will be grave for humanity. Surely there will be massive displacements of human populations, and these will be accompanied by the spread of famine. Further, various tropical diseases, including that global scourge malaria, will spread further and further away from the equator, and thus become even more pervasive threats.

As human activities continue to cause CO_2 levels to increase in the atmosphere, the question is not about the reality of global warming but rather "when and how much," and in particular, how extreme will the impacts be and will they be felt most fully in this or the succeeding generation. The weather patterns the authors of this book experience every spring come to mind. You probably all know from the famous movie *The Wizard of Oz* that in Kansas, USA, tornados and thunderstorms arrive every spring. Before any given spring season we do not know precisely how many tornados and thunderstorms we'll experience, nor do we know the precise dates they'll arrive. Still, we know they're coming. The situation is similar with global warming. Already, as CO_2 concentrations have begun to climb (Figure 18-4), temperatures have begun to rise (Figure 18-5). There is a potentially broad range of future temperatures that are forecasted, all higher than values typical of today. The variation in projected ranges partly depends on the details of the models

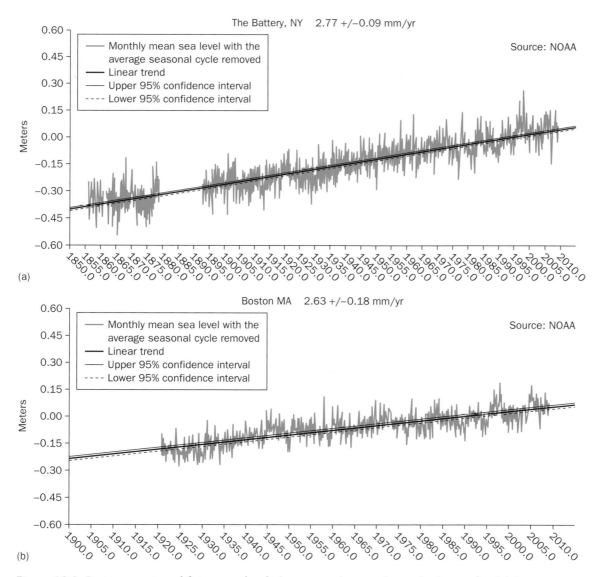

Figure 18.3 Past, present, and future sea-level. Average values and trends shown for (a) lower Manhattan, New York City, NY and (b) Boston, MA. Diagrams from NOAA.

used to estimate future temperatures, and it also partly depends on our ability to minimize future dramatic increases in CO_2 concentrations. From our perspective, the potential spread of tropical diseases and increasing incidences of famine sounds like things worth avoiding. And, in fact, avoiding such dire consequences is becoming more feasible. This is because there are readily available technological solutions that can serve to minimize future increases in CO_2 concentrations; additional investment in scientific research will improve upon these solutions and hasten their entrance into the marketplace.

The Current Biodiversity Crisis

Mass extinctions are other phenomena documented from the study of the history of life and relevant to our future prospects. Five times there have been episodes

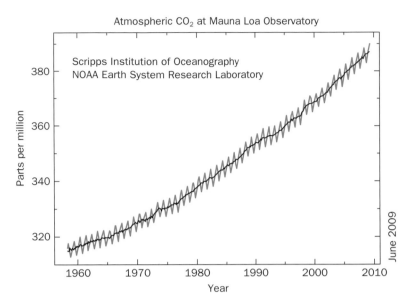

Figure 18.4 Atmospheric CO_2 levels. Average value and trend shown from a monitoring station in Hawaii. Diagram from NOAA.

when more than 50 percent of all the species on Earth died in a relatively short period of time. There were different causes for each of these extinctions: asteroids, carbon dioxide, global cooling, but all had the same end result. The habitat of the globe was degraded and this sent species spiraling towards extinction. Further, global ecosystems collapsed, and it took many millions of years for them to recover. We are currently in the midst of the biodiversity crisis, the 6th great mass extinction in the history of life. Extinction rates and levels are as high now as they have been in the past. Again, habitats are being degraded, but this time we are the cause. A mass extinction like the one happening now is not an untested theory, and instead can be compared to real, tangible events that have already happened on Earth. Indeed, the present day pattern is similar to earlier patterns. We can also see the byproduct of these mass extinctions in the fossil record: global ecosystem collapse and species depart like rivets in the wings of a plane. Enough rivets go from a plane and the passengers are in serious trouble. Ditto the case for species and our biosphere: either the quality of our lives will be seriously degraded or we will go extinct too.

Mapping a Course for Future Changes—Climate and Life

Our understanding of prehistoric life, and this is true for instances when it comes to climate change and the biodiversity crisis, shows that often it is not the present that is the key to the past, but rather it is the past that is the key to understanding the present. The concept of uniformitarianism turned upside down if you will.

We know from studies of the history of Earth's life and its climate that these systems do not respond gradually to repeated affronts. Instead, there is usually a lag and then sudden and dramatic change. Consider the case of the Earth's climate system. When we look at how long it takes for the climate system to

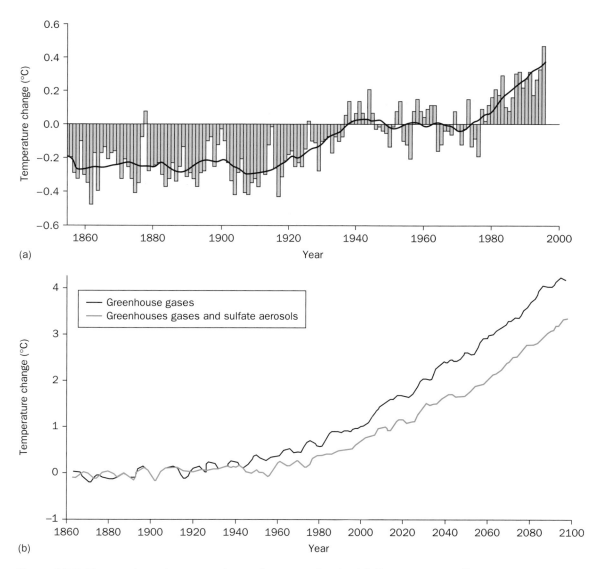

Figure 18.5 **Temperature changes and greenhouse gas levels.** (a) Past patterns of temperature change scaled to the average from 1951 to 1980. (b) Past, present, and future values of surface air temperature change modeled to greenhouse gas levels and greenhouse gas and sulfate levels; the green curve produces the most reliable values. Diagrams from C. B. Cox and P. D. Moore (2005) *Biogeography: An Ecological and Evolutionary Approach*, 7th Edn, Blackwell Publishing.

jump from one state, relatively cold, to another, relatively warm, it usually happens remarkably quickly, on the order of decades. In terms of the history of Earth, that is no time at all. A nice example comes from roughly 550 years ago when the Earth was plunged into a sudden cold snap called "the Little Ice Age." The Little Ice Age was preceded by a relatively warmer interval. In fact, there was a thriving colony of Vikings in Greenland (Figure 18-6); they made their living as farmers during this warmer interval. The shift from the warmer interval to the Little Ice Age was sudden. Within a generation or two, temperatures plunged, farms failed, and the colony in Greenland starved. This story is skillfully told in Jane Smiley's book "*The Greenlanders.*" Will we, for different reasons (warming instead of cooling), face the same fate as the colony in Greenland? Hopefully not, of course.

Figure 18.6 Ruined Norse church from Greenland. The Hvalsey Fjord Church in South Greenland. The church was built around AD 1300 and is the best preserved Norse ruin in Greenland. The last written record concerning Norse Greenland tells about a wedding between two Icelanders that took place in this church on September 16, 1408. Image and text from G. Nyegaard, Greenland National Museum & Archives, Nuuk, Greenland.

The lesson, although some of catastrophist thinking is bankrupt in its reliance on miraculous explanations, aspects of it provide a powerful means for understanding much of what we see in the history of life in terms of the relative suddenness of events on geological time scales. Modern experience, the fountainhead of uniformitarianism, is not enough to explain how every geological and biological mechanism operates. When changes in climate are abrupt, many species, including humans, may be unable to adapt and will face extinction.

One of the pervasive features of the fossil record relates to the phenomenon of extinction, and the history of life shows us that not only is extinction omnipresent, but sometimes there are truly massive extinction events. Each of these mass extinctions was precipitated by a huge, catastrophic event that eliminated much of the biota. In large measure, what survived during these time periods did so largely as a matter of chance to the extent that they could not see these types of changes coming and thereby be adapted to withstand them. Instead, they survived truly out of blind luck, or because they happened to have had a trait that evolved for another reason but allowed them to weather the proverbial storm. Mass extinctions thereby introduce an important element of chance, unpredictability, and contingency into the history of life. A classic example involves the large, nonflying dinosaurs and the mammals at the end of the Cretaceous. Before the extinction, mammals happened to be mostly small and in important respects competitively inferior to the large nonflying dinosaurs, yet it was the mammals that survived the mass extinction at the end of the Cretaceous. By just looking at Cretaceous life forms like mammals and dinosaurs and extrapolating we never would have predicted this. For similar reasons, it is hard to look at where we stand today and make predictions about long-term evolutionary prospects for any one species. Mass extinctions mean that each time period is characterized by its own contingent realities. Still, where we are headed is a topic that many people are interested in.

Throughout the history of life there is little evidence that more progressive or complex lineages are favored evolutionarily, even though human society may still largely think that evolution is about accumulating progress through time. Whether society likes it or not though, different species are not mounting a ladder of ever improving complexity and advancement. In an important sense some of the earliest and simplest lineages to evolve on Earth, which can be broadly classified as different types of bacteria, still dominate the biosphere. Complexity appears, and progress happens, due to the simple principle of diffusion: life started out as something minimally complex, and the only way to go was up. There was not an active force driving this upward climb; therefore, there should be no reason to expect that the origin of humans, admittedly very complex organisms, was manifestly predictable. Moreover, there is no reason our species should expect to persist indefinitely on this planet.

The reasons are partly tied up with the fact that evolution is not just about increasing complexity. This is because one of the important ways evolution happens is by natural selection, and adaptation is a very local process. Organisms within populations of species become adapted to a local environment and an adaptation to one environment cannot be treated as any better, or worse, than an adaptation to any other environment. Further, what works well in one environment may not work well in another environment. This is relevant since our species evolved during a time of unusually cold climate, at least when compared to the rest of Earth history. By changing the Earth's climate, we are removing ourselves from the very climatic conditions that favored our own lineage's origins and subsequent persistence. The adaptations that served humans well during one time of Earth history might not serve so well during a time when climate conditions are different. As complex as we and our technologies are, we cannot change our underlying evolutionary adaptations that ultimately evolved in a local environment. In short, because adaptation happens in local environments, and the most complex lineages are not evolutionarily favored, we cannot think that our own place on this Earth is anything but tenuous. (We should say, of course, that the habit of claiming that every human trait is an adaptation is a bad one. Many so-called human adaptations likely evolved for different reasons, or even by chance, and are akin to the tiny spaces or spandrels in Saint Mark's Cathedral in Venice, Italy.)

Concluding Remarks

It is true that by some criteria our existence on this planet seems very secure: in particular, human population size continues to increase dramatically, suggesting that for now the current human strategy is a successful one. However, a consideration of prehistoric life suggests that our long-term persistence is far more tenuous. Although our species evolved roughly 150,000 years ago, the Earth has been here for 4.5 billion years. Ultimately, the influence of our species, and each individual human to a lesser extent, amounts to the moisture of our breath on a mirror in the span of seconds mapped against the immensity of geological time: ephemeral, transient. But in terms of our children and our children's children, our impacts and the threats are very real. What we are taking part in is an emerging mass extinction, accomplished by human-mediated habitat degradation and climate change, and potentially disastrous for much of the life on this planet, including ourselves. With such an outlook it seems easy to become pessimistic. The comic Woody Allen remarked in his *My Speech to the Graduates* that now, "more than any other time in history, mankind faces a crossroads. One path leads to despair and utter hopelessness. The other, to total extinction. Let us pray we have the wisdom to choose correctly." Allen remarked that he was not being pessimistic but rather expressing the "panicky conviction of the absolute meaningless of existence" but hey, he was doing what good comics do: combining wit and humor. We firmly believe that in spite of the gravity of the situation there is cause for optimism; it is possible to choose wisely and ameliorate the environmental changes and the biological crisis we're causing. If so, there will be hope for the future of humanity, and the truth is we're hopeful. The hope lies in current and future generations. We know so much, perhaps more than enough from your perspective, about prehistoric life, and thus we have some idea of where things are going. Further, we know the root causes of the problems we face today and how to fix them. Thankfully, although the problems are large, they are identifiable. One real evolutionary trend that we identified in this book is the dramatic increase in brain size throughout primate history, culminating in our own species. We suspect that our intellects, the key to our evolutionary origins, are the key to our long-term survival. Each person, acting at the local level and using that intellect, really can make a difference. Just enough people have to stand up and say stop, and also do it.

Additional Reading

Allen, W. 1980. *Side Effects*. Ballantine, New York; 213 pp.

Diamond, J. 1992. *The Third Chimpanzee: The Evolution and Future of the Human Animal.* Harper, New York; 416 pp.

Gould, S. J. 1977. *Ontogeny and Phylogeny*. Harvard University Press, Cambridge, MA; 520 pp.

Smiley, J. 2005. *The Greenlanders*. Anchor, New York; 608 pp.

Weisman, A. 2007. *The World Without Us*. Thomas Dunne Books, New York; 336 pp.

Index

absolute time   10–13, 16
adaptations   92, 101, 115, 123, 133, 134, 144, 152, 163–168, 190, 191, 295, 379
albedo effect   228, *229*, 233, 234, 245, 249
algae   19, 40, 55, 58, 237, 257, 269, 274, *275*, 304, 307, 309, 313, 315–317
Allen, Woody   380
Alvarez, Luis   129
Alvarez, Walter   129
amber   26, 157, *158*
ammonites/ammonoids   47, 60, 127, *128*, 129, 201, 299, 313
amniote   351, 352
amphibian   3, 19, 48, 55, 104, *154*, 325, *326*, 329, 330, 350–352
anagenesis   *105*, 108, 109, 111, 112
anapsids   353, 355
Andrews, Roy Chapman   *199*, 201
angiosperms   *22*, 53, 129
animal embryos   277–279, 281, 295, 323, 324, 352
anomalocarid/*Anomalocaris*   287, *288*, 290–296
anoxic/anoxia   137, 139
Antarctica   102, 146, 189, 209, 210, 217, 218, 224, 229, 245, 366
  plants   *247*
anti-oxidants   262
aragonite   25, 34, 40
Archaeocyathans   *301*, 309
Archaeopteryx   88, *90*, 91, 107, 355
Aristotle   53, *54*
arthropods   47, 48, 58, *77*, 180, 273, 276, 279–287, 290, 294, 295, 337, 340, 350

artiodactyls   60–62
asteroids   123, 129–132, 134, 137, 143, 296

background extinction   125, 126
bacteria   17, 19, 26, 27, 34, 39, 88, 116–120, 144, 172–174, 237, 254, 257, 261–271, 288, 313, 324, 361, 368, 369, 379
Bambach, Richard   114
banded iron formation   261–263
bats   134, 154, *155*, 338, *339*, 352
beavers/*Daemonelix*   145, 147, 204, *205*
beetles   111, 290
Benton, Mike   104, 105
Berra, Yogi   117
biogeography   224
birds   19, 49, 54, 55, *77*, 78, 81–88, *90*, 104, 107, 126, 129, *154*, *155*, 160, *162*, 163, 167, 200–204, 206, 221, 224, 225, 325, *326*, 329–332, 338, 350, 355–358
bivalves/clams   10, 25, *28*, *39*, 42, 43, 45, 50, 51, 55, *60*, *102*, 103, *104*, 126, 140, 273, 274, *312*
blastula   *324*
blooded,
  cold   329, 330, 355–357
  warm   329, 352, 355–357
body size   167, 168, 276, 320, 327–335
brachiopods   *24*, 45, *102–104*, *106*, 140, 273, *280*, 310, *312*, 313
brain size   66, 67, 110, 325–336, *327*, 380
Briggs, Derek   283, 284, 291
Brontotheres   *206*
Brooks, Dan   120

Brown, Barnum *194*
bryozoans 49, 160, 280, 297
Burgess Shale 26, 29, 232, 276, 282–287, *284*,
    290–295, 339
Buss, Leo 321, 323

calcium carbonate 24, 34, 40, 301
Cambrian radiation 232, 273–281, 285–289,
    295, 296, 309, 320, 325
carbon dioxide/CO_2 4, 137–139, *140*, 144,
    189, 230–234, 238, 244, 251, 259, 267,
    307, 313, 314, 374, 375, *376*
Carboniferous 37, 180, 209, 210, 217, 243,
    352, 354, *plate 10.1*
catastrophist 8, 9, 125, 378
cathedral(s),
    Gothic 181, *182–184*, 185, 187, 190
    St Mark's 164, *165*, 349
cave bears 153, *202*
cephalochordates 341, *342*
chalicotheres 63, *64*
characters,
    derived 152–154
    primitive 153, 154, 156, 164
Cheetham, Alan 160
chelicerates 284
chemical fossils 257
chimpanzees 64–67, *65*, *78*, 91, 106, 152, *153*,
    156, 157, 328, *329*, 333, *335*, 338, 363,
    371, 372, *373*, 374
chloroplasts 266–268, *269*
chordates 125, 273, 294, 295, 339, 340–342
cladistics 154, 156, 158
cladogenesis 108, *109*, 111, 112
climate change 129, 209, 223, 224, 230, 244,
    251, 271, 313, 314, 318, 376, 380
cnidarians 273, 276, 279, 280, *281*, 291,
    301–306, 310, 313, 314, 323
coal 8, 31, 36, *37*, 139, 242–244
coal swamp *244*
colony/colonial 49, 303, 321, 322, 325
comets 10, 103, 136, 137, 143, 145
competition 71, 100–105, 110, 223, 224, 269,
    295, 324
continental drift 209–214, 216, 217
contingency 124, 134, 293–295, 339, 378
cooperation 268, 269, 272, 321, 323, 324
coprolites 18, 38, 50
corals, 25, 45, 49, *50*, 126, 140, 240, *300, 302*,
    303–307, *308, 311*, 313–317, 323
    rugosans 304, *306*, 310, 313
    tabulates 310, *311*
craters 129, *131*, 132, *136*, 137, 171, 185, *186*,
    187–191, 253
Cretaceous *133*, 234–236, 244, 313, 346, *Plate
    10.1*
crinoids 28, 45, *46*, 47, 49, 273, *297*
crocodiles 77, 78, 150, *151, 154*, 200, 245
Crustacea (crabs) 38, 39, 47, *48*, 49, 58, 273,

284, *285*, 293, 294
Cuvier, Georges 8, 63, 64, 124
cyanobacteria 257, *258*, 261, 264, 267, 270,
    308, 309
Cyclops 196, *197*

Darwin, Charles 4, 8, 18, 58, 64, 65, 74–87,
    91–99, 101, 105, 106, 115, 116, 133,
    146, 150, 163, 164, 165, 190, 212, 214,
    223, 224, 269, 274, 281, 321
De Candolle, Augustine 101
deforestation 234, *239*
development (animal) 45, 291, 295, 324, 341,
    350, 372, 374
diagenesis 40
diapsids 353–355
diatom *21*, 23, 134, *135*
dinosaurs, 3, 17, 19, 28, 88, 91, 104, 107, 124,
    127, 129, 134, 149, 154, 157–169,
    200–209, 217, 230, 235, 236, 279, 281,
    295–299, 313, 330, 352–358, 379
    Allosaurus 89
    Apatosaurus (*Brontosaurus*) 198, 206, 330, *331*
    Carcharodontosaurus 332
    Deinonychus 127
    dinosaur eggs 199
    embryos *278*
    extinction of 129–133
    feathers and 87, 162, 356–358, *162, 163*
    flatulence *235, 236*, 244
    footprints 203, *204*
    Giganotosaurus 332
    intelligence 330–333
    Oviraptor 200, *278*
    Triceratops 199, *200*
    Troodon 162, 332, *333*
    Tyrannosaurus rex 162, 166, *167*, 199, 202,
        331, 332, 338, 354, *355, 356*, 357
disparity 289–292, 296
DNA 26, 59, 64, 69, 150, 152–160, 169, 264,
    266, 268, 269, 369, 372
Dollo, Louis 14
dolomite 31, 34
Drake equation 361, 363–365, 370
dust bowl *240*

Earth *187, 214*
echinoderm 45, 47, 50, 273, *280*, 282
echinoid/sea urchin 45, 49, 315, *316*
Ediacarans *179*, 180, 231, 232, 274, 276–280,
    320
Eldredge, Niles 106–108, 111, 142, 155, 333
elephants 5, 8, 26, 120, 124, 145, 157, *176,
    177*, 193, 196, 198, 325, 338
eukaryotes 265–271
Europa *369*
evolution 4, 8, 14, 17, 21, 26, 49–53, 58–67,
    71–98, 101–110, 115–125, 134,
    150–154, 158–164, 167–169, 171–173,

177–180, 185, 190, 193, 201, 203, 209, 212, 214, 217–221, 223–225, 228, 232, 236, 238, 254, 264, 265–272, 273–276, 281, 290–297, 319–325, 329, 336, 338–345, 350–352, 358, 359, 361, 363, 372, 379, 380

facies   10, 13
Fischer, Al   245
Flessa, Karl   43
flight   88, 154, 338
flying buttresses   *184*, 185, 190
forams   *119*, 120, *121*
Fortey, Richard   284

Gaia   232, 233
Galapagos   81, 83, 84, 91, 224, 225
Galileo's principle   171–190
gamma-ray burst   140, *141*, 144
gastropods (snails)   25, 38, 45, 50, 55, *60*, 78, 107, 112, *113*, *114*, 115, 140, 152, *153*, 157, 273
geographic isolation   114, 218, 219, *220*, 276, 296
geological time scale   14–18, 28, 69, 120–126, 132, 133, 144, 148, 378
germ-line cells   321–323
gill arches   *344*
glaciation   140–142, 209, 217, 231, *250*
glass microspherules   *132*
global warming   139, 228, 229, 232, 234, 244, 245, 264, 313, 317, 374, *377*
Glossopteris   *211*, 217
Gondwana   205, 217, *plate 10.1*
gorilla   64, *77*, 106, 333, *334, 335*, 363
Gould, Stephen Jay   4, 9, 72, 91, 103–109, 111, 116, 120, 124, 125, 134, 164–168, 171, 180, 187, 190, 284, 289–296, 328–330, 333, 372
graptolites   *297*
gravity   76, 187, 188, 251, 305
Great American Interchange   219, 221, *223*, 224, 249
Greeks (Ancient)   4, 17, 53, 79, 193, 196–201, 206, 247, 361
greenhouse effect   139, 140, 229–234, 245, 251, 313, *377*
Greenland   *378*
griffins   *198*, 201
ground sloth (giant)   8, *9*, *145*, 157, *223*

habitat degradation   143, 144, 147, 148, 299–301, 314, 370, 380
Haeckel, Ernst   78, 85
hagfish   343, *348*
halite   *36*
Hallam, Tony   139
Hansen, Thor   112, 114
Harper, Dave   296

hemichordates   *340*
Hennig, Willi   154
Himalayas   216, 217, 219–221
hippos   59, 61, 250
HMS Beagle/Voyage of the   81, 83, 91, 93, 146
hominids   60, 64, 65, *66*, 67, 69, 124, 145, 333, 336, 372
   Australopithecus afarensis (Lucy)   66, 67, *68*
   Australopithecus africanus   66, 67, 332, *335*
   Australopithecus boisei and robusti   66, *67*
   Homo erectus   57, 58, 65, *66*, 67, *68*, 69, 110, 333, *335*
   Homo habilis   57, 58, 65, *66*, 67
   Homo neanderthalensis   57, 58, 65, *66*, 69, 70, 335
   Homo sapiens   55, 57, 58, *65*, *66*, 69, 110, 144, 151, *177, 332*, 333, *334, 335*, 336, 338, *373*
Hooke, Robert   3, 4
horses   1, 55, *61*, 62, 63, 147
horseshoe crabs   273, 284
Hutton, James   6–8, 28
Huxley, Thomas Henry   105, 116

ice ages   140, 168, 231, 250, 251, 377
icehouse   245
Ichthyostega   350, *351*
ichthyosaurs   338
iguana   *92*
insects   26, 47, 111, 156, 157, *158*, 180, *181*, 237, 284, 293, 320, 350, 354, 363
iridium   129–131, 136, 137, 143
Irish elk   167, *168*
isotopes   10, 11, 366

Jablonski, Dave   114
Jackson, Jeremy   160
Jamaica   314–317
jawless fish   *343, 348*
Jefferson, Thomas   8
jellyfish   179, 273, 274, 279, *280*, 281, 291, 303, *304*, 306, 323, 343
Jerison, Harry   328
Jupiter   187, 366, 369
Jurassic Park   157, 159

Kaesler, Roger   *194*
Knoll, Andy   120, 139, 140, 278
Knowlton, Nancy   160

Lagerstätten   282
Lamarck, Jean-Baptiste   58, *59*, 80, 81, 97, 115
lamprey   343, *344, 348*
larvae   27, 42, 112, 114, 350
lemur   333, *334*
Lieberman, Bruce Smith   140, 141, *195*
limestone   31, 34, *35*, 44, 253
Linnaean hierarchy   56–63, 85, 150
Linnaeus   55, *56*, 57–63

llama   81, *83*, 84
lobe-finned fish   347–349
lungfish   *348*, 349, 350
lungs   174, *175*
Lyell, Charles   8, 9, 85, 86

macroevolution   100–104
magnetic field   215
Malthus, Thomas   95
mammals   17, 19, 49, 55, 56, 59, 61–63, *77*,
        78, 105, 117, 124, 134, 144, 145–147,
        151–154, 164, 168, 196, 201, 206,
        221–224, 235, 279, 281, 282, 295, 297,
        323, 325, 328–335, 346, 350, 352, 355,
        358, 372, 379
mammoths   8, 26, *27*, 145, *147*, 148, 157, 196,
        197
Mantell, Gideon   *194*
Margulis, Lynn   267, 269
Marrella   286, 292, 295
Mars   185, *186*, 187, 188–190, 241, *362*, 363,
        366–369
    Martian meteorite   366, *367*, *368*, 369
mass extinctions,   4, 105, 122–126, 134, 143
    biodiversity crisis   124, 143, 144, 299, 313,
        371, 375, 376
    Cretaceous–Tertiary   123, 126–136
    Late Devonian   123, 143
    Ordovician–Silurian   123, 140–142, 144
    Permo-Triassic   137–140, 143, 149
    Late Triassic   104, 123, 143
mastodons   8, 193, 196, *223*
Mayor, Adrienne   193, 196, 198, 201, 203
McGhee, George   143
McLennan, Deborah   120
McShea, Dan   116
Melott, Adrian   136, 141
Mendel, Gregor   96
Mercury   185, *186*, 187–190, 361
Mesosaurus   210, *211*, 214, 217
Messinian Salinity Crisis   246–248
methane   139, 230, 234, *235*, *236*, 251
Milankovitch cycle   250, 251
millipede   180, 284, *352*
mitochondria   *266*, 267–269
Moas   *357*, 358
mollusc   45, 47, 55, 59, 102, 140, 191, 273,
        280, 282, 363
molting   47, 48
monkeys   333, *334*
Moon   4, 45, 171, 185, *186*, 187, *188*, 189,
        190, 303–306, 363
Morris, Simon Conway   125, 284
mosasaurs   *20*, 202, *203*, 338
multicellularity   269, 276, 319–325

NASA   363, 366–370
Native Americans,   193, 196, 201–206
    White Bear   193, *195*

Nautiloid   *47*, 50, *60*, *297*
Newell, Norman   126
Norris, Richard   118, 119

ocean currents   112, *138*, 225, 245, *246*, *249*
ocean ridges   *214*, 215, 216, 219
ocean trenches   215, 216
octopus   *364*
oldest body fossil   *257*
ontogeny   45, 372
Opabinia   *287*, 290, 291, 294
opossum   221, *223*, 224
Ordovician radiation   295–297
oreodonts   *62*
organelles   267, 268
Origin of Species   4, 76, 80, 91, 94, 96, 101,
        106, 115, 116, 164, 212, 226, 281
original horizontality   5
oxygen   3, *43*, 137, 144, 174–180, 185, 228,
        232, 242, 243, 257, 259–267, 271, 276,
        282, 317, 320, 344, 350
ozone layer   141–144, 264, 265, 271

panda   91, 92, *93*, 116, 124, 144
Pangaea   137, *138*, 230, 231–233
perissodactyls   56, 60–63
Permian   18, 104, 126, 137, *138*, 139, 140, 143,
        219, 230–232, 310–314, 352
phosphatization   27
photosynthesis   131, 237, 257, 259, 263, 267,
        269, 274, 307
Pikaia   *294*, 295, 339, 341, 342, 348
placoderms   344, *345*, 346, *348*
plants, oldest   *238*
plate tectonics   189, *190*, 191, 208–224,
        245–251
pollen   19, 21, *23*
polymerase chain reaction (PCR)   156
predation   38, *39*, 73, 316
progress   *97*, 98, 101, 115–122, 134, 293, 329,
        336, 368, 379
prokaryotes   *266*, 268–270
Proterospongia   321, *322*
pterosaurs   127, *128*, *154*, *162*, 202, 338
punctuated equilibria   107–111, 160, 372
pyrite   260, *261*

radioactive decay   10, *11*, 71–73, 190
radiometric dating   71, 132, 256, 366
Rap Music   *75*, 76
Raup, Dave   126, 134
ray-finned fish   *326*, 346–349
Ray, John   54
reefs   259, 299–304, *305*, 307–317
relative time   10, 13–16
reproduction,   78, 95, 190, 259, 268, 321, 322,
        350, 351
    asexual   271
    sexual   271, 272

Rhea   81, *84*
rivers,
  braided   240, *241*
  meandering   241, *242*
rocks,
  igneous   30, 254
  metamorphic   30, 37, 254, 257
  oldest   *12*, 256, 257
  sedimentary   5, *6*, 10, 12, 13, 29–31, 36, 37,
    52, 241, 243, 254–257
  siliciclastic   31, *32–34*
Rodinia   230, 231
rudists   *312*, 313

saber-toothed cats   145, *146*, 147
Sagan, Carl   361, 365
Schopf, J. W.   259
seafloor spreading   215–219
sea level   245, 374, *375*
Seilacher, Dolf   179
selection,
  gene   110, 111
  natural   85–101, 110–115, 133, 164–169,
    190, 209, 214, 269, 321, 379
  species   103, 111–115
Sepkoski, J. J.   126, 161, 163
Serial Endosymbiont Theory   268, *269*, 270
sharks   325, *326*, 329–331, 343, 344, *345*, *346*,
    348
shocked quartz   132
Simpson, George G.   2, 3, 221
Smith, William   8
snakes   *154*, 202, *326*, 329, 330, 350
Snowball Earth   231, 233, 276
South America   70, 81, *82*, 84–86, 91, 145,
    146, 154, 189, 196, 209, 210, 216, 217,
    221–226, 249, 349, 358
spandrels (pendentives)   164, 165, *166*, 290
speciation,   70, 106, 108–112, 114, 143, 218,
    220, 276, 295, 296
  allopatric   108, *218*, 219
species concepts   55–59, 71
spiders   284, 293, 350
sponges   23, 39, 42, *43*, 49, 178–180, 276, 277,
    279, *280*, 281, 309, 310, 323
Stanley, Steven   137
stasis   *106*, 107, 109, 110

Steno, Nicholas   5, 10
Stigall, Alycia   143
stromatolites   257, *258*, 259, *260*, 308, 309
stromatoporoids   309, *310*, 313
superposition   5
synapsids   352–355

tapeworm   *178*, 180, 320
taphonomy   38–42
terror birds   221, *222*
tetrapods   104, 105, *348*, 349–351
Tiktaalik   350, *351*
Time's arrow/Time's cycle   4, 14, 171, 372
titanotheres   *63*
tools   58, 64, 67, 67, 69, 332, 335, 358
trace fossils   18, 49–51, *51*, *204*, 256
trees   21, 26, 45, 53, 66, 88, 116, 129, 157, 238
  oldest   *239*
trilobites   14, *15*, *20*, *41*, 47, 76, *107*, *108*, 126,
    154, *155*, *181*, 210, 212, *213*, 214, *255*,
    256, 273, 274, 279–284, *283*, *285*, 287,
    291, *292*, 295, *297*, 299
turtles   *89*, 107, 245, 329, 353, 354

ultraviolet (UV) radiation   264, 265
unconformity   6, *7*, 30
uniformitarian(ism)   8, 9, 85, 109, 132, 376
uraninite   261
urochordates   340, *341*, *342*, 372

Valentine, James   219
Venus   139, 185, *186*, 187–190, 361
vertebrates   19, *20*, 49, 127, 175, 196, 237, 273,
    274, 282, 325–358, 372
Vrba, Elisabeth   119, 164, 223, 224

Walcott, Charles   283
Wallace, Alfred Russel   *86*, 94–96, 101, 110,
    214
Wegener, Alfred   209
whale,   55, 59, *77*, 106, 134, 325, 346, 352,
    363
  blue whale   116, *338*
Wignall, Peter   139
wolf   *57*, 145

Xiao, Shuhai   278